"Information through Innovation"

MICROCOMPUTER DATABASE MANAGEMENT
USING
Microsoft Access
Version 2.0

PHILIP J. PRATT
PAUL M. LEIDIG
Grand Valley State University

boyd & fraser publishing company

 An International Thomson Publishing Company

Danvers • Albany • Bonn • Boston • Cincinnati • Detroit • London • Madrid • Melbourne
Mexico City • New York • Paris • San Francisco • Singapore • Tokyo • Toronto • Washington

Acquisitions Editor: Anne E. Hamilton
Production Coordinator: Patty Stephan
Production Services: Ruttle, Shaw & Wetherill, Inc.
Compositor: Ruttle, Shaw & Wetherill, Inc.
Interior Design: Becky Herrington
Cover Design: Diana Coe
Cover Photo: Joseph Drivas / The Image Bank
Manufacturing Coordinator: Tracy Megison

I(T)P The ITP™ logo is a trademark under license.

Printed in the United States of America

This book is printed on recycled, acid-free paper that meets Environmental Protection Agency Standards.

For more information, contact boyd & fraser publishing company:

boyd & fraser publishing company
One Corporate Place • Ferncroft Village
Danvers, Massachusetts 01923, USA

International Thomsom Publishing Europe
Berkshire House 168-173
High Holborn
London, WC1V 7AA, England

Thomas Nelson Australia
102 Dodds Street
South Melbourne 3205
Victoria, Australia

Nelson Canada
1120 Birchmont Road
Scarborough, Ontario
Canada MlK 5G4

International Thomson Editores
Campose Eliseos 385, Piso 7
Col. Polanco
11560 Mexico D.F. Mexico

International Thomson Publishing GmbH
Konigswinterer Strasse 418
53227 Bonn, Germany

International Thomson Publishing Asia
221 Henderson Road
#05-10 Henderson Building
Singapore 0315

International Thomason Publishing Japan
Hirakawacho Kyowa Building, 3F
2-2-1 Hirakawacho
Chiyoda-Ku, Tokyo, 102, Japan

1 2 3 4 5 6 7 8 9 10 KE 8 7 6 5 4

Library of Congress Cataloging-in-Publication Data

Pratt, Philip J., 1945–
    Microcomputer database management using Microsoft Access 2.0 /
  Philip J. Pratt, Paul M. Leidig.
        p.   cm.
    Includes bibliographical references and index.
    ISBN 0-87709-386-5
    1. Database management.   2. Microsoft Access.   3. Microcomputers—Programming.
I. Leidig, Paul M.   II. Title.
QA76.9.D3P72934   1994
005.75′65—dc20

94-23262
CIP

Text with 3.5" Student Data Disk: ISBN 0-87709-560-4

CONTENTS

Database courses in leading computer science programs have been offered comprehensively since the 1970s to students aspiring to computer-related careers. Once they become professionals, students often become involved in the design, development, implementation, and maintenance of large, mainframe, database-oriented application systems. Some work in related areas, such as database administration.

Until recently, these professionals were the only people who had direct contact with databases and database management systems (DBMSs). With the advent of microcomputer DBMSs, however, the picture has changed dramatically. Virtually anyone can now be considered potential users of such systems, including such diverse groups as home computer owners, owners of small businesses, and end users in large organizations. Recently the spreadsheet was the tool that most microcomputer owners felt was essential; however, many are now turning to DBMSs.

The major microcomputer database systems have continually added ease-of-use features so that users can begin to apply the products relatively quickly. Truly effective operation of such products, however, requires more than just product knowledge. It requires a general knowledge of the database area, including such topics as database design, database administration, and application development using these systems. Although the depth of understanding required is certainly not as great for the majority of users as it is for the data-processing professional, a lack of any understanding in these areas precludes effective use of the product in all but the most limited applications.

ABOUT THIS BOOK

This book is intended for anyone interested in becoming familiar with database management. It is appropriate for students in introductory database classes in computer science or information systems programs. It is also appropriate for students in database courses in related disciplines, such as business at either the undergraduate or graduate levels, since such students require a general understanding of the database environment. In addition, *Microcomputer Database Management Using Microsoft Access* is ideal for courses introducing students of any discipline to database management, which is becoming increasingly popular. This book is also aimed at potential purchasers of a microcomputer database package who want to make effective use of such a package.

This book assumes that students have some familiarity with personal computers. A single introductory course is the only required background. Although students need not have any background in programming to effectively use this text, those with a little programming background will be able to explore certain topics in more depth than students without such a background.

Although database management on mainframes is discussed in the book, the main thrust is on microcomputer database management.

DISTINGUISHING FEATURES

Emphasis on the Relational Model   The main emphasis of the text is on the relational model. Not only is the relational model becoming the dominant model for mainframe systems, but microcomputer systems are almost exclusively relational.

Coverage of Other Models	Although other models—specifically the hierarchical and network models—are typically not found on microcomputer systems, students should be exposed to these models. The other models shed light on alternative approaches to a problem. A brief study of them also gives students a better perspective on the relational model, its relationship to other approaches, and its place in database management. Thus, these models are presented in the text, but only at a very intuitive level.
Data Manipulation within the Relational Model	Clearly, the most important approach to data manipulation in the years to come is SQL. This language has been adopted as a standard. Because many systems already support it, familiarity with SQL is crucial. This book contains a discussion of SQL as well as many examples and exercises through which students can gain needed familiarity. This book also covers other approaches to data manipulation within the model, including the three most important: Query-by-Example (QBE), the relational algebra, and natural languages.
Database Design	The important process of database design is given detailed treatment in the text. A methodology is presented that represents a highly useful subset of the full-scale database design methodology. The design methodology presented in this text will be more than enough to allow most microcomputer users to develop a correct design for whatever requirements they face. Those few who become involved in the design of a major mainframe database or an especially complex microcomputer database can use reference [14] (see p. 446) to quickly and easily step into the material on database design.
Functions Provided by a DBMS	Because current microcomputer DBMSs offer such a wide variety of features, students must know the functions that such systems *should* provide. These functions are presented and discussed and the methods by which one DBMS, Microsoft Access, furnishes these functions are examined.
Database Administration (DBA)	Although the office of database administration (DBA) is essential in the mainframe environment, it is also important in a microcomputer environment, especially if several users are to share the database. Thus, this text includes a detailed discussion of the DBA function.
DBMS Selection	The process of selecting a DBMS is important, given the myriad systems available. Unfortunately, this is not an easy task. To help students make such selections effectively, the text includes a detailed discussion of the process and offers a comprehensive checklist.
Applications Generators	The text includes a detailed discussion of what an application system is, what an applications generator is, and how an applications generator is used in the creation of an application system. The relationships between the terms *applications generator, fourth-generation language,* and *fourth-generation environment* are presented as part of the discussion of applications generators.
Glossary	The glossary contains definitions of the important terms, shown in **boldface** in the text.
Numerous Realistic Examples	The book contains numerous examples to illustrate each of the concepts discussed. The examples are realistic and represent the kinds of probems encountered in the design, manipulation, and administration of databases.
Questions and Exercises	The book contains a wide variety of questions. At key points within the chapters, the students are asked questions to ensure that they understand the material before proceeding. The answers to these questions are given immediately following the questions. At the end of each chapter, there are exercises that test students' recall of the important

points in the chapter as well as their ability to apply what they have learned. Finally, the modules include computer assignments in which the students use Microsoft Access to solve a number of problems related to what they have been learning.

Instructor's Manual

The companying *Instructor's Manual* contains detailed teaching tips, instructions on integrating Microsoft Access into the course, answers to exercises in the text, test questions (and answers), and detailed project assignments.

Projects

Projects in which students actually use a DBMS are essential in a database course. Several such projects are presented in the *Instructor's Manual,* together with tips for administering the projects. These projects vary in size and can be implemented by students working either individually or in teams using Microsoft Access. Students can also use these projects as the basis for database design exercises.

ORGANIZATION OF THE TEXTBOOK

The textbook consists of nine chapters, nine modules, and an appendix. The chapters deal with general database topics and are not geared to any specific DBMS. The modules are devoted specifically to Microsoft Access. The recommended arrangement of coverage is to intersperse chapters and modules: That is follow Chapter 1 with Module 1, follow Chapter 2 with Module 2, and so on. The appendix contains a procedure summary for Microsoft Access.

Introduction

Chapter 1 provides a general introduction to the field of database management.

Data Models

Chapter 2 presents the concept of data models. The relational, network, and hierarchical models are covered at an intuitive level.

The Relational Models

The relational model is detailed in Chapters 3 and 4. Chapter 3 covers the data definition and manipulation aspects of the model using SQL, QBE, the relational algebra, and natural languages. Chapter 4 discusses some advanced aspects of the model such as views, the use of indexes, the catalog, and the relational integrity rules. It also includes a discussion of the question: *What does it take to be relational?*

Database Design

Chapters 5 and 6 are devoted to database design. Chapter 5 covers the normalization process, which provides a mechanism to find and correct bad design. In Chapter 6, a methodology for database design is presented and illustrated through a number of examples.

Functions of a DBMS

Chapter 7 discusses the features that a full-functional microcomputer DBMS should provide.

Database Administration

Chapter 8 is devoted to the role of database administration (DBA). Also included in this chapter is a discussion of the process of selecting a DBMS.

Applications Generators

Chapter 9 deals with the process of applications generation. It covers applications generators, their features, and their use in the process. It also describes the term *fourth generation* and how it relates to the term *applications generator.*

Introduction to Microsoft Windows

Module 1 introduces the graphical user interface (GUI) of Microsoft Windows. The concepts of multiple windows, using a mouse, and pull-down menus are covered.

Introduction to Access	Module 2 introduces Microsoft Access menus and windows as well as the process of creating a table and printing a simple report.
Queries	Module 3 covers the Access implementation of Queries, the Query Wizard, and Query-by-Example (QBE). Both single- and multi-table queries are included.
Advanced Relational Features	Module 4 discusses the way in which some of the advanced features of the relational model are implemented in Access. The use of indexes, searching, and views are covered.
Data-Entry Support	Module 5 covers the creation and use of custom forms for both single- and multi-table forms. It also includes the creation of validity checks.
Advanced Reports	Module 6 covers some advanced report techniques including the ReportWizard, tabular reports, sorting and grouping, and embedding subreports.
Microsoft Access and Functions of a DBMS	Module 7 discusses the manner in which Microsoft Access furnishes the functions of a DBMS, such as the use of a catalog, shared updates, recovery, security, integrity, and data independence.
Advanced Topics	Module 8 covers special topics such as OLE links between other Windows applications, advanced form techniques including one-to-many relationships, adding option groups to forms, and the use of macros.
Graphing in Access	Module 9 covers the process of including graphs, embedded OLE graph objects using Microsoft Graphs, and graphing Access data.

GENERAL NOTES TO THE STUDENT

Tutorial Material	Some of the material in the text is tutorial in nature. To get maximum benefit from the text, it is important that you work through this tutorial material. To assist you in this process, each step you are actually to perform is preceded by a bullet (■).
Embedded Questions	Special questions have been embedded in a number of places in the text. Sometimes, these questions are meant to ensure that you understand some crucial material before you proceed. In other cases, the questions are designed to give you the chance to consider some special concept before it is actually presented. In all cases, the answer to the question immediately follows the question. You could simply read the question and its answer. You will receive maximum benefit from the text, however, if you take the time to work out the answers to the questions and then check your answers against the ones given in the text before you proceed with your reading.
End-of-Chapter Material	The end-of-chapter material consists of a summary and exercises. In addition, at the end of each module there are computer assignments. The summary briefly describes the material covered in the chapter. The exercises require that you recall and apply the important material in the chapter. Answers to the odd-numbered exercises are given in the text. The computer assignments involve using Microsoft Access to solve a number of problems related to what you have been learning.

ACKNOWLEDGMENTS

We would like to acknowledge the individuals who made contributions in the preparation of this book.

We appreciate the efforts of the following individuals, who reviewed the text and gave many helpful suggestions:

Judy Adamski, Jenison, Michigan
Michael Alessio, Borland International
Cindy L. Bonfini-Hotlosz, West Virginia Northern Community College
Linda Denny, Sinclair Community College
John M. Lloyd, Montgomery County Community College
Michael Michaelson, Palomar College
Robert E. Norton, San Diego Mesa College
David Seymour, Ludington, Michigan
Hung-Lian Tang, Bowling Green State University
Connie Wells, Georgia State University
Francis Whittle, Dutchess Community College

Thanks also to Larry Malloy and Jerry Kinsel of Oakton Community College, who class tested this text.

The efforts of the following members of the staff of boyd & fraser have been invaluable: Anne Hamilton, Acquisitions Editor, and Patty Stephan, Production Coordinator. We would also like to thank Peg Markow and Kathy Barrett and the staff of Ruttle, Shaw & Wetherill for all their hard work and editing help.

Introduction to Database Management

OBJECTIVES

1. Provide a general introduction to the field of database management.
2. Introduce some basic terminology.
3. Describe the advantages and disadvantages of database processing.

1.1 INTRODUCTION

Chris, Henry, Pat, and Maxine are all taking a microcomputer database course at a local college. Each of the four is interested in learning how to use a microcomputer to manage data effectively, and each one has come to the class with a special project in mind.

Case 1: Chris Chris is the director of volunteers for a large, service-oriented organization. When volunteers are needed for a project, it's his responsibility to locate and enlist them. Currently, he keeps the information he needs on cards in a filing cabinet. For each volunteer, he has a card that contains pertinent information. In a separate drawer, he has a card for each project in which his organization is involved. Some of his responsibilities, such as determining the skills of a particular volunteer, involve locating a single card. This aspect of his job is not very complicated. Other responsibilities, such as gathering the names of all the volunteers who have a particular skill, are much more difficult, since they involve examining every card in the cabinet. Preparing reports is also a time-consuming process.

What Chris would really like is an easy way to enter all this information into his computer and to get it back when he needs it. In particular, he would like to be able to ask the computer such questions as the following:

- What interests and/or skills are required for a given project?
- Which volunteers have a particular combination of skills and have not worked on any projects in the last two years?

He also would like to be able to group volunteers from the same family and ask the following question:

- Which families have more than one person who is suited to work on a given project?

Case 2: Henry Henry owns a chain of four bookstores. Henry has used a computerized file-oriented system to organize the data he uses to run his bookstores.

He gathers and organizes information about publishers, authors, and books. Each book has a code that uniquely identifies the book. In addition, he records the title, the publisher, the type of book, the price, and whether the book is paperback. He also records the author or authors of the books along with the number of units of the book that are in stock in each of the branches.

Henry uses this information in a variety of ways. For example, a customer may be interested in books written by a certain author or of a certain type. Henry wants to be able to tell the customer which books by the author or of that type he currently has in stock. If the customer wants a book that is not in stock at one branch, Henry needs to be able to determine if any of the other branches currently have it.

Case 3: Pat Pat is the office manager for a dental practice. She has been asked by one of the dentists in the practice to look into acquiring a computer to manage all of the appointment scheduling, billing, and preparation of insurance forms. This is a complex task that requires a wide range of information. The following summary of terms used in the dental practice illustrates the extent of the information that is needed.

There are two types of *providers* of services: *dentists* and *hygienists*. *Patients* have *appointments*. Each appointment is *scheduled* with a given provider on a given *date* at a given *time* in a given *room* and for some specific collection of *services*. Patients are grouped into *households* (families), which are ultimately responsible for paying for any services rendered. Patients in a given household may all be covered by the same dental *insurance* policy or they may be covered by different policies. Because patients have this insurance through their *employers*, it is important to relate patients to employers and employers to the insurance companies whose policies cover their employees. Also, if more than one patient in a household is covered by an insurance policy, it is important to distinguish between *primary* and *secondary insurance carriers*. Within a household, a given insurance policy may be the primary carrier for one patient and the secondary carrier for another patient.

Case 4: Maxine Maxine schedules sections of courses at a college. For the purpose of simplicity, we will call these sections *classes*. She determines which faculty member will teach which class and when, and assigns each class to a room. Each room has certain characteristics, such as the number of seats, the types of lab and video display equipment present, and so on. Each class has its own set of requirements for a room in which it may be held. Each faculty member has certain courses that he or she is capable of teaching as well as specific courses that he or she would prefer to teach. In making her schedules, Maxine needs to match rooms, classes, and faculty in such a way that there are no conflicts, while making sure that faculty members are capable of teaching the classes to which they are assigned. In addition, she tries to assign faculty members to the courses that they prefer to teach.

With regard to the microcomputer database course they are taking, what is it that Chris, Henry, Pat, and Maxine all have in common? They all need to be able to store and retrieve data in an efficient and organized way. Further, all of them are interested in more than one category of information. In database management we call these categories **entities**.

Chris is interested in such entities as volunteers, projects, and households. Henry is interested in books, authors, publishers, and branches. Pat is interested in patients, providers (dentists and hygienists), households, services, appointments, insurance companies, rooms, and so on. Maxine is interested in classes, faculty members, rooms, and times.

Besides wanting to store data that pertains to more than one entity, the four students are also interested in relationships between the entities. Chris, for example, is interested not only in volunteers, projects, and households but also in the relationship between volunteers and projects (which volunteers are assigned to which projects), between volunteers and households (which volunteers are a part of which households), and so on. Henry is interested in relating books to the authors who wrote them, to the publishers who published them, and to the branches that have them in stock.

What these students most want to know is how to maintain and use a database. A **database** is a structure that contains information about many kinds of entities and about the relationships between the entities. Henry's database, for example, would contain information about books, authors, branches, and publishers. It would provide facts that related authors to the books they directed and branches to the books they currently have in stock. With the use of a database, Henry would be able to start with a particular book and find out who wrote it as well as which branches have it. Alternatively, he could start with an author and find all the books he or she wrote, together with the publishers of these books.

Approach 1: Writing Programs to Maintain a Database

One possible approach to the problem would be to use a high-level language like BASIC to write programs. Some of the programs would allow users to enter and modify data. Others would be used to produce reports. Still others might be used to calculate statistics about the data that has been stored. Some programs could produce special documents. To handle customer billing, a collection of programs would include a program that printed invoices. Such a collection for payroll would have to contain programs to print paychecks and W-2 forms. Other programs could be devised to answer such questions as, "Do we currently have available any volunteers with skills X, Y, and Z?"

Such a collection of programs is often called a **system of programs** or a **software system**; sometimes it is simply called a **system**. Commercial developers often call these systems of programs **software packages**. They may also be called an **application system** or an **application package**. The programs that make up the collection are often called **application programs**.

Case 2: Henry and BASIC   Using BASIC, Henry had created such a system to maintain and report on books, publishers, authors, and branches several years ago. The system that Henry created used several **files**. (If you have written programs before, you are familiar with the concept of a file. You may have referred to this type of file as a **data file**. For those of you who are not familiar with this concept, some sample files, represented in the form of tables, are shown in Figure 1.1a. The rows in the table are called **records**, and the columns are called **fields**.)

Figure 1.1a

BRANCH,
PUBLSHR, and
AUTHOR files

BRANCH

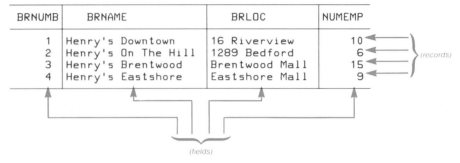

PUBLSHR

PUBCODE	PUBNAME	PUBCITY
AH	Arkham House Publ.	Sauk City, Wisconsin
AP	Arcade Publishing	New York
AW	Addison-Wesley	Reading, Mass.
BB	Bantam Books	New York
BF	Boyd and Fraser	Boston
JT	Jeremy P. Tarcher	Los Angeles
MP	McPherson and Co.	Kingston
PB	Pocket Books	New York
RH	Random House	New York
RZ	Rizzoli	New York
SB	Schoken Books	New York
SI	Signet	New York
TH	Thames and Hudson	New York
WN	W.W. Norton and Co.	New York

AUTHOR

AUTHNUMB	AUTHNAME
1	Archer, Jeffrey
2	Christie, Agatha
3	Clarke, Arthur C.
4	Francis, Dick
5	Cussler, Clive
6	King, Stephen
7	Pratt, Philip
8	Adamski, Joseph
10	Harmon, Willis
11	Rheingold, Howard
12	Owen, Barbara
13	Williams, Peter
14	Kafka, Franz
15	Novalis
16	Lovecraft, H. P.
17	Paz, Octavio
18	Camus, Albert
19	Castleman, Riva
20	Zinbardo, Philip
21	Gimferrer, Pere
22	Southworth, Rod
23	Wray, Robert

Figure 1.1b

Description of
BRANCH,
PUBLSHR, and
AUTHOR files

BRANCH

NAME	DESCRIPTION
BRNUMB	Branch number
BRNAME	Branch name
BRLOC	Branch location
NUMEMP	Number of employees

PUBLSHR

NAME	DESCRIPTION
PUBCODE	Publisher code
PUBNAME	Publisher name
PUBCITY	Publisher city

AUTHOR

NAME	DESCRIPTION
AUTHNUMB	Author number
AUTHNAME	Author name

The first three of Henry's files are shown in Figure 1.1a. Descriptions of the fields are given in 1.1b. The files are called *BRANCH, PUBLSHR,* and *AUTHOR.*

(**Note:** There are many cases where names are limited to eight characters. To make sure that we would not have problems in such cases, names in this book will adhere to the eight-character limit. This is why the name for the publisher file has been shortened to *PUBLSHR.*)

The *BOOK* file is shown in Figure 1.2a with a description of its fields appearing in Figure 1.2b on the next page.

Figure 1.2a

BOOK file

BOOK

BK CODE	BKTITLE	PUB CODE	BK TYPE	BK PRICE	PB
0180	Shyness	BB	PSY	7.65	T
0189	Kane and Abel	PB	FIC	5.55	T
0200	The Stranger	BB	FIC	8.75	T
0378	The Dunwich Horror and Others	PB	HOR	19.75	F
079X	Smoke-screen	PB	MYS	4.55	T
0808	Knockdown	PB	MYS	4.75	T
1351	Cujo	SI	HOR	6.65	T
1382	Marcel Duchamp	PB	ART	11.25	T
138X	Death on the Nile	BB	MYS	3.95	T
2226	Ghost from the Grand Banks	BB	SFI	19.95	F
2281	Prints of the 20th Century	PB	ART	13.25	T
2766	The Prodigal Daughter	PB	FIC	5.45	T
2908	Hymns to the Night	BB	POE	6.75	T
3350	Higher Creativity	PB	PSY	9.75	T
3743	First Among Equals	PB	FIC	3.95	T
3906	Vortex	BB	SUS	5.45	T
5163	The Organ	SI	MUS	16.95	T
5790	Database Systems	BF	CS	54.95	F
6128	Evil Under the Sun	PB	MYS	4.45	T
6328	Vixen 07	BB	SUS	5.55	T
669X	A Guide to SQL	BF	CS	23.95	T
6908	DOS Essentials	BF	CS	20.50	T
7405	Night Probe	BB	SUS	5.65	T
7443	Carrie	SI	HOR	6.75	T
7559	Risk	PB	MYS	3.95	T
7947	dBASE Programming	BF	CS	39.90	T
8092	Magritte	SI	ART	21.95	F
8720	The Castle	BB	FIC	12.15	T
9611	Amerika	BB	FIC	10.95	T

Figure 1.2b

Description of
BOOK file

Q & A

```
BOOK

NAME      DESCRIPTION

BKCODE    Book code
BKTITLE   Book title
PUBCODE   Book publisher code
BKTYPE    Book type
BKPRICE   Price
PB        Paperback? (T or F)
```

Question:

To check your understanding of the relationship between publishers and books, answer the following questions: Who published *Knockdown*? Which books did Signet publish?

Answer:

The publisher code (*PUBCODE*) in the row in the *BOOKS* table for *Knockdown* is PB. Examining the *PUBLSHR* table, we see that PB is the code assigned to Pocket Books.

To find the books published by Signet, we look up its code in the *PUBLSHR* table and see that it is SI. Next, we look for all records in the *BOOK* table for which the publisher code is SI and find that Signet published *Cujo, Carrie, The Organ,* and *Magritte.*

Figure 1.3a

WROTE and
INVENT files

The table called *WROTE* in Figure 1.3a is used to relate books and authors. The sequence number indicates the order in which the authors of a particular text should be listed. The table called *INVENT* in the same figure is used to indicate the number of units of a particular book that are currently on hand at a particular branch. The first row, for example, indicates that there are two units of the book whose code is 0180 currently on hand at Branch 1. Descriptions of these tables are shown in Figure 1.3b.

WROTE

BK CODE	AUTH NUMB	SEQ NUMB
0180	20	1
0189	1	1
0200	18	1
0378	16	1
079X	4	1
0808	4	1
1351	6	1
1382	17	1
138X	2	1
2226	3	1
2281	19	1
2766	1	1
2908	15	1
3350	10	1
3350	11	2
3743	1	1
3906	5	1
5163	12	2
5163	13	1
5790	7	1
5790	8	2
6128	2	1
6328	5	1
669X	7	1
6908	22	1
7405	5	1
7443	6	1
7559	4	1
7947	7	1
7947	23	2
8092	21	1
8720	14	1
9611	14	1

INVENT

BK CODE	BR NUMB	OH
0180	1	2
0189	2	2
0200	1	1
0200	2	3
079X	2	1
079X	3	2
079X	4	3
1351	1	1
1351	2	4
1351	3	2
138X	2	3
2226	1	3
2226	3	2
2226	4	1
2281	4	3
2766	3	2
2908	1	3
2908	4	1
3350	1	2
3906	2	1
3906	3	2
5163	1	1
5790	4	2
6128	2	4
6128	3	3
6328	2	2
669X	1	1
6908	2	2
7405	3	2
7559	2	2
7947	2	2
8092	3	1
8720	1	3
9611	1	2

Figure 1.3b	WROTE			INVENT		

WROTE

NAME	DESCRIPTION
BKCODE	Book code
AUTHNUMB	Author number
SEQNUMB	Sequence number. Used to indicate order in which multiple authors should be listed.

INVENT

NAME	DESCRIPTION
BKCODE	Book code
BRNUMB	Branch number
OH	Number of units on hand

Description of *WROTE* and *INVENT* files

Question: To check your understanding of the relationship between authors and books, answer the following questions: Who wrote *The Organ*? (Be sure to list the authors in the correct order.) Which books did Jeffrey Archer write?

Answer: To determine who wrote *The Organ*, we first examine the *BOOK* table to find its book code (5163). Next we look for all rows in the *WROTE* table in which the book code (*BKCODE*) is 5163. There are two such rows. In one of them the author number (*AUTHNUMB*) is 12, and in the other, it is 13. All that is left is to look in the *AUTHOR* table to find the authors who have been assigned the numbers 12 and 13. The answer is Barbara Owen (12) and Peter Williams (13). The sequence number for author 12 is 2, however, and the sequence number for author 13 is 1. Thus, listing the authors in the proper order, the authors are Peter Williams and Barbara Owen.

To find the books written by Jeffrey Archer, we look up his number in the *AUTHOR* table and find that it is 1. Then we look for all rows in the *WROTE* table for which the author number is 1. There are three such rows. The corresponding book codes are 0189, 2766, and 3743. Looking up these codes in the *BOOK* table, we find that Jeffrey Archer wrote *Kane and Abel*, *The Prodigal Daughter*, and *First Among Equals*.

Question: A customer in Branch 1 wishes to purchase *The Vortex*. Is it currently in stock in Branch 1?

Answer: Looking up the code for *The Vortex* in the *BOOK* table, we find it is 3906. To find out how many copies are in stock in Branch 1, we look for a row in the *INVENT* table with 3906 in the *BKCODE* column and 1 in the *BRNUMB* column. Since there is no such row, Branch 1 doesn't have any copies of *The Vortex*.

Question: We would like to obtain a copy of *The Vortex* for this customer. Which other branches currently have it in stock and how many copies do they have?

Answer: We already know that the code for *The Vortex* is 3906. (If we didn't, we would simply look it up in the *BOOK* table.) To find out the branches that currently have copies, we look for rows in the *INVENT* table with 3906 in the *BKCODE* column. There are two such rows. The first one indicates that Branch 2 currently has one copy. The second indicates that Branch 3 currently has two copies.

Henry wrote several programs to allow him to update the data in his files. Separate programs were required to allow for the addition, correction, and deletion of books, publishers, and authors. Henry also had to create programs to produce any of the reports he wanted. Further, the logic that we just discussed for relating books and publishers and for relating authors and books had to be built into these programs.

**Approach 2:
Using a
Database
Management
System**

In the system that Henry developed in BASIC, he effectively implemented a database. He certainly maintained data on several entities (books, publishers, and authors). He also maintained relationships between these entities. To the computer, however, Henry was dealing with nothing more than a collection of isolated files. It was only through the efforts of Henry's programs that the crucial relationships were maintained. Another way of stating this would be to say that Henry's programs *managed the database*. As Henry discovered, this can be a *very complex task*.

Fortunately, we no longer necessarily have to create our own programs, because the computer is now able to assist in managing the database for us. The tool it uses is called a **database management system**, or **DBMS**. A DBMS is a program or collection of programs whose function is to manage a database on behalf of the people who use it. It greatly simplifies the task of manipulating and using a database. In fact, had Henry used a DBMS, he might not have had to write *a single program*! (*Note:* You will have a chance to see exactly how you can use one of the most popular DBMSs to manage a database in the module that follows this chapter and in subsequent modules.)

Case 1: Chris and a DBMS   Chris decided that a microcomputer DBMS could fulfill the needs of his service organization. He determined the structure of the database he needed (this is called **designing a database**), following the procedure discussed in the course that he took. His database contained the information he needed about volunteers, projects, and households. He then communicated this design to the DBMS and began to enter data. He found that through the use of the DBMS, all of the reports he needed were easy to produce. Since then, he has had no need to resort to any programming.

Case 2: Henry and a DBMS   Even though he already had the system he developed in BASIC, Henry was so impressed with what he learned about databases and database management systems that he decided to purchase a DBMS. He then designed his database and communicated the design to the DBMS. He entered his data and was easily able to produce a number of reports. He could now perform many important tasks without writing any programs at all.

However, the DBMS did not completely satisfy Henry's needs. He found, for example, that when he had entered PA as the publisher code for a particular book, the DBMS simply let him do it, *even though there was no publisher with code PA on file*. In the programs that he had written in BASIC, he had included a feature to prevent this sort of thing from happening. He also discovered that some of his most important reports could not be produced with the built-in facilities of the DBMS.

**Approach 3:
Programming
with a DBMS**

Fortunately, many DBMSs allow users to write programs to supplement the built-in features of the DBMS. Some allow programs to be written in existing languages such as BASIC, Pascal, or COBOL. These languages are expanded to include additional commands that allow for the accessing and updating of a database. Other DBMSs include their own language. These languages usually contain all of the typical structures one expects in a modern programming language, along with a number of special features that are geared toward manipulation of the database. Many DBMSs also offer features that provide screen and printer management. The languages furnished by the better DBMSs provide features that are superior to those found in many non-DBMS languages (for example, BASIC, Pascal, COBOL, and FORTRAN).

Case 2: Henry and a DBMS   Henry was pleased to discover that his DBMS included its own programming language. He found that with the DBMS he could write programs to overcome the deficiencies he had spotted. Using a program, he could ensure that a book could not be entered if the book's publisher was not already in the database. This programming capability enabled him to produce reports that were beyond the basic capabilities of the DBMS.

Henry was also pleasantly surprised to find that he could develop programs with the DBMS much more rapidly than he had developed the BASIC programs in his previous system. Simple commands could be used in his programs to accomplish various important tasks, such as accessing the database, interacting with the screen, printing lines on reports, copying data from one file to another, and so on. He found tools within the system to allow the development of large portions of some of the important programs. He is convinced that he has now found the perfect environment for his system.

Approach 4: Using a Commercial Software Package

Case 3: Pat, a Software Package, and a DBMS   For a variety of reasons, Pat decided not to use a DBMS herself to develop the system for the dental practice. First, the many entities and complicated relationships made the requirements so complex that she felt the problem should be handled by a computer professional. Pat had another reason for not wanting to undertake the task. She was very leery of attempting to develop a system on which the whole billing operation of the dental practice would depend. She believed there were many issues involved in the development and use of such a system that she didn't understand, and that seemed to be yet another reason for putting such an undertaking in the hands of professionals.

That's the direction that the dental practice took. They purchased a software package from an organization that specialized in software for dental offices, and it seems to have fit their needs quite well. The package is geared specifically toward appointment scheduling, billing, and the preparation of insurance forms.

Pat did decide she could do certain things with a DBMS, however. In fact, even though the dental practice purchased the software package, she determined that her office had some needs that more than justified the purchase of a DBMS. She has a number of things in mind to do with it. She plans to track the history of various types of treatment by household and by family. She is also going to track the use of supplies in various types of treatment. The package that the practice purchased does not do these things, and she feels comfortable doing them herself with the DBMS. The purposes for which she uses the DBMS will be completely separate from those of the package that the practice purchased.

Approach 5: Solving Part of the Problem with a DBMS

Case 4: Maxine and a DBMS   Maxine originally intended to purchase a software package to handle her scheduling needs. In her estimation, however, none of the packages she looked at were really suitable. Some would require her to change the way she carried out the scheduling process; others did not supply certain reports that she felt were essential. At this point, she happened to be taking a database course and wondered if she might be able to develop the system herself with a DBMS.

Maxine talked to the professor who was teaching the course. In the process, she discovered that simply using a DBMS would not solve her problem. The DBMS could be used to maintain crucial information about rooms, faculty, and classes, but the actual process of scheduling was beyond its built-in capabilities.

This meant that someone would have to write programs to do the scheduling. The scheduling process itself is very intricate, and the task of writing programs to implement it seemed complicated as well. Maxine felt she was not in a position to undertake such a formidable task. Happily, the professor decided to have a team of students try to develop the system Maxine needed as their project for a course they would enroll in during the next semester.

The professor also pointed out to Maxine that she could use the DBMS to carry out certain productive activities even before the full system was developed. Even if she made out the schedule entirely by hand without the benefit of the computer, she could still derive benefits from the DBMS. She could begin by manually entering her schedule into a database. Although that might sound like extra work, once it had been done, she could easily get a variety of reports that would be very useful. She could print out the schedule of each faculty member and the schedule of all the classes in a given room, or the times and locations at which the various sections of a given course were offered — that is, a time schedule. If someone needed to schedule a room for a special offering, such as a seminar, she could have the system search for rooms that would be open at the desired time, would be capable of seating the number of people expected to attend, and would contain any special equipment that was required.

Maxine is excited both by the possibility of the students at the college developing the complete system she needs and by the amount of useful work she can accomplish even before this has been done. She is eager to begin.

The interests and needs of these four students are typical of those of today's micro-computer users. The foregoing approaches to their problems are all being used by people around the world today. Ever increasing numbers of people are finding database management systems the ideal tool for solving a wide variety of problems.

Section 1.2 presents some background material on database management and some of the most commonly used terms. In sections 1.3 and 1.4, we will examine some of the advantages and disadvantages of database management systems.

1.2 BACKGROUND

This section introduces some terminology and concepts that are very important in the database environment. Some of the terms will be familiar to you from the material in the preceding section.

Entities, Attributes, and Relationships

The most fundamental terms are entity, attribute, and relationship. An **entity** is really just like a noun; it is a person, place, or thing. The entities of interest to Henry, for example, are such things as publishers, branches, and books. The entities that are of interest to Chris include members and projects. In her dental office, Pat is interested in such entities as patients, households, services, providers of services, appointments, and so on.

An **attribute** is a property of an entity. The term is used here exactly as it is used in everyday English. For the entity *person*, for example, the list of attributes might include such things as eye color and height. For Henry, the attributes of interest for the entity *book* are such things as code, title, type of book, price, and so on.

Figure 1.4 shows two entities, *PUBLSHR* and *BOOK*. It also shows a number of attributes. The *PUBLSHR* entity has three attributes: publisher code (*PUBCODE*), publisher name (*PUBNAME*), and publisher city (*PUBCITY*). The attributes are really just the columns in the table. The *BOOK* entity has six attributes: book code (*BKCODE*), book title (*BKTITLE*), publisher code (*PUBCODE*), book type (*BKTYPE*), book price (*BKPRICE*), and paperback (*PB*). (The last attribute simply indicates whether or not the book is paperback.)

The final key term is relationship. When we speak of a **relationship**, we really mean an association between entities. There is an association between publishers and books, for example. A publisher is associated with all of the books that it publishes, and a book is associated with its publisher. Technically, we say that a publisher is *related to* all of the books it publishes, and a book is *related to* its publisher.

This particular relationship is called a **one-to-many** relationship. *One* publisher is associated with *many* books, but each book is associated with only *one* publisher. (In this type of relationship, the word *many* is used differently than in everyday English; it may not always literally mean a large number. In this context, it would mean that a publisher can be associated with *any number* of books.)

A one-to-many relationship is often represented pictorially in the fashion shown in Figure 1.5. In such a diagram, entities and attributes are represented in precisely the same way as they are shown in Figure 1.4. The relationship is represented by an arrow. The "one" part of the relationship, in this case *PUBLSHR*, is indicated by a single-headed arrow, and the "many" part of the relationship, in this case *BOOK*, is indicated by a double-headed arrow.

Figure 1.4

Entities and attributes

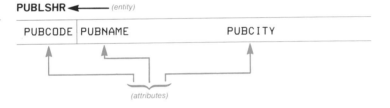

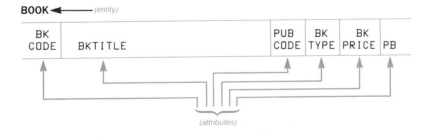

Figure 1.5

One-to-many relationship

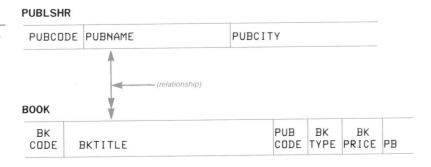

Files and Databases

You encountered the word *file* earlier in this chapter. If you have done some programming yourself, you are probably already familiar with the word. Basically, a file used to store data, which is often called a *data file*, is the computer counterpart to an ordinary paper file you might keep in a filing cabinet. You will recall that Chris kept such a file on the members in his organization. His filing cabinet was filled with cards, one for each member. The crucial aspect of this type of file is that it houses information on a *single entity* and the attributes of that entity. In Chris's case, the single entity was *member*. Each card kept information on the crucial attributes of one member of the organization.

A database is much more than a file, however. Unlike a typical data file, a database can house information about more than one entity. And there is another difference. A database also holds information about the relationships among the various entities. Not only would Henry's database have information about both books and publishers, for example, it would also hold information relating publishers to the books they had produced. Formally, the definition of a database is as follows:

Definition: A **database** is a structure that can house information about multiple types of entities, the attributes of these entities, and the relationships among the entities.

Database Management Systems

Managing a database is inherently a complicated task. Fortunately, software packages called **database management systems** can do the job of manipulating databases for us. A database management system, or **DBMS**, is a software product through which users interact with a database. The actual manipulation of the underlying database is handled by the DBMS. In some cases, users may interact with the DBMS directly, as shown in Figure 1.6a. In other cases, users may interact with programs; these programs in turn interact with the DBMS (see Figure 1.6b). In either case, it is only the DBMS that actually accesses the database.

Figure 1.6a

Using a database management system directly

Figure 1.6b

Using a database management system from a program

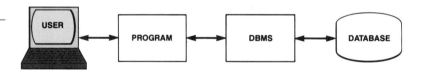

Using a DBMS, for example, Henry can request the system to find publisher PB and the system will either locate this publisher and give us the data or tell us that no such publisher exists in the database. All the work involved in this task is performed by the DBMS. If publisher PB is in the database, Henry can then ask for the books this publisher has published and again the system will perform all the work involved in locating these books. Likewise, when Henry stores a new book in the database, the DBMS performs all the tasks necessary to ensure that the book is related to the appropriate publisher.

Mainframe DBMSs have been in use since the 1960s. They have continually been enhanced over the years, gaining in selection of features and in performance. Recently, microcomputer DBMSs that possess many of the features of their mainframe counterparts have become available. The leaders in this field, such as dBASE™ from Ashton-Tate and R:BASE® from Microrim, are also improved on a continuous basis. They make the power of database management available to large numbers of microcomputer users. The focus of this text is microcomputer database management systems. For a discussion of database management on mainframes, see [9], [10], [11], [13], [14], and [15] in the references at the end of the book.

Database Processing

When we use the term **database processing**, we mean that the data to be processed is stored in a database and the data in the database is being manipulated by a DBMS. We have seen how database processing will benefit Chris, Henry, Pat, and Maxine with their individual systems. Still greater benefit is obtained by combining the activities of several users and allowing them to share a common database.

Let's first consider the nondatabase approach illustrated in Figure 1.7. Mary, Jeff, and Joan are three separate users at the same college. Mary is involved in enrolling students in courses, in producing class lists, and so on. She has her own system of programs and files that she uses to perform this activity on the computer. Her files contain information on classes, on faculty members who teach these classes, and on the students who are enrolled in them.

Figure 1.7

Nondatabase approach

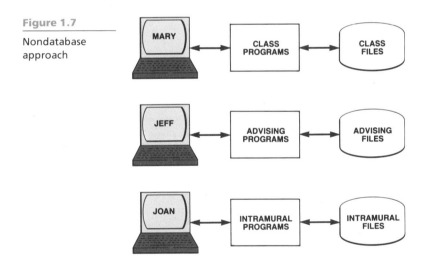

Jeff is involved in the advising process, that is, the process of advising students about their programs and their progress toward a degree. He has his own system of programs and files. His files, which are totally separate from Mary's, contain information on faculty members, on the students who are advised by them, and on the requirements that have already been fulfilled by these students.

Joan is in charge of maintaining information on the intramural athletic program. She too has her own system of programs and files. Her files contain information on the various sports that are available, the teams that participate in these sports, the students who belong to these teams, and the current records of the teams.

Two major problems arise with this nondatabase approach. The first problem is duplication of data. Mary, Jeff, and Joan are each keeping information about students, for example. Presumably, each of them will need the address of all the students. Thus the address of each student is stored in at least three separate places in the computer. Not only is this wasteful of space, but it causes a real headache when a student moves and his or her address must be changed.

The second problem is that it is extremely difficult to fulfill requirements that involve data from more than one system. The format of the files might not even be compatible from one system to another. The following question and answer segment demonstrates such a requirement.

Q & A

Question:

Suppose we want to list the number, name, and address of a particular student. We also want to list the classes in which the student is currently enrolled, the name of the student's advisor, and the intramural sports in which the student is participating. Where would we find the necessary data?

Answer:

The student's number, name, and address could come from any one of the three systems. The classes in which the student is enrolled could be found in Mary's system. The name of the student's advisor would be found in Jeff's system. The sports in which the student is participating could be found in Joan's system. Thus, this requirement involves data from all three systems.

By contrast, in a database approach, instead of having separate collections of files, Mary, Jeff, and Joan would be able to share a common database managed by a DBMS (see Figure 1.8).

Figure 1.8

Database approach (using a database management system)

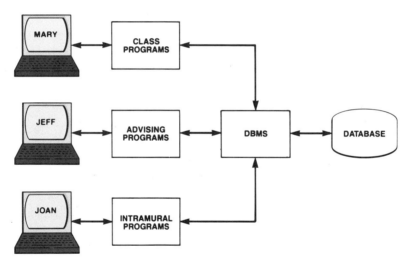

Each student would appear only once, so his or her address would likewise appear only once. No space would be wasted, and changing a student's address would be a very simple procedure. Further, since all the data would be in a single database, listing the information on a student, the student's classes, the student's advisor, and the sports in which the student was participating would now be quite possible. In fact, with a good DBMS, it should be a simple task.

1.3 ADVANTAGES OF DATABASE PROCESSING

The database approach to processing, using a microcomputer DBMS, offers ten clear advantages. They are listed in Figure 1.9 and are discussed below.

Figure 1.9	**Advantages of Database Processing**
Advantages of database processing	1. Lower Cost
	2. Getting More Information from the Same Amount of Data
	3. Sharing of Data
	4. Balancing Conflicting Requirements
	5. Controlled or Eliminated Redundancy
	6. Consistency
	7. Integrity
	8. Security
	9. Increased Productivity
	10. Data Independence

1. **Lower cost.** This is an interesting advantage. In discussions about mainframe DBMSs, cost is usually listed as a disadvantage (see [14]). By the time all the appropriate components have been purchased, the total price can easily run between $100,000 and $400,000. The size and complexity of a DBMS may also necessitate the use of more hardware. Purchasing (or leasing) this additional hardware represents another cost.

   On microcomputers, the opposite is true. Some of the more limited (but still very useful) systems are priced under $100. Even for the most powerful and most sophisticated of these microcomputer DBMSs, the price ranges between $500 and $800. These prices make the features of the DBMS available to a wide range of users.

2. **Getting more information from the same amount of data.** The primary goal of a computer system is to turn data (recorded facts) into information (the knowledge gained by processing these facts). In a nondatabase environment, data is often partitioned into several disjointed systems, each system having its own collection of files. Any request for information that would necessitate accessing data from more than one of these collections can be extremely difficult. In many cases, for all practical purposes, it is considered impossible. Thus, the desired information is unavailable, not because it is not stored in the computer, but because of the way it has been broken down into the various collections of files. When, instead, all the data for the various systems is stored in a single database, the information becomes available. Given the power of a modern DBMS, not only is the information available, but the process of getting it can be a quick and easy one.

3. **Sharing of data.** The data of various users can be combined and shared among authorized users, allowing all users access to a greater pool of data. Several users can have access to the same piece of data, for example, a customer's address, and still use it in a variety of ways. When an address is changed, however, the new address immediately becomes available to all users. In addition, new applications can be developed through the use of the existing data in the database without the burden of having to create separate collections of files.

4. **Balancing conflicting requirements**. For the database approach to function adequately within an organization, a person or group should be in charge of the database itself, especially if it is to serve a number of users. This person or group is often called **Database Administration (DBA)**. By keeping the overall needs of the organization in mind, DBA can structure the database in such a way that it benefits the entire organization, not just a single group. While this may mean that an individual user group is served less well than it would have been if it had its own isolated system, the organization as a whole is better off. And ultimately, when the organization benefits, so do the individual groups of users.

5. **Controlled or eliminated redundancy**. With database processing, data that was formerly kept separate in a nondatabase system is integrated into a single database, so multiple copies of the same data no longer exist. With the nondatabase approach, Mary, Jeff, and Joan each had a copy of the address of each student, but with the database approach, each student's address would occur only once, thus eliminating the duplication (technically called **redundancy**).

   Eliminating redundancy not only saves space but makes the process of updating much simpler. With the database approach, changing the address of a student would mean making *one* single change. With the nondatabase approach, in which each student happened to be stored in three different places, the same change of address would mean that *three* changes had to be made in the computer.

   Although eliminating redundancy is the ideal, it is not always possible. Sometimes, for reasons having to do with performance, we might choose to introduce a limited amount of redundancy into a database. But, even in these cases, we would be able to keep the redundancy under tight control, thus obtaining the same advantages. This is why it is technically better to say that we *control* redundancy rather than *eliminate* it.

6. **Consistency**. Suppose an individual student's address were to appear in more than one place. Student 176, for example, might be listed at 926 Meadowbrook at one spot within our data and 2856 Wisner at another. The data in the computer would then be *inconsistent*. Since the potential for this sort of problem is a direct result of redundancy, and since the database approach eliminates (or at least controls) redundancy, there is much less potential for the occurrence of this sort of inconsistency with the database approach.

7. **Integrity**. An **integrity constraint** is a rule that must be followed by data in the database. Here is an example of an integrity constraint: The director number given for any movie must be that of a director who is *already in the database*. A database has **integrity** if the data in it satisfies all integrity constraints that have been established. A good DBMS should provide an opportunity for users to articulate these integrity constraints when they describe the database. The DBMS should then ensure that these constraints are never violated. According to the integrity constraint just articulated, the DBMS should *not allow* us to store data about a given movie if the director number that we enter is not the number of a director whose name is already in the database.

8. **Security**. **Security** is the prevention of access to the database by unauthorized users. A good DBMS has a number of features that help ensure the enforcement of security measures.

9. **Increased productivity.** A DBMS frees the programmers who are writing programs to access a database from having to engage in mundane data manipulation activities thus making the programmers more productive. A good DBMS comes with many features that allow users to gain access to data in the database without having to do any programming at all. This increases the productivity both of programmers, who may not need to write complex programs in order to perform certain tasks, and of nonprogrammers, who may be able to get the results they seek from the data in the database without waiting for a program to be written for them.

10. **Data independence.** The structure of a database often needs to be changed. For example, new user requirements may necessitate the addition of an entity, an attribute, or a relationship, or a change may be required to improve performance. A good DBMS provides **data independence**, which is the property that the structure of a database can change without the programs that access the database having to change. Without data independence, a lot of unnecessary effort can be expended in changing programs to match the new structure of the database. The presence of many programs in the system may make this effort so prohibitive that a decision is made not to change the database. With data independence, the effort of changing all the programs is unnecessary. Thus, when the need arises to change the database, the decision to do so is more likely to be made.

For other perspective on the advantages of database processing, see [9], [11], [12], [13], [14], and [15].

1.4   DISADVANTAGES OF DATABASE PROCESSING

As you would expect, if there are advantages to doing something in a certain way, there are also disadvantages. The area of database processing is no exception. In terms of numbers alone, the advantages outweigh the disadvantages, but the latter are listed in Figure 1.10 and explained below.

Figure 1.10

Disadvantages of database processing

Disadvantages of Database Processing

1. Larger Size
2. Greater Complexity
3. Greater Impact of a Failure
4. Recovery More Difficult

1. **Larger size.** In order to support all the complex functions that it provides to users, a database management system must be a large program that occupies megabytes of disk space as well as a substantial amount of internal memory.

2. **Greater complexity.** The complexity and breadth of the functions furnished by a DBMS make it a complex product. Users of the DBMS must understand the features of the system in order to take full advantage of it, and there is a great deal for them to learn. In design and implementation of a new system that uses a DBMS, many choices have to be made, and it is possible to make incorrect choices, especially with an insufficient understanding of the system. Unfortunately, a few incorrect choices can spell disaster for the whole project. This is especially true for a large mainframe project that serves many users, but it can also apply to microcomputer projects.

3. **Greater impact of a failure**. If each user has a completely separate system, the failure of any single user's system does not necessarily affect any other user. If, on the other hand, several users are sharing the same database, a failure on the part of any one user which damages the database in some way may affect *all of the other users*.

4. **Recovery more difficult**. Because a database is inherently more complex than a simple file, the process of recovering it in the event of a catastrophe is also more complicated than the process of recovering a simple file. This is particularly true if the database is being updated by a large number of users at the same time. It must first be restored to the condition it was in when it was last known to be correct; any updates made by users since that time must be redone. The greater the number of users involved in updating the database, the more complicated this task becomes.

For other perspectives on the disadvantages of database processing, see [9], [11], [12], [13], [14], and [15].

SUMMARY

1. An entity is a person, place, or thing. An attribute is a property of an entity. A relationship is an association between entities.
2. A database is a structure that can house information about many different entities and about the relationships between these entities.
3. A database management system is a software package whose function is to manipulate a database on behalf of users.
4. Database processing offers a number of advantages, including the following:
   a. lower cost
   b. getting more information from the same amount of data
   c. sharing of data
   d. balancing conflicting requirements
   e. controlled or eliminated redundancy
   f. consistency
   g. integrity
   h. security
   i. increased productivity
   j. data independence
5. The disadvantages to database processing include the following:
   a. larger size
   b. greater complexity
   c. greater impact of a failure
   d. recovery more difficult

KEY TERMS

application package

application program

attribute

column

data independence

database

database design

database management
    system (DBMS)

entity

field

file

integrity

integrity constraint

record

redundancy

relational model

relationship

security

software package

table

EXERCISES

1. What is a software package? What is another name for it?
2. What is the difference between an application package and an application program?
3. What is a file? A record? A field?
4. Was Henry implementing a database with the programs he wrote in BASIC? Explain your answer.
5. Give two reasons why Pat did not want to use a DBMS by herself to develop a system for managing her office.
6. What is an entity? An attribute?
7. When we speak of relationship, what exactly do we mean?
8. What is a one-to-many relationship?
9. What is a database?
10. What is a DBMS?
11. Why is the cost of a DBMS listed as an advantage in the microcomputer environment and a disadvantage on mainframes?
12. How is it possible to get more information from the same amount of data through using a database approach as opposed to a file approach?
13. What is meant by "sharing of data"?
14. What is DBA? What kinds of things does DBA do in a database environment?
15. What is redundancy? What are the problems associated with redundancy?
16. How does consistency result from controlling or eliminating redundancy?
17. What is meant by integrity as it is used in this chapter?
18. What is meant by security? What does the DBMS have to do with security?
19. What is meant by data independence? Why is it desirable?
20. How can the size of a DBMS be one of its disadvantages?
21. How can the complexity of a DBMS be a disadvantage?
22. Why can a failure in a database environment be more serious than one in a file environment?
23. Why can recovery be more difficult in a database environment?

2

Data Models

OBJECTIVES

1. Introduce Premiere Products, the company that is used as a basis for many of the examples throughout the text.
2. Introduce the concept of a data model.
3. Describe the relational model.
4. Introduce the network and CODASYL models.
5. Introduce the hierarchical model.

2.1   INTRODUCTION

A database management system (DBMS) must furnish a method for storing and manipulating information about a variety of entities, the attributes of the entities, and the relationships between the entities. Several approaches can be taken by a DBMS to do this, and DBMSs are often categorized by the general approach that they take. This chapter discusses the general categories of DBMSs. These categories are usually called data models.

Since a DBMS must both store and manipulate data, both of these facets must be addressed in a data model. Thus, a **data model** has two components, usually called structure and operations. The **structure** refers to the way in which the system constructs, or structures, the data. More properly, it refers to the way in which the users *perceive* that the system structures data. It really doesn't matter what the DBMS does with the data behind the scenes; it just matters how it appears to the user. The **operations** are the facilities given to the users of the DBMS for the purpose of manipulating data within the database.

The vast majority of DBMSs follow one of three models: the relational model, the network model, or the hierarchical model. In this chapter, we will discuss the basic ideas behind these three models as well as the relative strengths and weaknesses of each one. In subsequent chapters, we will focus only on the relational model. There are two reasons for this. First, the main focus of the book is microcomputer DBMSs, which are almost exclusively relational. Second, even on mainframes, relational model systems are becoming more prevalent all the time. At the very least, systems that follow one of the other two models are typically adding relational characteristics to their systems.

Before introducing the three models, we will examine the requirements of a company called Premiere Products, which will be referred to in many examples throughout this chapter and in the rest of the text. After this examination is completed in section 2.2, we will move on to study the models. We will look at the basic concepts of the relational model in section 2.3, of the network model in section 2.4, and of the hierarchical model in section 2.5.

2.2 PREMIERE PRODUCTS

Premiere Products, a distributor of appliances and sporting goods, needs to maintain the following information:

1. For sales reps, it needs to store the sales rep's number, name, address, total commission, and commission rate.
2. For customers, it needs to store the customer's number, name, address, current balance, and credit limit, and the number of the sales rep who represents the customer.
3. For parts, it needs to store the part's number, its description, the units on hand, the item class, the number of the warehouse in which the item is stored, and the unit price.

Premiere Products also must store information on orders. A sample order is shown in Figure 2.1. Note that there are three parts to the order. The heading (top) of the order contains the order number; the date; the customer's number, name, and address; the sales rep number; and the sales rep name. The body of the order contains a number of *order lines*, sometimes called *line items*. Each order line contains a part number, a part description, the number of units of the part that were ordered, and the quoted price for the part. It also contains a total (usually called an *extension*) which is the product of the number ordered and the quoted price. Finally, the footing (bottom) of the order contains the order total. The additional items that Premiere Products must store with respect to orders are as follows.

Figure 2.1

Premiere Products
order

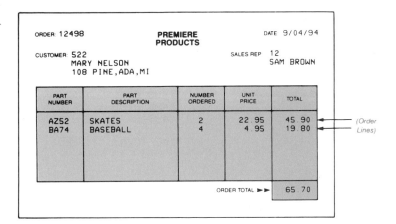

1. For the orders themselves, it needs to store the order number, the date the order was placed, and the number of the customer who placed the order. Note that the customer's name and address and the number of the sales rep who represents the customer are stored with customer information. The name of the sales rep is stored with sales rep information.
2. For each order line, it needs to store the order number, the part number, the number of units ordered, and the quoted price. The part description, you will recall, is stored with information on parts. The product of the number of units ordered and the quoted price is not stored, since it can easily be computed when needed.
3. The overall order total is not stored but will be computed when the order is produced.

Figure 2.2 shows sample data for *PREMIERE PRODUCTS*. We see that there are three sales reps whose numbers are 3, 6, and 12. The name of sales rep 3 is Mary Jones. Her address is 123 Main Street, Grant, MI. Her total commission is $2,150.00, and her commission rate is 5 percent (.05).

Figure 2.2

PREMIERE PRODUCTS sample data

SLSREP

SLSRNUMB	SLSRNAME	SLSRADDR	TOTCOMM	COMMRATE
3	MARY JONES	123 MAIN,GRANT,MI	2150.00	.05
6	WILLIAM SMITH	102 RAYMOND,ADA,MI	4912.50	.07
12	SAM BROWN	419 HARPER,LANSING,MI	2150.00	.05

CUSTOMER

CUSTNUMB	CUSTNAME	ADDRESS	BALANCE	CREDLIM	SLSRNUMB
124	SALLY ADAMS	481 OAK,LANSING,MI	418.75	500	3
256	ANN SAMUELS	215 PETE,GRANT,MI	10.75	800	6
311	DON CHARLES	48 COLLEGE,IRA,MI	200.10	300	12
315	TOM DANIELS	914 CHERRY,KENT,MI	320.75	300	6
405	AL WILLIAMS	519 WATSON,GRANT,MI	201.75	800	12
412	SALLY ADAMS	16 ELM,LANSING,MI	908.75	1000	3
522	MARY NELSON	108 PINE,ADA,MI	49.50	800	12
567	JOE BAKER	808 RIDGE,HARPER,MI	201.20	300	6
587	JUDY ROBERTS	512 PINE,ADA,MI	57.75	500	6
622	DAN MARTIN	419 CHIP,GRANT,MI	575.50	500	3

ORDERS

ORDNUMB	ORDDTE	CUSTNUMB
12489	90294	124
12491	90294	311
12494	90494	315
12495	90494	256
12498	90594	522
12500	90594	124
12504	90594	522

ORDLNE

ORDNUMB	PARTNUMB	NUMBORD	QUOTPRCE
12489	AX12	11	14.95
12491	BT04	1	402.99
12491	BZ66	1	311.95
12494	CB03	4	175.00
12495	CX11	2	57.95
12498	AZ52	2	22.95
12498	BA74	4	4.95
12500	BT04	1	402.99
12504	CZ81	2	108.99

PART

PARTNUMB	PARTDESC	UNONHAND	ITEMCLSS	WREHSENM	UNITPRCE
AX12	IRON	104	HW	3	17.95
AZ52	SKATES	20	SG	2	24.95
BA74	BASEBALL	40	SG	1	4.95
BH22	TOASTER	95	HW	3	34.95
BT04	STOVE	11	AP	2	402.99
BZ66	WASHER	52	AP	3	311.95
CA14	SKILLET	2	HW	3	19.95
CB03	BIKE	44	SG	1	187.50
CX11	MIXER	112	HW	3	57.95
CZ81	WEIGHTS	208	SG	2	108.99

We also see that there are ten customers, numbered 124, 256, 311, 315, 405, 412, 522, 567, 587, and 622. The name of customer 124 is Sally Adams. Her address is 481

Oak Street, Lansing, MI. Her current balance is $418.75, and her credit limit is $500. The number 3 in the column entitled *SLSRNUMB* indicates that Sally is represented by sales rep 3 (Mary Jones).

Skipping down for a moment to the table labeled *PART*, we see that there are ten parts, whose part numbers are AX12, AZ52, BA74, BH22, BT04, BZ66, CA14, CB03, CX11, and CZ81. Part AX12 is an iron, and the company has 104 units of this part on hand. These parts are in item class HW (housewares) and are stored in warehouse 3. The price of an iron is $17.95.

Moving back up to the table labeled *ORDER*, we see that there are seven orders, numbered 12489, 12491, 12494, 12495, 12498, 12500, and 12504. Order 12489 was placed on September 2, 1994, by customer 124 (Sally Adams).

The table labeled *ORDLNE* may seem strange at first glance. Why do we need a separate table for the order lines? Couldn't they somehow be included in the *ORDER* table? The answer is yes, they could. The table *ORDER* could potentially be structured in the manner shown in Figure 2.3. Examining this table, we see that the same orders as those shown in Figure 2.2 are present, with the same dates and the same customer numbers. In addition, each row contains all of the order lines for a given order. Examining the fifth row, for example, we see that order 12498 has two order lines. One of these order lines is for two AZ52's at $22.95 each; the other is for four BA74's at $4.95 each.

Figure 2.3

ORDER and *ORDLNE* combined

ORDERS

ORDNUMB	ORDDTE	CUSTNUMB	PARTNUMB	NUMBORD	QUOTPRCE
12489	90294	124	AX12	11	14.95
12491	90294	311	BT04	1	402.99
			BZ66	1	311.95
12494	90494	315	CB03	4	175.00
12495	90494	256	CX11	2	57.95
12498	90594	522	AZ52	2	22.95
			BA74	4	4.95
12500	90594	124	BT04	1	402.99
12504	90594	522	CZ81	2	108.99

Q & A

Question:

How is the same information represented in Figure 2.2?

Answer:

Take a look at the table in Figure 2.2 labeled *ORDLNE* and examine the sixth and seventh rows. The sixth row indicates that there is an order line on order 12498 for two AZ52's at $22.95 each. The seventh row indicates that there is an order line on order 12498 for four BA74's at $4.95 each. Thus, the same information that we find in Figure 2.3 is represented here, although in two separate rows rather than one.

It may seem that it would be better not to take two rows to represent the same information that can be represented in one row. There is a problem with the arrangement shown in Figure 2.3, however: the table is more complicated. In Figure 2.2, there is a single entry at each location in the table. In Figure 2.3, some of the individual positions within the table contain multiple entries, and, further, there is a correspondence between these entries (in the row for order 12498, it is crucial to know that the AZ52 corresponds to the 2 in the *NUMBORD* column, not the 4, and the 22.95 in the *QUOTPRCE* column, not the 4.95). There are practical issues to worry about, such as:

1. How much room do we allow for these multiple entries?
2. What if an order has more order lines than we have allowed room for?
3. Given a part, how do we determine which orders contain order lines for that part?

Certainly, none of these problems is unsolvable. They do add a level of complexity, however, that is not present in the arrangement shown in Figure 2.2. In the structure shown in that figure, there are no multiple entries to worry about; it does not matter how many order lines exist for any order; and finding all of the orders that contain order lines for a given part is easy (just look for all order lines with the given part number in the *PARTNUMB* column). In general, this simpler structure is preferable, and that is why order lines have been placed in a separate table.

To test your understanding of the *PREMIERE PRODUCTS* data, answer the following questions, using the data in Figure 2.2.

Q & A

Question: Give the numbers of all the customers represented by MARY JONES.

Answer: 124, 412, and 622. (Look up the number of MARY JONES in the *SLSREP* table and obtain the number 3. Then find all customers in the *CUSTOMER* table that have the number 3 in the *SLSRNUMB* column.)

Question: Give the name of the customer who placed order 12491, then give the name of the sales rep who represents this customer.

Answer: DON CHARLES, SAM BROWN. (Look up the customer number in the *ORDER* table and obtain the number 311. Then find the customer in the *CUSTOMER* table who has customer number 311. Using this customer's sales rep number, which is 12, find the name of the sales rep in the *SLSREP* table.)

Question: List all of the parts that appear on order 12491. For each part, give the description, number ordered, and quoted price.

Answer: *PARTNUMB*: BZ66, *PARTDESC*: WASHER, *NUMBORD*: 1, *QUOTPRCE*: 311.95. Also *PARTNUMB*: BT04, *PARTDESC*: STOVE, *NUMBORD*: 1, *QUOTPRCE*: 402.99. (Look up each *ORDLNE* table row in which the order number is 12491. Each of these rows contains a part number, the number ordered, and the quoted price. The only thing missing is the description of the part. Use the part number to look up the corresponding description in the *PART* table.)

Q & A

Question: Why is the column *QUOTPRCE* a part of the *ORDLNE* table? Can't we just take the part number and look up the price in the *PART* table?

Answer: If we do not have the *QUOTPRCE* column in the *ORDLNE* table, the price for a part on an order line must be obtained by looking up the price in the *PART* table. While this may not be bad, it does prevent Premiere Products from charging different prices to different customers for the same part. Since Premiere Products wants the flexibility to quote different prices to different customers, we include the *QUOTPRCE* column in the *ORDLNE* table. If you examine the *ORDLNE* table, you will see cases in which the quoted price matches the actual price in the *PART* table and cases in which it differs.

2.3   THE RELATIONAL MODEL

Structure Within the Relational Model

You have actually already seen a relational database. A relational database is essentially just a collection of tables like the ones we just looked at for Premiere Products in Figure 2.2. A **relational model** database is perceived by the user as being just such a collection. (Notice again the phrase "perceived by the user," indicating as before that what matters is how things appear to the user, not what is taking place behind the scenes.) You might wonder why this model is not called the "table" model, or something along that line if a database is a collection of tables. Formally, these **tables** are called **relations**, and this is where the model gets its name.

How does a DBMS that follows the relational model handle entities, attributes of entities, and relationships between entities? Entities and attributes are fairly simple. Each entity gets a table of its own. Thus, in the *PREMIERE PRODUCTS* database, there is a table for sales reps, a separate table for customers, and so on. The attributes of an entity become the columns in the table. In the table for sales reps, for example, there is a column for the sales rep number, a column for the sales rep name, and so on.

What about relationships? At Premiere Products there is a one-to-many relationship between sales reps and customers (each sales rep is related to the many customers he or she represents, and each customer is represented to the one sales rep who represents the customer). How is this relationship implemented in a relational model database? The answer is through common columns in two or more tables. Consider again Figure 2.2. The column *SLSRNUMB* of the sales rep table and the column *SLSRNUMB* of the customer table are used to implement the relationship between sales reps and customers; that is, given a sales rep, we can use these columns to determine all of the customers he or she represents, and, given a customer, we can use these columns to find the sales rep who represents the customer.

Let us now attempt to be a little more precise in our description of a relation. As we discussed, a relation is essentially just a two-dimensional table. If we consider the tables in Figure 2.2, however, we can see that there are certain restrictions we would probably want to place on relations. Each column should have a unique name, and entries within each column should all "match" this column name; that is, if the column name is *CREDLIM*, all entries in that column should in fact be credit limits. Also, each row should be unique. After all, if two rows are absolutely identical, the second row doesn't give us any information that we don't already have. In addition, for maximum flexibility, the ordering of the columns and the rows should be immaterial. Finally, the table will be simplest if each position is restricted to a single entry, that is, if we do not allow multiple entries (often called **repeating groups**) in an individual location in the table. These ideas lead to the following definitions:

Definition: A **relation** is a two-dimensional table in which

1. the entries in the table are single-valued, that is, each location in the table contains a single entry;
2. each column has a distinct name (technically called the attribute name);
3. all of the values in a column are values of the same attribute (that is, all entries must match the column name);
4. the order of columns is immaterial;
5. each row is distinct; and
6. the order of rows is immaterial.

Definition: A **relational database** is a collection of relations.

Note: Later in the text, we will encounter situations in which a structure satisfies all the properties of a relation *except for property 1;* that is, some of the entries contain repeating groups and thus are not single-valued. Such a structure is called an **unnormalized relation**. This jargon is certainly a little strange, in that an unnormalized relation is thus not a relation at all. It is the term that is used for such a structure, however. The table shown in Figure 2.3 is an example of an unnormalized relation.

Each row of a relation is technically called a **tuple**, and each column is technically called an **attribute**. Thus, we have two different sets of terms: relation, tuple, attribute, and table, row, column. There is even a third set. The table could be viewed as a file (in fact, this is how relational databases are often stored, with each relation in a separate file). In this case, we would call the rows records and the columns fields. We now have *three* different sets of terms! Their correspondence is shown below.

Formal Terms	Alternative One	Alternative Two
relation	table	file
tuple	row	record
attribute	column	field

Of these three sets of choices, the one that is becoming the most popular is alternative one: tables, rows, and columns. One reason for its popularity is that it seems the most natural to the nontechnical user. A second reason is that many (if not most) of the commercial relational DBMSs use this set of terms. All the terms are often used interchangeably. In this text, we will use tables, rows, and columns.

It would be nice to have a concise way of indicating the tables and columns in a relational database without having to draw the tables themselves as we did in Figure 2.2. Perhaps we could draw some empty tables, such as those shown in Figure 2.4. This seems rather cumbersome, however. Fortunately, there is a commonly accepted shorthand representation of the structure of a relational database. We merely write the name of the table and then within parentheses list all of the columns in the table. Thus, this sample database consists of:

```
SLSREP (SLSRNUMB, SLSRNAME, SLSRADDR,
           TOTCOMM, COMMRATE)

CUSTOMER (CUSTNUMB, NAME, ADDRESS, BALANCE,
           CREDLIM, SLSRNUMB)

ORDER (ORDNUMB, DATE, CUSTNUMB)

ORDLNE (ORDNUMB, PARTNUMB, NUMBORD,
           QUOTPRCE)

PART (PARTNUMB, PARTDESC, UNONHAND,
           ITEMCLSS, WREHSENM, UNITPRCE)
```

Figure 2.4

PREMIERE PRODUCTS sample structure

SLSREP

SLSRNUMB	SLSRNAME	SLSRADDR	TOTCOMM	COMMRATE

CUSTOMER

CUSTNUMB	CUSTNAME	ADDRESS	BALANCE	CREDLIM	SLSRNUMB

ORDERS

ORDNUMB	ORDDTE	CUSTNUMB

ORDLNE

ORDNUMB	PARTNUMB	NUMBORD	QUOTPRCE

PART

PARTNUMB	PARTDESC	UNONHAND	ITEMCLSS	WREHSENM	UNITPRCE

Notice that there is some duplication of names. The column *SLSRNUMB* appears in *both* the *SLSREP* table *and* the *CUSTOMER* table. Suppose a situation existed wherein the two might be confused. If we merely wrote *SLSRNUMB*, how would the computer know which *SLSRNUMB* we meant? How would a person looking at what we had written know which one we meant, for that matter? We need a mechanism for indicating the one to which we are referring. One common approach to this problem is to write both the table name and the column name, separated by a period. Thus, the *SLSRNUMB* in the *CUSTOMER* table would be written *CUSTOMER.SLSRNUMB*, whereas the *SLSRNUMB* in the *SLSREP* table would be written *SLSREP.SLSRNUMB*. Technically, when we do this we say that we **qualify** the names. It is *always* acceptable to qualify data names, even if there is no possibility of confusion. If confusion may arise, however, it is *essential* to do so.

There is one other important topic to discuss before we leave the structure part of the relational model. The **primary key** of a table (relation) is the column or collection of columns that uniquely identifies a given row. In the *SLSREP* table, for example, the sales rep's number uniquely identifies a given row. (Sales rep 6 occurs in only one row of the table, for instance.) Thus, *SLSRNUMB* is the primary key. Primary keys are typically indicated in the shorthand representation by underlining the column or collection of columns that comprises the primary key. Thus, the complete shorthand representation for the *PREMIERE PRODUCTS* database would be:

```
SLSREP (SLSRNUMB, SLSRNAME, SLSRADDR,
               TOTCOMM, COMMRATE)

CUSTOMER (CUSTNUMB, NAME, ADDRESS, BALANCE,
               CREDLIM, SLSRNUMB)

ORDER (ORDNUMB, DATE, CUSTNUMB)

ORDLNE (ORDNUMB, PARTNUMB, NUMBORD,
               QUOTPRCE)

PART (PARTNUMB, PARTDESC, UNONHAND,
               ITEMCLSS, WREHSENM, UNITPRCE)
```

Q & A

Question: Why does the primary key to the *ORDLNE* table consist of two columns, not just one?

Answer: No single column uniquely identifies a given row. It requires two: *ORDNUMB* and *PARTNUMB*.

Operations Within the Relational Model

There are many approaches to manipulating a relational database. One of the most prevalent is a language called **SQL** (**Structured Query Language**), which was developed by IBM. The basic form of an SQL command is simply

```
SELECT ...
     FROM ...
     WHERE ...
```

We list the columns that we wish to see printed after the word SELECT. After the word FROM, we list all tables that contain these columns. Finally, we list any restrictions to be applied after the word WHERE. For example, if we wanted to print the name of the customer whose number was 124, we would type:

```
SELECT NAME
     FROM CUSTOMER
     WHERE CUSTNUMB = 124
```

and the computer would respond with:

```
NAME
----------
SALLY ADAMS
```

Suppose we wish to see the number and name of each customer together with the number and name of the sales rep who represents the customer. This task involves the use of the two tables, *CUSTOMER* and *SLSREP*, together with the relationship between them. In the SELECT statement, we list all of the columns that we wish included on the report, *CUSTNUMB*, *NAME*, *SLSRNUMB*, and *SLSRNAME*. We have a slight problem, however. There is a *SLSRNUMB* column in both tables, so we must indicate which one we want by qualifying *SLSRNUMB*. (Actually, in this case, since we are looking for combinations of customers and sales reps where the sales rep numbers match, the value of *CUSTOMER.SLSRNUMB* and the value of *SLSREP.SLSRNUMB* will be the same. Thus, it would seem that we shouldn't have to bother with qualifying *SLSRNUMB*. The computer doesn't know that the two will be the same, however, and thus will insist that *SLSRNUMB* be qualified. We would get an error message if we did not do so.) The SELECT clause is thus:

```
SELECT CUSTNUMB, NAME, SLSREP.SLSRNUMB, SLSRNAME
```

After the word FROM, we list both of the tables involved, *CUSTOMER* and *SLSREP*. The FROM clause is:

```
FROM CUSTOMER, SLSREP
```

Finally, we list any restrictions in the WHERE clause. If we did not list any restrictions in the query, we would get *all possible combinations of customers and sales reps*. Therefore, we need a condition that will restrict the output to only those combinations which *match*, that is, the combinations in which *CUSTOMER.SLSRNUMB = SLSREP.SLSRNUMB*. The WHERE clause to accomplish this is:

```
WHERE CUSTOMER.SLSRNUMB = SLSREP.SLSRNUMB
```

The complete query is:

```
SELECT CUSTNUMB, NAME, SLSREP.SLSRNUMB, SLSRNAME
    FROM CUSTOMER, SLSREP
    WHERE CUSTOMER.SLSRNUMB = SLSREP.SLSRNUMB
```

to which the computer would respond:

CUSTNUMB	NAME	SLSRNUMB	SLSRNAME
124	SALLY ADAMS	3	MARY JONES
256	ANN SAMUELS	6	WILLIAM SMITH
311	DON CHARLES	12	SAM BROWN
315	TOM DANIELS	6	WILLIAM SMITH
405	AL WILLIAMS	12	SAM BROWN
412	SALLY ADAMS	3	MARY JONES
522	MARY NELSON	12	SAM BROWN
567	JOE BAKER	6	WILLIAM SMITH
587	JUDY ROBERTS	6	WILLIAM SMITH
622	DAN MARTIN	3	MARY JONES

Advantages and Disadvantages

One big advantage of relational model systems is that they are generally easier to use than systems that follow the other models. A variety of methods are available for manipulating relational databases. These methods are simpler than the ones available for manipulating network and hierarchical model systems. Users of relational systems do not need the in-depth knowledge of the underlying database structure which is required in these other systems.

Another advantage concerns data independence. Data independence, which is the ability to make changes in the database structure without having to make changes in programs that access the database, is one of the advantages of all types of DBMSs. Relational model systems offer a degree of data independence that is much higher than that of network or hierarchical systems. A far wider range of types of changes can be made without affecting programs that access the database. In addition, even those types of changes which can be made using network or hierarchical systems can usually be made much more easily in relational systems.

At the present time, relational systems have two problems, both of which should be corrected in the not-too-distant future. The first problem concerns efficiency. In spite of claims to the contrary, current relational model systems are not as efficient as some of the top network and hierarchical systems. Basically, this means that relational model systems may not be appropriate for developing some large-scale application systems. When there is a great deal of activity involving a very large database, a relational model system may not be able to provide the required performance.

You will probably be surprised at the second problem. It concerns **integrity**, which is the property of ensuring that the data in the database satisfies certain restrictions. These restrictions, which are often called **integrity constraints**, could include such things as the following:

- No two sales reps can have the same number.
- A credit limit must be $300, $500, $800, or $1000.
- The sales rep number on any row in the *CUSTOMER* table must be the number of a sales rep who actually exists, that is, who appears on some row in the *SLSREP* table.

The problem is that many current relational model systems *do not contain any facilities to enforce such integrity constraints.*

For additional perspectives on the basic relational model, see [8], [9], [11], [12], [13], [14], and [15].

2.4   *THE NETWORK MODEL*

Structure Within the Network Model

A **network model** database is perceived by the user as a collection of record types (which represent the entities), fields within these record types (which represent the attributes), and *explicit* relationships between these record types. Such a structure is called a **network**, and it is from this term that the model takes its name. The fact that the relationships are explicit distinguishes the network model from the relational model, in which relationships are *implicit* (derived from matching columns in the tables).

The CODASYL approach, also called the **CODASYL model**, falls within the general network model. While some systems that are network model systems do not follow the CODASYL model, they are the exception. To many people, the CODASYL model and the network model are synonymous, and that is how we will treat them here. In fact, in the rest of the discussion of the network model, we will use CODASYL terminology.

Consider the database in Figure 2.5, for example. The rectangles represent the record types in the database. There is one for each of the entities: sales reps, customers, orders, parts, and order lines.

Figure 2.5

PREMIERE PRODUCTS network database structure

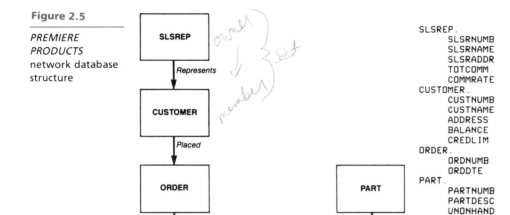

```
SLSREP.
    SLSRNUMB
    SLSRNAME
    SLSRADDR
    TOTCOMM
    COMMRATE
CUSTOMER.
    CUSTNUMB
    CUSTNAME
    ADDRESS
    BALANCE
    CREDLIM
ORDER.
    ORDNUMB
    ORDDTE
PART.
    PARTNUMB
    PARTDESC
    UNONHAND
    ITEMCLSS
    WREHSENM
    UNITPRCE
ORDLNE.
    NUMBORD
    QUOTPRCE
```

The arrows represent the relationships. In particular, they represent the familiar one-to-many relationships. The arrow goes from the "one" part of the relationship, called the **owner**, to the "many" part, called the **member**. Since one sales rep is related to many customers but each customer is related to exactly one sales rep, we have a one-to-many relationship, and thus an arrow, from *SLSREP* to *CUSTOMER*. In this relationship, *SLSREP* is the owner and *CUSTOMER* is the member. There is a term for the relationship in CODASYL systems; it is called a **set**. (Note that this use of the term differs from the one usually associated with it in mathematics.)

Sets are given names. The set from *SLSREP* to *CUSTOMER* is called *REPRESENTS*, indicative of the fact that a sales rep "represents" a customer. The set from *CUSTOMER* to *ORDER* is called *PLACED*, to indicate that a customer "placed" a number of orders. Similarly, an order "contains" order lines and a part "is on" a number of order lines, so the set from *ORDER* to *ORDLNE* is called *CONTAINS* and the set from *PART* to *ORDLNE* is called *IS_ON*.

Operations Within the Network Model

As we will see in the following examples, we manipulate a network model database essentially by *following the arrows*. Arrows may be followed in either direction. Suppose we locate a given sales rep, for example. If we follow the arrow from *SLSREP* to *CUSTOMER*, we obtain the many customers represented by this sales rep. If we start with a customer, however, and follow the same arrow only in the reverse direction, we obtain the single sales rep who represents the given customer, that is, the single owner.

The sample data shown in Figure 2.6 illustrates these processes. This figure shows three sales reps and ten customers. The relationship between them is indicated by the lines in the figure. Thus, sales rep 3 represents customers 124, 412, and 622; sales rep 6 represents customers 256, 315, 567, and 587; sales rep 12 represents customers 311, 405, and 522. (While we will not go into the manner in which these relationships are actually implemented, this is a reasonable way to picture it.) The following three examples demonstrate the process of manipulating a network model database.

Example 1: Given the data shown in Figure 2.6, print a list of all of the customers of sales rep 12.

We first ask the DBMS to FIND sales rep 12. Assuming there is such a sales rep in the database, we then repeatedly ask to FIND the NEXT customer within the collection of customers represented by this sales rep, until reaching the end of the list of such customers. (The exact command would be *FIND NEXT WITHIN REPRESENTS*.) The first time we made the request, we would obtain customer 311; the second time, we would obtain customer 405; the next time, we would obtain customer 522. When we tried to find another customer related to this sales rep, the DBMS would indicate that there were no more. At this point, our task would be complete.

Figure 2.6

Sales reps, customers, and the relationship between them

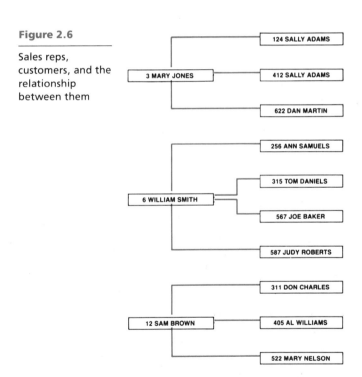

Example 2: Find the sales rep who represents customer 315.

We first ask the DBMS to FIND customer 315. Having accomplished this, we then ask the DBMS to FIND the OWNER of customer 315 in the relationship between sales reps and customers. The owner is a *single* sales rep, in this example, sales rep 6. (The exact command is *FIND OWNER WITHIN REPRESENTS*.)

Note that we find the sales rep who represents a customer by following the arrow, not by looking at a sales rep number in a customer record, as we would if we were using the relational model. Therefore, in a network structure, there is *no need for such a field*.

Some processing involves following more than a single arrow. Consider Figure 2.7, for example. This figure contains the same data and relationships that were shown in Figure 2.6. Additionally, it contains orders, together with the relationships between customers and orders. Thus we see that customer 124 placed orders 12489 and 12500; customer 256 placed order 12495; and so on.

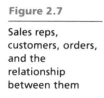

Figure 2.7

Sales reps, customers, orders, and the relationship between them

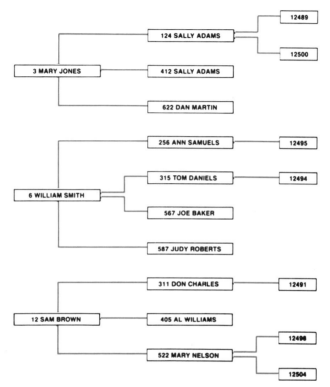

Example 3: List all the orders placed by customers of sales rep 12.

As in Example 1, we first ask the DBMS to FIND sales rep 12. We also repeatedly ask to FIND the NEXT customer within the collection of customers represented by this sales rep (*FIND NEXT WITHIN REPRESENTS*) until reaching the end of the list of such customers. The difference here is that once we have found a customer, we repeatedly ask the DBMS to FIND the NEXT order owned by the customer (*FIND NEXT WITHIN PLACED*). We do this until reaching the end of the list of orders placed by the indicated customer. Only at this point do we proceed on to the next customer owned by the sales rep. In our example, after finding customer 311, we would obtain the single order 12491. Since this is the only order for this customer, we would then be able to proceed to customer 405. Since this customer has no orders, we would then be able to proceed on to customer 522. At this point, we would find the first order for this customer, order 12498. After displaying details about this order, we would move on to the next order placed by this customer, order 12504. Since this is the last order for this customer, once we had displayed details about this order, we would be ready to move on to the next customer. There are, however, no more customers, and thus the whole process would then be complete.

Question: How would you handle the following query: List the number and name of the sales rep who represents customer 124. In addition, list the numbers of all orders placed by this customer.

Answer: Ask the DBMS to FIND customer 124. Assuming this step is successfully completed, ask the DBMS to FIND the OWNER of this customer in the relationship between sales reps and customers (*FIND OWNER WITHIN REPRESENTS*). Finally, ask the DBMS to repeatedly FIND the NEXT order owned by this customer (*FIND NEXT WITHIN PLACED*).

Advantages and Disadvantages

The biggest advantage of network model systems is efficiency. Network model DBMSs are known for their ability to handle huge databases with large amounts of activity. Network model systems have also existed for some time now, and a great deal of software has been developed which uses them. Network systems also provide some of the types of integrity support which are lacking in current relational systems. For example, in network systems, it is easy to ensure that each sales rep in the database will have a different number; if sales rep 3 is already in the database, an attempt to add a second sales rep 3 will be reflected by the DBMS. Similarly, a network system guarantees that the data for a sales rep will be placed in the database before any customer he or she represents. Thus, we know that if Dan Martin is assigned to Mary Jones, who is sales rep 3, he will not be placed in the database unless she has already been placed there.

Network systems, on the other hand, are more difficult to use than relational systems. Getting the desired information out of a network database involves following the proper sequence of arrows in the correct order. We need to be very familiar with the underlying database structure in order to determine how best to work our way through the database in order to satisfy a given requirement in an efficient manner. If we do not do this correctly, we will not obtain the additional efficiency the network model can provide.

It is also more difficult to make changes in the database structure in a network system than in a relational system. Many types of changes that could be made in relational systems without affecting application programs cannot be made in any easy way in network systems, and, if they are made, changes are required in all of the programs that access the affected part of the database.

For additional perspectives on the general network model and the CODASYL model, see [9], [10], [11], [13], [14], and [15].

2.5   THE HIERARCHICAL MODEL

Structure and Operations Within the Hierarchical Model

Where the structure used in the network model is the network, the structure used in the **hierarchical model** is the **hierarchy**, or **tree**. A tree can be thought of as a network with an added restriction: no box can have more than one arrow entering it. (It doesn't matter how many arrows leave a box.) A tree is thus a more restrictive structure than a network.

Some of the terminology used for the two models differs as well. Rather than the "one" and "many" ends of a relationship being called *owners* and *members*, as they are in the network model, they are called **parents** and **children** in the hierarchical model.

There is no special name for a relationship in the hierarchical model; that is, there is no term that is analogous to the term *set* in the network model. A different collection of commands is used to manipulate hierarchical model databases, although the basic ideas are similar. There are differences in the underlying strategies used by hierarchical and network DBMSs to store and manipulate the database. Perhaps the biggest difference, however, arises when the database to be implemented is not a tree.

Consider the network shown in Figure 2.8, for example. In this network, a box called *PAYMENT* has been added to the diagram shown in Figure 2.5. This new box will be used to keep track of the payments made by each customer. There is a one-to-many relationship between *CUSTOMER* and *PAYMENT*, which is represented by an arrow from *CUSTOMER* to *PAYMENT*. Thus, in this new diagram, there are two arrows *leaving* the *CUSTOMER* box. There are also two arrows *entering* the *ORDLNE* box. The two arrows leaving the *CUSTOMER* box do not prevent this structure from being a tree. The fact that there is a box, namely *ORDLNE*, with two arrows entering it prevents it from being a tree. The structure shown in Figure 2.8, which is thus not a tree, *cannot* be implemented directly by means of a hierarchical system.

Figure 2.8

PREMIERE PRODUCTS network database structure with additional record type (*PAYMENT*)

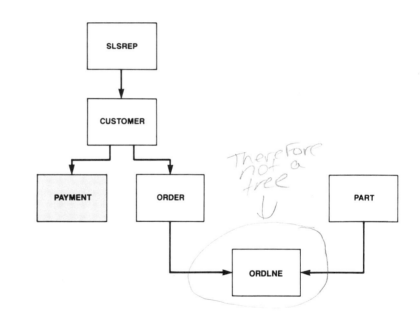

Fortunately, the structure can be implemented, although not in quite as appealing a fashion as is possible in a network system. Although many of the issues involved in such an implementation are rather tricky and are beyond the scope of our discussion, the basic idea is relatively simple. The idea is to create two separate trees (each called a **physical database**), and to create a relationship in which the parent is in one of the physical databases and the child is in the other. Such a relationship is called a **logical child relationship**.

Figure 2.9 shows how this technique could be applied to the structure of Figure 2.8. One physical database contains sales reps, customers, payments, orders, and order lines, as well as all of the relationships between them. The other contains only parts. In addition, the relationship between parts and order lines is implemented as a logical child relationship, in which the *PART* record is the parent and the *ORDLNE* record is the child.

Advantages and Disadvantages

Relative to relational model systems, hierarchical systems offer the same advantages and suffer from the same disadvantages that network systems do. In many ways, hierarchical systems and network systems are comparable. If the underlying database design is really a tree, then a hierarchical model system may outperform a network system. On the other hand, if the underlying structure is not a tree, the reverse may very well be true. In general, though, the systems are quite comparable, so the advantages and disadvantages discussed for the network model apply equally well here.

Figure 2.9

PREMIERE PRODUCTS database split into two separate physical databases with a logical child relationship between them

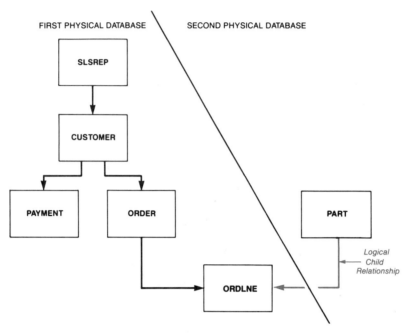

For additional perspectives on the hierarchical model, see [9], [10], [11], [13], [14], and [15].

SUMMARY

1. DBMSs are categorized by the data model that they follow. A data model has two components:
   a. structure (the way the system structures the data)
   b. operations (the facilities provided to users for manipulating data in the database)
2. The three main data models are as follows:
   a. the relational model
   b. the network model
   c. the hierarchical model
3. Premiere Products is an organization whose requirements include the following entities:
   a. sales reps   d. parts
   b. customers   e. order lines
   c. orders
4. A relation is a two-dimensional table in which
   a. the entries are single-valued;
   b. each column has a distinct name;
   c. all of the values in a column are values of the same attribute (the one identified by the column name);
   d. the order of columns is immaterial;
   e. each row is distinct; and
   f. the order of rows is immaterial.

5. A relational database is a collection of relations.

6. An unnormalized relation is a structure in which entries need not be single-valued but which satisfies all of the other properties of a relation.

7. The terms relation, tuple, and attribute correspond to the terms table, row, and column, respectively.

8. Relation, tuple, and attribute also correspond to file, record, and field, respectively.

9. A column name is qualified by preceding it with the table name and a period, for example, *SLSREP.SLSRNUMB*.

10. The primary key is the column or columns that uniquely identify a given row within the table.

11. SQL is a language used to manipulate relational databases. The basic form of an SQL command is SELECT-FROM-WHERE.

12. The advantages of relational model systems are
   a. ease of use and
   b. a high degree of data independence.

13. The disadvantages of relational model systems are that
   a. they are less efficient than systems that follow the other models, and
   b. they are weak in integrity support.

14. A network model database is a collection of record types and explicit one-to-many relationships between these record types.

15. The CODASYL model is a subset of the network model, in which
   a. relationships are called sets;
   b. the "one" part of the set is called the owner; and
   c. the "many" part is called the member.

16. The advantages of network model systems are that
   a. they tend to be more efficient than relational systems, and
   b. they provide better integrity support than relational systems.

17. The disadvantages of network model systems are that
   a. they tend to be more difficult to use than relational systems, and
   b. they provide less data independence than relational systems.

18. The structure in the hierarchical model is the hierarchy.

19. A hierarchy (also called a tree structure) is a network structure with an added restriction: no box can have more than one arrow entering it.

21. If a network that is *not* a tree is to be implemented using a hierarchy system, the network is split into trees with logical child relationships between them.

22. Relative to relational model systems, hierarchical systems have the same advantages and disadvantages that the network systems do.

23. If the structure to be implemented is actually a tree, a hierarchical system may outperform a network system. If it is not a tree, the reverse may very well be the case.

KEY TERMS

children	network model	relational model
CODASYL model	owner	repeating group
data model	parent	row
hierarchical model	physical database	set
hierarchy	primary key	table
logical child	qualification	tree
member	relation	tuple
network	relational database	unnormalized relation

EXERCISES

1. What is a data model?
2. What are the components of a data model?
3. What are the three main data models?
4. Using the data for *PREMIERE PRODUCTS* as shown in Figure 2.2, give an answer for each of the following problems.
    a. Give the names of all customers represented by William Smith.
    b. How many customers have a balance that is over their credit limit?
    c. Which sales reps represent customers whose balance is over their credit limit?
    d. Which customers placed orders on 9/05/94?
    e. Which sales reps represent any customers who placed orders on 9/05/94?
    f. Which customers currently have a STOVE on order?
    g. List the number and description of all parts that are currently on order by any customer represented by William Smith.
5. Why are order lines in the *PREMIERE PRODUCTS* database in a separate table rather than being part of the *ORDERS* table?
6. What is a relation?
7. What is a relational database?
8. What is an unnormalized relation? Is it a relation according to the definition of the word *relation*?
9. What is a tuple? What is a more common name for it?
10. How is the term *attribute* used in the relational model? What is a more common name for it?
11. Describe the shorthand representation of the structure of a relational database. Illustrate this technique by representing Henry's database as shown in Figures 1.1, 1.2, and 1.3.
12. What does it mean to qualify the name of a column? How is this done?
13. What is a primary key? What is the primary key for each of the tables in Henry's database? (See Exercise 11.)
14. What is SQL? What is it used for?
15. What are the advantages of the relational model? What are the disadvantages?
16. What is a network?
17. What is the relationship between the network model and the CODASYL model?
18. What is a set? An owner? A member?
19. Why might there be columns in a relational model table for which there are no corresponding fields in a network model record?
20. How would you handle the following query in a CODASYL model database for *PREMIERE PRODUCTS*: Print a list of all orders placed by customers of sales rep 6.
21. What are the advantages of the network model compared to the relational model? What are the disadvantages?
22. What is another name for hierarchy? What is the difference between a hierarchy and a network?
23. Describe briefly how a network that is not a hierarchy can be implemented by means of a hierarchical model DBMS.
24. What are the advantages of the hierarchical model as compared to the other two models? What are the disadvantages?

The Relational Model: Data Definition and Manipulation

OBJECTIVES

1. Present the language SQL and illustrate data definition and manipulation in SQL.
2. Describe the relational algebra and demonstrate the select, project, and join operations.
3. Present QBE (Query-by-Example) and illustrate its use in retrieving data.
4. Demonstrate the concept of a natural language through the use of a hypothetical natural language called "NL".

3.1 INTRODUCTION

In the last chapter, we discussed the basic structure of the relational model. We saw that data is stored in the form of tables. Each entity in the database has a table of its own. The attributes (properties) of the entity are the columns in the table. Relationships between entities are affected through common columns.

In this chapter, we examine a variety of ways of manipulating data in a relational database. In section 3.2, we discuss the language SQL (Structured Query Language). Originally developed by IBM under the name SEQUEL, SQL is perhaps the most important language designed to manipulate relational databases. SQL is not only the language used in DB2, IBM's mainframe relational DBMS; it is also used in many other DBMSs as well, both on mainframes and on microcomputers. A myriad of systems use it already, and there are constant rumors about other systems furnishing this support in the not-too-distant future.

In section 3.3, we will study a very visual approach to the process, called QBE (Query-by-Example). In section 3.4, we will investigate the relational algebra, one of the first methods proposed for retrieving data from relational databases. Finally, in section 3.5, we discuss the use of natural languages to retrieve data from relational databases. The examples used throughout this chapter all refer to the PREMIERE PRODUCTS database (see Figure 3.1 on the next page).

Figure 3.1

*PREMIERE
PRODUCTS*
sample data

SLSREP

SLSRNUMB	SLSRNAME	SLSRADDR	TOTCOMM	COMMRATE
3	MARY JONES	123 MAIN,GRANT,MI	2150.00	.05
6	WILLIAM SMITH	102 RAYMOND,ADA,MI	4912.50	.07
12	SAM BROWN	419 HARPER,LANSING,MI	2150.00	.05

CUSTOMER

CUSTNUMB	CUSTNAME	ADDRESS	BALANCE	CREDLIM	SLSRNUMB
124	SALLY ADAMS	481 OAK,LANSING,MI	418.75	500	3
256	ANN SAMUELS	215 PETE,GRANT,MI	10.75	800	6
311	DON CHARLES	48 COLLEGE,IRA,MI	200.10	300	12
315	TOM DANIELS	914 CHERRY,KENT,MI	320.75	300	6
405	AL WILLIAMS	519 WATSON,GRANT,MI	201.75	800	12
412	SALLY ADAMS	16 ELM,LANSING,MI	908.75	1000	3
522	MARY NELSON	108 PINE,ADA,MI	49.50	800	12
567	JOE BAKER	808 RIDGE,HARPER,MI	201.20	300	6
587	JUDY ROBERTS	512 PINE,ADA,MI	57.75	500	6
622	DAN MARTIN	419 CHIP,GRANT,MI	575.50	500	3

ORDERS

ORDNUMB	ORDDTE	CUSTNUMB
12489	90294	124
12491	90294	311
12494	90494	315
12495	90494	256
12498	90594	522
12500	90594	124
12504	90594	522

ORDLNE

ORDNUMB	PARTNUMB	NUMBORD	QUOTPRCE
12489	AX12	11	14.95
12491	BT04	1	402.99
12491	BZ66	1	311.95
12494	CB03	4	175.00
12495	CX11	2	57.95
12498	AZ52	2	22.95
12498	BA74	4	4.95
12500	BT04	1	402.99
12504	CZ81	2	108.99

PART

PARTNUMB	PARTDESC	UNONHAND	ITEMCLSS	WREHSENM	UNITPRCE
AX12	IRON	104	HW	3	17.95
AZ52	SKATES	20	SG	2	24.95
BA74	BASEBALL	40	SG	1	4.95
BH22	TOASTER	95	HW	3	34.95
BT04	STOVE	11	AP	2	402.99
BZ66	WASHER	52	AP	3	311.95
CA14	SKILLET	2	HW	3	19.95
CB03	BIKE	44	SG	1	187.50
CX11	MIXER	112	HW	3	57.95
CZ81	WEIGHTS	208	SG	2	108.99

3.2 SQL

In this section, we examine the SQL language through a number of examples. Like most modern languages, SQL is basically free format. This means that there are no special rules for spacing when one is typing an SQL command. Commas, however, are essential. There is one little problem that is not unique to SQL. The word *ORDER* has special meaning in SQL and cannot be used as the name of a table. In any system with such a restriction, we must pick another name for the *ORDER* table. In our examples, we will use the name *ORDERS*.

In the examples that follow, we will investigate the manner in which tables may be described, data may be retrieved, new data may be added, data may be changed, and data may be deleted.

Data Definition

Example 1: Creation of a database.

Statement: Describe the layout of the sales rep table to the DBMS.

```
CREATE TABLE SLSREP
    (SLSRNUMB        DECIMAL(2),
     SLSRNAME        CHAR(15),
     SLSRADDR        CHAR(25),
     TOTCOMM         DECIMAL(7,2),
     COMMRATE        DECIMAL(2,2))
```

This illustrates the manner in which, using SQL, we describe a new table to the DBMS. We name the table, the columns, and the physical characteristics of the columns. In this example, we are describing a table that will be called *SLSREP*. It contains five columns: *SLSRNUMB*, *SLSRNAME*, *SLSRADDR*, *TOTCOMM*, and *COMMRATE*. The word *DECIMAL* indicates that *SLSRNUMB* is a numeric field; that is, it can contain only numbers. The number 2 in parentheses indicates that *SLSRNUMB* is two digits in length. The *CHAR* indicates that *SLSRNAME* is a character field (also called an *alphanumeric* field); that is, it can contain any type of character. The number 15 indicates that *SLSRNAME* is fifteen characters in length. Similarly, *SLSRADDR* is a character field that is twenty-five characters in length. *TOTCOMM* is numeric and is seven digits long. The 2 that follows the comma indicates that the last two of the seven digits represent two decimal places. Similarly, *COMMRATE* is two digits long, and both of the two digits are decimal places. We can visualize this statement as setting up for us a blank table with appropriate column headings (see Figure 3.2).

Figure 3.2

Blank *SLSREP* table

SLSREP

SLSRNUMB	SLSRNAME	SLSRADDR	TOTCOMM	COMMRATE

We now consider the data manipulation features of the language, beginning with those aspects of the language which are devoted to retrieving information from the database.

Simple Retrieval

The basic form of an SQL expression is quite simple. It is merely SELECT-FROM-WHERE. After the SELECT, we list those columns which we wish to have displayed. After the FROM, we list the table or tables that are involved in the query. Finally, after the WHERE, we list any conditions that apply to the data we wish to retrieve.

Example 2: Retrieval of certain columns and all rows.

Statement: List the number, name, and balance of all customers.

Since we want all customers listed, there is no need for the WHERE clause (we have no restrictions). The query is thus:

```
SELECT CUSTNUMB, CUSTNAME, BALANCE
     FROM CUSTOMER
```

The computer will respond with:

```
CUSTNUMB CUSTNAME        BALANCE
     124 SALLY ADAMS      418.75
     256 ANN SAMUELS       10.75
     311 DON CHARLES      200.10
     315 TOM DANIELS      320.75
     405 AL WILLIAMS      201.75
     412 SALLY ADAMS      908.75
     522 MARY NELSON       49.50
     567 JOE BAKER        201.20
     587 JUDY ROBERTS      57.75
     622 DAN MARTIN       575.50
```

Example 3: Retrieval of all columns and all rows.

Statement: List the complete *ORDERS* table.

We could certainly use the same structure that is shown in Example 2. However, a shortcut is available. Instead of listing all of the column names after the SELECT, we can use the symbol *. This indicates that we want all columns listed (in the order in which they have been described to the system during data definition). If we want all columns listed but in a different order, we would have to type the names of the columns in the order in which we wanted them to appear. In this case, assuming that the normal order is appropriate, the query would be:

```
SELECT *
     FROM ORDERS
```

Result:

```
ORDNUMB ORDDTE CUSTNUMB
  12489  90294      124
  12491  90294      311
  12494  90494      315
  12495  90494      256
  12498  90594      522
  12500  90594      124
  12504  90594      522
```

Example 4: Use of the WHERE clause.

Statement: What is the name of customer 124?

We use the WHERE clause to restrict the output of the query to customer 124 as follows:

```
SELECT CUSTNAME
     FROM CUSTOMER
     WHERE CUSTNUMB = 124
```

Result:

```
CUSTNAME

SALLY ADAMS
```

The condition in the WHERE clause need not involve equality. We can also use > (greater than), > = (greater than or equal to), < (less than), < = , (less than or equal to), or ~ = (not equal). (In some settings, < > is used for not equal.)

Example 5: Use of a compound condition within the WHERE clause.

Statement: List the descriptions of all parts that are in warehouse 3 and for which there are more than 100 units on hand.

Compound conditions are possible within the WHERE clause using AND, OR, and NOT. In this case, we have:

```
SELECT PARTDESC
     FROM PART
     WHERE WREHSENM = 3
     AND UNONHAND > 100
```

Result:

```
PARTDESC
IRON
MIXER
```

Example 6: Use of computed columns.

Statement: List the number, name, and available credit for all customers who have at least an $800 credit limit.

This statement poses a problem for us. There is no *available credit* column in our database. It is, however, computable from two columns that are present, *CREDLIM* and *BALANCE* (*available credit = CREDLIM – BALANCE*). Fortunately, SQL permits us to specify computations within the SQL expression. In this case, we would have:

```
SELECT CUSTNUMB, CUSTNAME, CREDLIM - BALANCE
     FROM CUSTOMER
     WHERE CREDLIM >= 800
```

Result:

```
CUSTNUMB CUSTNAME      3
     256 ANN SAMUELS 789.25
     405 AL WILLIAMS 598.25
     412 SALLY ADAMS  91.25
     522 MARY NELSON 750.50
```

Note that the heading for the new column is simply the number 3. Since this column does not exist in the *CUSTOMER* table, the computer does not know how to label the column and instead uses the number 3 (for the *third* column). There is a facility within SQL to change any of the column headings to whatever we desire. For now, though, we will just accept the headings that SQL will produce automatically. (Some variation exists among different versions of SQL regarding column headings for computed columns. Your version may very well treat them differently. Some versions, for example, will label the column as *CREDLIM – BALANCE*; that is, the computation will be used as the column heading.)

Example 7: Use of a wild card.

Statement: List the number, name, and address of all customers who live in Grant.

The only problem posed by this query results from a single column containing the street address, the city, and the state. If we had a separate column for the city, the query would not be difficult (the WHERE clause would be *WHERE CITY = 'GRANT'*). In this case, however, all we can say is that anyone living in Grant has GRANT somewhere within his or her address, but *we don't know where*. Fortunately, SQL provides a facility that we can use in this situation, as the following illustrates:

```
SELECT CUSTNUMB, CUSTNAME, ADDRESS
     FROM CUSTOMER
     WHERE ADDRESS LIKE '%GRANT%'
```

Result:

CUSTNUMB	CUSTNAME	ADDRESS
256	ANN SAMUELS	215 PETE,GRANT,MI
405	AL WILLIAMS	519 WATSON,GRANT,MI
622	DAN MARTIN	419 CHIP,GRANT,MI

The symbol % is used as what is termed a *wild card*. Using this feature, we are asking for all customers whose address is "like" some collection of characters, followed by GRANT, followed by some other characters. Note that this query would also pick up a customer whose address was 123 Grantview Street, Ada, Michigan. We would probably be safer to ask for addresses like '%,GRANT,%'.

Sorting

In any language that is used to access a database, an essential feature is the ability to sort the data that is retrieved in whatever manner is desired. In SQL, this is accomplished by the ORDER BY clause, as the following example demonstrates.

Example 8: Sorting.

Statement: List the number, name, and address of all customers. The report should be ordered by name.

To have the output sorted, indicate the key on which the data is to be sorted in an ORDER BY clause. In this example, the formulation is:

```
SELECT CUSTNUMB, CUSTNAME, ADDRESS
     FROM CUSTOMER
     ORDER BY CUSTNAME
```

Result:

CUSTNUMB	CUSTNAME	ADDRESS
405	AL WILLIAMS	519 WATSON,GRANT,MI
256	ANN SAMUELS	215 PETE,GRANT,MI
622	DAN MARTIN	419 CHIP,GRANT,MI
311	DON CHARLES	48 COLLEGE,IRA,MI
567	JOE BAKER	808 RIDGE,HARPER,MI
587	JUDY ROBERTS	512 PINE,ADA,MI
522	MARY NELSON	108 PINE,ADA,MI
412	SALLY ADAMS	16 ELM,LANSING,MI
124	SALLY ADAMS	481 OAK,LANSING,MI
315	TOM DANIELS	914 CHERRY,KENT,MI

(Note that since the names are stored in a single column and we have stored the names in the form FIRST LAST, the list is alphabetized on the basis of *first* names.)

Example 9: Sorting with multiple keys, descending order.

Statement: List the customer number, name, and credit limit of all customers, ordered by decreasing credit limit and by customer number within credit limit.

This example calls for sorting on multiple keys (customer number within credit limit) as well as the use of descending order for one of the keys. This is accomplished as follows:

```
SELECT CUSTNUMB, CUSTNAME, CREDLIM
    FROM CUSTOMER
    ORDER BY CREDLIM DESC, CUSTNUMB
```

Result:

CUSTNUMB	CUSTNAME	CREDLIM
412	SALLY ADAMS	1000
256	ANN SAMUELS	800
405	AL WILLIAMS	800
522	MARY NELSON	800
124	SALLY ADAMS	500
587	JUDY ROBERTS	500
622	DAN MARTIN	500
311	DON CHARLES	300
315	TOM DANIELS	300
567	JOE BAKER	300

Built-in Functions

Many languages include built-in functions to allow the calculation of the number of entries, the sum or average of all of the entries in a given column, and the largest or smallest of the entries in a given column. In SQL, these functions are called COUNT, SUM, AVG, MAX, and MIN, respectively. The manner in which they are used is demonstrated in the next two examples.

Example 10: Use of the built-in function COUNT.

Statement: How many parts are in item class HW?

In this query, we are interested in the number of rows in the table for which the item class is HW. We could count the number of part numbers in this table or the number of descriptions or the number of entries in any other column; it doesn't make any difference. (Rather than requiring us to arbitrarily pick one of these, some versions of SQL allow us to use the symbol *.) The query could thus be formulated as follows:

```
SELECT COUNT(PARTNUMB)
    FROM PART
    WHERE ITEMCLSS = 'HW'
```

Result:

1
4

Example 11: Use of SUM.

Statement: Count the number of customers and find the total of their balances.

The only real difference between COUNT and SUM (other than the obvious one that they are computing different statistics) is that in the case of SUM, we *must* specify the column for which we want a total. (It doesn't make sense to use the *, even if this is permitted for COUNT.) This query is thus:

```
SELECT COUNT(CUSTNUMB), SUM(BALANCE)
    FROM CUSTOMER
```

Result:

1	2
10	2944.8

Querying Two Tables

Example 12: Joining two tables together.

Statement: For each customer, list the number and name of the customer together with the number and name of the sales rep who represents the customer.

Here we indicate all columns we wish displayed in the SELECT clause. In the FROM clause, we list all of the tables involved in the query. Finally, in the WHERE clause, we give the condition that will restrict the data to be retrieved to only those rows from the two tables that match. We have a problem, however. There is a column in *SLSREP* called *SLSRNUMB*, as well as a column in *CUSTOMER* called *SLSRNUMB*. In this case, if we merely mention *SLSRNUMB*, SQL will not know which one we mean. It is necessary to **qualify** *SLSRNUMB*, that is, to specify which one we are referring to. You will recall that we do this by preceding the name of the column with the name of the table, followed by a period. (The *SLSRNUMB* column in the *SLSREP* table is *SLSREP.SLSRNUMB*, and the *SLSRNUMB* column in the *CUSTOMER* table is *CUSTOMER.SLSRNUMB*.) The query is as follows.

```
SELECT CUSTNUMB, CUSTNAME, SLSREP.SLSRNUMB, SLSRNAME
    FROM CUSTOMER, SLSREP
    WHERE CUSTOMER.SLSRNUMB = SLSREP.SLSRNUMB
```

Result:

CUSTNUMB	CUSTNAME	SLSRNUMB	SLSRNAME
124	SALLY ADAMS	3	MARY JONES
256	ANN SAMUELS	6	WILLIAM SMITH
311	DON CHARLES	12	SAM BROWN
315	TOM DANIELS	6	WILLIAM SMITH
405	AL WILLIAMS	12	SAM BROWN
412	SALLY ADAMS	3	MARY JONES
522	MARY NELSON	12	SAM BROWN
567	JOE BAKER	6	WILLIAM SMITH
587	JUDY ROBERTS	6	WILLIAM SMITH
622	DAN MARTIN	3	MARY JONES

Example 13: Restricting the rows in a JOIN.

Statement: For each customer whose credit limit is $800, list the number and name of the customer together with the number and name of the sales rep who represents the customer.

In the previous example, the condition in the WHERE clause served only to relate a customer to a sales rep. While relating a customer to a sales rep is essential in this example as well, we also want to restrict the output to only those customers whose credit limit is $800. This is accomplished by a compound condition, as follows:

```
SELECT CUSTNUMB, CUSTNAME, SLSREP.SLSRNUMB, SLSRNAME
    FROM CUSTOMER, SLSREP
    WHERE CUSTOMER.SLSRNUMB = SLSREP.SLSRNUMB
    AND CREDLIM = 800
```

Result:

CUSTNUMB	CUSTNAME	SLSRNUMB	SLSRNAME
256	ANN SAMUELS	6	WILLIAM SMITH
405	AL WILLIAMS	12	SAM BROWN
522	MARY NELSON	12	SAM BROWN

Subqueries

Example 14: Nesting queries.

Statement: List the number and name of any sales rep who represents any customer whose balance exceeds his or her credit limit.

The only real difference between this query and the preceding one is that in this one, we are not required to list any information about the customer. Customer information still enters the picture, however, since it is used in determining which sales reps should be listed. For this reason, the *CUSTOMER* table is still listed in the FROM clause. The query is thus:

```
SELECT SLSREP.SLSRNUMB, SLSRNAME
    FROM CUSTOMER, SLSREP
    WHERE CUSTOMER.SLSRNUMB = SLSREP.SLSRNUMB
    AND BALANCE > CREDLIM
```

Result:

SLSRNUMB	SLSRNAME
3	MARY JONES
6	WILLIAM SMITH

There is another way in which we could logically approach the problem. We could do it in two steps. We could first find the numbers of all the sales reps who represent such customers by entering:

```
SELECT SLSRNUMB
    FROM CUSTOMER
    WHERE BALANCE > CREDLIM
```

and then somehow ask for the number and name of all sales reps whose numbers are in this collection. Actually, these two can be accomplished in a single step by using a feature of SQL called subqueries, as follows:

```
SELECT SLSRNUMB, SLSRNAME
    FROM SLSREP
    WHERE SLSRNUMB IN
        (SELECT SLSRNUMB
            FROM CUSTOMER
            WHERE BALANCE > CREDLIM)
```

In algebra, expressions within parentheses are evaluated first. The same holds true in SQL. The expression within parentheses, which is called a **subquery**, is evaluated first, producing a collection (actually a table) of those sales rep numbers that appear on some row in the *CUSTOMER* table on which the balance is greater than the credit limit. Once this has been accomplished, the remainder of the query can be executed, producing a list of sales rep numbers and names from those rows in the *SLSREP* table for which the sales rep number is in this collection.

Joining Multiple Tables

Example 15: Joining more than two tables.

Statement: For each part that is on order, list the part number, the number ordered, the order number, the number and name of the customer who placed the order, and the name of the sales rep who represents the customer.

A part is on order if it occurs on any row in the *ORDLNE* table. The part number, number ordered, and order number are all found within the *ORDLNE* table. If these were the only things required for the query, we could just enter:

```
SELECT PARTNUMB, NUMBORD, ORDNUMB
    FROM ORDLNE
```

This is not all we need, however. We also need the customer number, which is in the *ORDERS* table; the customer name, which is in the *CUSTOMER* table; and the sales rep name, which is in the *SLSREP* table. Thus, we really need to join *four* tables: *ORDLNE, ORDERS, CUSTOMER,* and *SLSREP.* The mechanism for doing this is essentially the same as the mechanism for joining two tables. The only difference is that the condition in the WHERE clause will be a compound condition. The WHERE clause in this case would be the following.

```
WHERE ORDERS.ORDNUMB = ORDLNE.ORDNUMB
AND    CUSTOMER.CUSTNUMB = ORDERS.CUSTNUMB
AND    SLSREP.SLSRNUMB = CUSTOMER.SLSRNUMB
```

The first condition relates an order line to an order with a matching order number. The second condition relates this order to the customer with a matching customer number. The final condition relates this customer to a sales rep based on a matching sales rep number.

For the complete query, we list all of the desired columns after the SELECT, qualifying any that appear in more than one table. After the FROM, we list all four tables that are involved in the query. The complete formulation is thus:

```
SELECT ORDLNE.PARTNUMB, NUMBORD, ORDLNE.ORDNUMB, ORDERS.CUSTNUMB,
    CUSTNAME, SLSRNAME
    FROM ORDLNE, ORDERS, CUSTOMER, SLSREP
    WHERE ORDERS.ORDNUMB = ORDLNE.ORDNUMB
    AND    CUSTOMER.CUSTNUMB = ORDERS.CUSTNUMB
    AND    SLSREP.SLSRNUMB = CUSTOMER.SLSRNUMB
```

Result:

PARTNUMB	NUMBORD	ORDNUMB	CUSTNUMB	CUSTNAME	SLSRNAME
AX12	11	12489	124	SALLY ADAMS	MARY JONES
AZ52	2	12498	522	MARY NELSON	SAM BROWN
BA74	4	12498	522	MARY NELSON	SAM BROWN
BT04	1	12491	311	DON CHARLES	SAM BROWN
BT04	1	12500	124	SALLY ADAMS	MARY JONES
BZ66	1	12491	311	DON CHARLES	SAM BROWN
CB03	4	12494	315	TOM DANIELS	WILLIAM SMITH
CX11	2	12495	256	ANN SAMUELS	WILLIAM SMITH
CZ81	2	12504	522	MARY NELSON	SAM BROWN

Q & A

Question: Could *CUSTOMER.CUSTNUMB* be used in place of *ORDERS.CUSTNUMB* in the query?

Answer: Yes, since the values for these two columns must match by virtue of the condition *CUSTOMER.CUSTNUMB = ORDERS.CUSTNUMB*. Thus, we could choose either one.

Question: Why didn't we have to say *ORDLNE.PARTNUMB*, since *PARTNUMB* also appears as a column in the *PART* table?

Answer: If the *PART* table were used as one of the tables in the query, we would have to qualify *PARTNUMB*. Since it is not, such qualification is unnecessary. (Among the tables listed in the query, only one column is labeled *PARTNUMB*.)

Certainly this last query is more involved than the previous ones. You may be thinking that SQL is not such an easy language to use after all. If you take it one step at a time, however, the query really isn't all that bad. To construct it in a step-by-step fashion we should do the following:

1. List all of the columns that we want printed on the report after the word SELECT. If the name of any column appears in more than one table, precede the column name with the table name (that is, qualify it).
2. List all of the tables involved in the query after the word FROM. This will usually be the tables that contain the columns listed in the SELECT clause. Occasionally, however, there might be a table that does not contain any columns used in the SELECT clause but does contain columns used in the WHERE clause. It must also be listed. (In the last query, if there had been no need to list a customer number or name, but we had needed to list the sales rep name, no columns from the *CUSTOMER* table would be used in the SELECT clause. The *CUSTOMER* table would still have been required, however, since columns from it must have been used in the WHERE clause.)
3. Taking the tables involved one pair at a time, put the condition that relates the tables in the WHERE clause. Join these conditions with AND. If there are any other conditions, they should also be included in the WHERE clause and should be connected to the others with the word AND. If, in the last example, we had only wanted parts present on orders placed by customers with a $500 credit limit, for example, one more condition would have been added to the WHERE clause, giving:

```
SELECT ORDLNE.PARTNUMB, NUMBORD, ORDLNE.ORDNUMB, ORDERS.CUSTNUMB,
    CUSTNAME, SLSRNAME
    FROM ORDLNE, ORDERS, CUSTOMER, SLSREP
    WHERE ORDERS.ORDNUMB = ORDLNE.ORDNUMB
    AND   CUSTOMER.CUSTNUMB = ORDERS.CUSTNUMB
    AND   SLSREP.SLSRNUMB = CUSTOMER.SLSRNUMB
    AND   CREDLIM = 500
```

Update

The remainder of the SQL examples involve the update features of SQL.

Example 16: Changing existing data in the database.

Statement: Change the name of customer 256 TO ANN JONES.

The SQL command to make changes in existing data is the UPDATE command. For this example, the formulation would be:

```
UPDATE CUSTOMER
    SET CUSTNAME = 'ANN JONES'
    WHERE CUSTNUMB = 256
```

Example 17: Adding new data to the database.

Statement: Add sales rep 14, (Name: ANN CRANE, Address: 123 RIVER,ADA,MI, Total Commission: 0, Commission rate: 0.05) to the database.

Addition of new data is accomplished through the INSERT command. If we have specific data, as in this example, we can use the INSERT command as follows:

```
INSERT INTO SLSREP
    VALUES
    (14,'ANN CRANE','123 RIVER,ADA,MI',0.00,0.05)
```

Example 18: Deleting data from the database.

Statement: Delete from the database the customer whose name is AL WILLIAMS.

To delete data from the database, the DELETE command is used, as in the following:

```
DELETE CUSTOMER
    WHERE CUSTNAME = 'AL WILLIAMS'
```

Note that this type of deletion can be very dangerous. If there happened to be another customer with the name AL WILLIAMS, this customer would also be deleted in the process. The safest type of deletion occurs when the condition involves the customer number. In such a case, since the customer number is unique, we can be certain that we will not cause accidental deletion of other rows in the table.

Example 19: Changing data in the database on the basis of a compound condition.

Statement: For each customer with a $500 credit limit whose balance does not exceed his or her credit limit, increase the credit limit to $800.

The only difference between this and the previous update example is that here the condition is compound. Thus, the formulation would be:

```
UPDATE CUSTOMER
    SET CREDLIM = 800
    WHERE CREDLIM = 500
    AND BALANCE < CREDLIM
```

Example 20: Creating a new table with data from an existing table.

Statement: Create a new table called *SMALLCUST* that contains the same columns as *CUSTOMER* but only the rows for which the credit limit is $500 or less.

The first thing we need to do is to describe this new table by means of the data definition facilities of SQL, as follows:

```
CREATE TABLE SMALLCUST
    (CUSTNUMB        DECIMAL(4),
     CUSTNAME        CHAR(15),
     ADDRESS         CHAR(25),
     BALANCE         DECIMAL(7,2),
     CREDLIM         DECIMAL(4),
     SLSRNUMB        DECIMAL(2))
```

Once this is done, we can use the same INSERT command that we encountered in Example 17. Here, however, we use a SELECT command to indicate what is to be inserted into this new table. The exact formulation is:

```
INSERT INTO SMALLCUST
    SELECT *
    FROM CUSTOMER
    WHERE CREDLIM <= 500
```

It should be clear from these examples that SQL is a very powerful language that allows comprehensive queries to be satisfied with brief formulations. The use of SQL is widespread. It is employed with many relational DBMSs, on everything from the largest mainframe to some of the smaller microcomputers, so it's a good idea to become familiar with it.

For other examples of SQL queries, see [8], [9], [11], and [14].

3.3 QBE

In this section, we will investigate a visual approach to manipulating relational databases. It is called **Query-by-Example** (or **QBE** for short) and was proposed by M. M. Zloof (see [16]). It is intended for use on a visual display terminal. Not only are results displayed on the screen in tabular form, but users enter their requests by filling in portions of the displayed tables.

In using QBE, we are first presented with a blank form on the screen:

We indicate which table we wish to manipulate by typing the name of the table in the first box:

ORDERS					

At this point, we can fill in the column headings for those columns we wish to have included in our queries. If we wish to include all columns from the table, we can employ a shortcut. We merely type:

ORDERS P.					

The **P.** stands for print. The system will respond with:

ORDERS	ORDNUMB	ORDDTE	CUSTNUMB		

In the following examples, we will give the QBE formulation but leave it to you to determine the results that the computer would produce. For the first few examples, let's assume that we have requested the part table with the following columns:

PART	PARTNUMB	PARTDESC	UNONHND	ITEMCLSS	WREHSENM

Example 1: Retrieving certain columns and all rows.

Statement: List the part number and description of all parts.

Using the same **P.** that we encountered earlier, we request all part numbers to be printed by putting a **P.** in the *PARTNUMB* and *PARTDESC* columns.

PART	PARTNUMB	PARTDESC	UNONHND	ITEMCLSS	WREHSENM
	P.	P.			

The system would then respond by filling in the part number column with all of the part numbers currently on file. If there are more part numbers than will fit on the screen, which will probably be the case, they will be displayed one screen at a time.

Example 2: Retrieving all columns and all rows.

Statement: List the complete part table.

We could certainly put a **P.** in each column in the table to obtain the desired result. A simpler method is available, however, and that is to put a single **P.** in the first column:

PART	PARTNUMB	PARTDESC	UNONHND	ITEMCLSS	WREHSENM
P.					

This indicates that we want the full table printed.

Example 3: Retrieval with a simple condition.

Statement: List the part numbers of all parts in item class HW.

We use the **P.**, as before, to indicate the columns that are to be printed. We can also place a specific value in a column:

PART	PARTNUMB	PARTDESC	UNONHND	ITEMCLSS	WREHSENM
	P.			HW	

This indicates that the part numbers to be printed should only be those for which the item class is HW.

Example 4: Retrieval with a compound condition involving AND.

Statement: List the part numbers for all parts which are in item class HW and are located in warehouse 3.

As you might expect, we can put specific values in more than one column. Further, we can also use the comparison operators =, >, > =, <, and < =, as well as ~ = (NOT EQUAL). It is common in QBE to omit the = symbol in "equal" and "not equal" comparisons, although it may be used if desired.

PART	PARTNUMB	PARTDESC	UNONHND	ITEMCLSS	WREHSENM
	P.			HW	3

In this case, we are requesting those parts for which the item class is HW *and* the warehouse is 3.

Example 5: Retrieval with a compound condition involving OR.

Statement: List the part numbers for those parts that are in item class HW or warehouse 3.

What we essentially have in this query is two queries. We want all parts that are in class HW. We also want all parts that are in warehouse 3. This is effectively how we enter our request, as two queries:

PART	PARTNUMB	PARTDESC	UNONHND	ITEMCLSS	WREHSENM
	P.			HW	
	P.				3

The first row indicates that we want all parts that are in class HW. The second row indicates that we also want all parts that are in warehouse 3.

Example 6: Retrieval using NOT.

Statement: List the part numbers of all parts that are not in item class HW.

We use the symbol ~ for NOT and enter:

PART	PARTNUMB	PARTDESC	UNONHND	ITEMCLSS	WREHSENM
	P.			~HW	

In each of the prior examples, we could have used a feature of the QBE language from which it draws its name: an example. The prior queries were simple enough, so there was no real need to use an example, but it would certainly have been legitimate to do so. To use an example, we pick a sample response that the computer could give to the query and enter it in the table. To indicate that it is merely an example, we underline it. Thus, in the previous query, we could have entered:

PART	PARTNUMB	PARTDESC	UNONHND	ITEMCLSS	WREHSENM
	P.XYZ			~HW	

We are indicating that XYZ is an example of the response we are expecting. It does *not* have to be an actual response that would be generated in response to this query. In this case, in fact, there is not even a part XYZ in the database. Notice the difference between the part number XYZ and the item class HW. The part number is underlined, indicating that it is strictly an example. The item class is not underlined, indicating that it is an actual value.

Working with a single table is thus fairly simple. But what if we need to use two tables and the relationship between them in order to obtain the desired information? This is also possible in QBE. To illustrate the method, let's assume that we will use the following two tables.

CUSTOMER	CUSTNUMB	CUSTNAME	BALANCE	CREDLIM	SLSRNUMB

and

SLSREP	SLSRNUMB	SLSRNAME	SLSRADDR	TOTCOMM	COMMRATE

Example 7: Retrieval using more than one table.

Statement: For each customer whose credit limit is $800, list the number and name of the customer together with the number and name of the sales rep who represents the customer.

This query cannot be satisfied by using a single table. The customer name is in the customer table, whereas the sales rep name is in the sales rep table. We need the equivalent of a **join** operation. This is accomplished by having two tables on the screen and filling them in as follows:

CUSTOMER	CUSTNUMB	CUSTNAME	BALANCE	CREDLIM	SLSRNUMB
	P.	P.		800	123

SLSREP	SLSRNUMB	SLSRNAME	SLSRADDR	TOTCOMM	COMMRATE
	P.123	P.			

In this case, there is a **P.** in both the *CUSTNUMB* and *CUSTNAME* columns of the *CUSTOMER* table, indicating that these are the columns in which results are to be printed. Further, there is an example, 123, in the sales rep number column of the *CUSTOMER* table, as well as *the same* example in the sales rep number column of the *SLSREP* table. These examples are necessary, and the fact that they are the same is *crucial*. This is what tells the system how the tables are to be joined. Finally, since the 800 is not underlined, it is *not* an example but rather a specific restriction. In other words, we are telling the system to PRINT the number and name of any customers in the

CUSTOMER table for whom the value in the credit limit column is 800 and for whom there is a row in the *SLSREP* table where the sales rep number matches this sales rep number in the *CUSTOMER* table.

For other examples of QBE queries, see [8], [9], [11], and [14].

3.4   THE RELATIONAL ALGEBRA

The relational algebra is one of the original approaches proposed for manipulating relational databases. In the relational algebra, operations act on tables to produce new tables, just as the operations of $+$, $-$, and so on, act on numbers to produce new numbers in the algebra with which you are familiar. Retrieving data from a relational database through the use of the relational algebra involves using these operations on existing tables to form a new table that contains the desired information. It may be necessary to use successive commands to form intermediate tables before the final result is obtained, as some of the following examples demonstrate. As you will notice in these examples, each command ends with a clause that reads GIVING, followed by a table name. This clause is requesting that the result of the execution of the command is to be placed in a table with the name we have specified.

There are eight commands that make up the relational algebra. In this text, we will examine three of the most important of them: SELECT, PROJECT, and JOIN.

Select

The relational algebra SELECT command takes a horizontal subset of a table, that is, it causes only certain rows to be included in the new table. (It should not be confused with the SQL SELECT command, which actually includes the power to accomplish all of the relational algebra commands.)   This SELECT causes the creation of a new table that consists of only those rows from the indicated table which meet some specified criteria.

Statement: List all information from the *CUSTOMER* table concerning customer 256.

In relational algebra:

```
SELECT CUSTOMER WHERE CUSTNUMB = 256 GIVING ANSWER.
```

In SQL:

```
SELECT * FROM CUSTOMER WHERE CUSTNUMB = 256
```

Project

The PROJECT command within the relational algebra takes a vertical subset of a table, that is, it causes only certain columns to be included in the new table.

Statement: List the number and name of all customers.

In relational algebra:

```
PROJECT CUSTOMER OVER (CUSTNUMB, CUSTNAME) GIVING ANSWER
```

In SQL:

```
SELECT CUSTNUMB, CUSTNAME FROM CUSTOMER
```

We can combine SELECT and PROJECT, as the following example shows:

Statement: List the number and name of all customers who have an $800 credit limit.

This is accomplished in a two-step process. We first use a SELECT command to create a new table that contains only those customers who have the appropriate credit limit. Then we project that table to restrict the result to only the indicated columns.

In relational algebra:

```
SELECT CUSTOMER WHERE CREDLIM = 800 GIVING TEMP
PROJECT TEMP OVER (CUSTNUMB, CUSTNAME) GIVING ANSWER
```

In SQL:

```
SELECT CUSTNUMB, CUSTNAME FROM CUSTOMER
    WHERE CREDLIM = 800
```

Join

The JOIN operation is at the heart of the relational algebra. It is the command that allows us to pull together data from more than one table. In the most usual form of the JOIN, two tables are **join**ed together on the basis of a common attribute. A new table is formed, containing the columns of both the tables that have been joined. Rows in this new table will be the concatenation (combination) of a row from the first table and a row from the second that match on the common attribute (often called the JOIN column). For example, suppose we wanted to JOIN the following tables, *CUSTOMER* and *SLSREP*, on *SLSRNUMB* (the join column) as follows.

CUSTOMER

CUSTNUMB	CUSTNAME	ADDRESS	SLSRNUMB
124	SALLY ADAMS	481 OAK,LANSING,MI	3
256	ANN SAMUELS	215 PETE,GRANT,MI	6
311	DON CHARLES	48 COLLEGE,IRA,MI	12
315	TOM DANIELS	914 CHERRY,KENT,MI	6
405	AL WILLIAMS	519 WATSON,GRANT,MI	12
412	SALLY ADAMS	16 ELM,LANSING,MI	3
522	MARY NELSON	108 PINE,ADA,MI	12
567	JOE BAKER	808 RIDGE,HARPER,MI	6
587	JUDY ROBERTS	512 PINE,ADA,MI	6
622	DAN MARTIN	419 CHIP,GRANT,MI	3

SLSREP

SLSRNUMB	SLSRNAME
3	MARY JONES
6	WILLIAM SMITH
12	SAM BROWN

Suppose that we wish to call the result of this join *TEMP*. It would look like:

TEMP

CUSTNUMB	CUSTNAME	ADDRESS	SLSRNUMB	SLSRNAME
124	SALLY ADAMS	481 OAK,LANSING,MI	3	MARY JONES
256	ANN SAMUELS	215 PETE,GRANT,MI	6	WILLIAM SMITH
311	DON CHARLES	48 COLLEGE,IRA,MI	12	SAM BROWN
315	TOM DANIELS	914 CHERRY,KENT,MI	6	WILLIAM SMITH
405	AL WILLIAMS	519 WATSON,GRANT,MI	12	SAM BROWN
412	SALLY ADAMS	16 ELM,LANSING,MI	3	MARY JONES
522	MARY NELSON	108 PINE,ADA,MI	12	SAM BROWN
567	JOE BAKER	808 RIDGE,HARPER,MI	6	WILLIAM SMITH
587	JUDY ROBERTS	512 PINE,ADA,MI	6	WILLIAM SMITH
622	DAN MARTIN	419 CHIP,GRANT,MI	3	MARY JONES

Note that the column on which the tables are joined appears only once. Other than that, all columns from both tables are present in the result.

The output from the join can be restricted to include only desired columns by using the PROJECT command, as the following example illustrates.

Statement: List the number and name of all customers together with the number and name of the sales rep who represents each customer.

In relational algebra:

```
JOIN CUSTOMER SLSREP
    WHERE CUSTOMER.SLSRNUMB = SLSREP.SLSRNUMB
    GIVING TEMP
PROJECT TEMP OVER (CUSTNUMB, CUSTNAME, SLSRNUMB,
    SLSRNAME) GIVING ANSWER
```

In SQL:

```
SELECT CUSTNUMB, CUSTNAME, SLSRNUMB, SLSRNAME
    FROM CUSTOMER, SLSREP
    WHERE CUSTOMER.SLSRNUMB = SLSREP.SLSRNUMB
```

This should give you the flavor of the relational algebra. Other types of joins and other commands are present in the relational algebra, but they are beyond the scope of this text. For information on these topics, see [9] and [14].

3.5 NATURAL LANGUAGES

Another approach to manipulating databases for the purpose of retrieval involves the use of so-called **natural languages**. When employing these languages, users type their queries as normal English questions or requests. These languages possess a built-in dictionary of words that they can interpret. Users can add words to this dictionary in order to tailor it to their particular application. In addition, these languages recognize the names of tables and columns in the database. Finally, they can recognize alternate names, or aliases, for table and column names; for example, *account number* may be another name for *customer number*.

These languages are discussed here because they represent an important way of retrieving data from a relational database. (They can be used with databases that follow other models, but for them to reach their full potential, the flexibility of the relational model is crucial.) Let's look at a few examples of querying a database through a fictitious (but representative) natural language called NL; in each case, an SQL formulation for the same query will also be presented.

Statement: Find the name of customer 256.

In NL:

```
User: What is the nme of customer 256?
Computer: I don't recognize "nme."  Did you mean "name"?
User: Yes.
Computer:      CUSTNAME
              ANN SAMUELS
```

In SQL:

```
SELECT CUSTNAME
     FROM CUSTOMER
     WHERE CUSTNUMB = 256
```

Statement: Find the names of all of the customers of sales rep 6.

In NL:

```
User: Give me the names of the customers of sales rep 6.
Computer:      CUSTNAME
              ANN SAMUELS
              TOM DANIELS
              JOE BAKER
              JUDY ROBERTS
```

In SQL:

```
SELECT CUSTNAME
     FROM CUSTOMER
     WHERE SLSRNUMB = 6
```

Statement: Of the names found in the previous query, which ones have a credit limit of $300?

In NL:

```
User: Which ones have a credit limit of 300?
Computer:      CUSTNAME
              TOM DANIELS
              JOE BAKER
```

In SQL:

```
SELECT CUSTNAME
     FROM CUSTOMER
     WHERE SLSRNUMB = 6
     AND CREDLIM = 300
```

One final note: if the user had placed the term *small customer* in the dictionary and had defined it as a customer whose credit limit is $300, then that last query could have been formulated as "Which ones are small customers?"

SUMMARY

1. SQL (Structured Query Language) is a language that is used to manipulate relational databases. The basic form of an SQL command is SELECT-FROM-WHERE.
   a. The CREATE command is used to describe the layout of a table.
   b. Computed columns can be used in SQL commands.
   c. Sorting is accomplished through the use of the ORDER BY clause.
   d. SQL has the built-in functions COUNT, SUM, AVG, MAX, and MIN.
   e. Joining two tables is accomplished in SQL through the use of a condition that relates matching rows in the tables to be joined, for example
      CUSTOMER.SLSRNUMB = SLSREP.SLSRNUMB.
   f. One query nested within another query is called a subquery and is executed first.
   g. The INSERT command is used to add new data.
   h. The UPDATE command is used to change existing data.
   i. The DELETE command is used to delete existing data.
2. QBE (Query-by-Example) is another language that is used to manipulate relational databases. QBE queries are indicated by filling in tables on the screen.
   a. P. is included in the columns to be printed.
   b. Values that are not underlined indicate specific restrictions.
   c. Underlined values are examples. They are optional unless they are used to relate two tables.
3. The relational algebra is an approach to manipulating relational databases in which there are operations that act on tables to produce new tables. The three most important operations are SELECT, PROJECT, and JOIN.
   a. The SELECT operation selects only certain rows from a table.
   b. The PROJECT operation selects only certain columns from a given table.
   c. The JOIN operation combines data from two tables.
4. Natural languages form yet another way of retrieving data from a relational database. Users type their requests in ordinary English when they use natural languages.

KEY TERMS

built-in function	relational algebra
computed column	relational model
join	select
natural language	SQL
project	subquery
Query-by-Example (QBE)	

EXERCISES

Questions 1 through 17 are based on the sample database of Figure 3.1 and deal with the language SQL. For each question, give both the appropriate SQL formulation and the result that would be produced.

1. Find the part number and description of all parts.
2. List the complete sales rep table.

3. Find the names of all the customers who have a credit limit of at least $800.
4. Give the order numbers of those orders placed by customer 124 on September 5, 1994.
5. Give the part number, description, and on-hand value (units on hand * price) for each part in item class AP. (On-hand value is really units on hand * cost but we do not have a cost column in the *PART* table.)
6. Find the number and name of all customers whose last name is NELSON.
7. List all details about parts. The output should be sorted by unit price.
8. Find out how many customers have a balance that exceeds their credit limit.
9. Find the total of the balances for all the customers represented by sales rep 12.
10. For each order, list the order number, the order date, the customer number, and the customer name.
11. For each order placed on September 5, 1994, list the order number, the order date, the customer number, and the customer name.
12. Find the number and name of all sales reps who represent any customer with a credit limit of $1000. Do this in two different ways: in one solution use a subquery; in the other, do not use a subquery.
13. For each order, list the order number, the order date, the customer number, the customer name, together with the number and name of the sales rep who represents the customer.
14. Change the description of part BT04 to OVEN.
15. Add order 12600 (date: 90694, customer number: 311) to the database.
16. Delete all customers whose balance is $0.00 and who are represented by sales rep 12.
17. Describe a new table to the database called *SPGOOD*. It contains only part number, description, and price. Once this has been done, insert the part number description and price of all parts whose item class is SG into this new table.

Questions 18 through 24 are also based on the sample database of Figure 3.1 but deal with QBE. For each question, give the appropriate QBE formulation.

18. List the number and name of all sales reps.
19. List the complete *CUSTOMER* table.
20. List the number and name of all customers represented by sales rep 3.
21. List the number and name of all customers represented by sales rep 3 and whose credit limit is $500.
22. List the number and name of all customers represented by sales rep 3 or who have a credit limit of $500.
23. List the number and name of all customers who are not represented by sales rep 3.
24. List the number and name of all customers who are represented by MARY JONES.

Questions 25 through 28 are also based on the sample database of Figure 3.1 but deal with the relational algebra. For each question, give both the appropriate relational algebra formulation and an equivalent SQL formulation.

25. List all information from the part table concerning part BT04.
26. List the number and name of all sales reps.
27. List the order number, order date, customer number, and customer name for each order.
28. List the order number, order date, customer number, and customer name for each order placed by any customer represented by MARY JONES.

Relational Model II: Advanced Topics

OBJECTIVES

1. Discuss views: what they are, how they are described, and how they are used.
2. Discuss the use of indexes for improving performance.
3. Define the catalog and explain its use.

4. Explain entity and referential integrity.
5. Discuss the manner in which the structure of a relational database can be changed.
6. Describe the characteristics a system must possess in order to be relational.

4.1 INTRODUCTION

In the last chapter, we examined data definition and manipulation within the relational model. In this chapter, we will investigate some other aspects of the model. In section 4.2, we will look at views, which represent a way of giving each user his or her own picture of the database to work with. In section 4.3, we will look at indexes and their use in improving performance. In section 4.4, we will discuss the catalog that is furnished by many relational DBMSs to provide users with access to information about the structure of a database. Two critical integrity rules are presented in section 4.5. Section 4.6 covers one of the real strengths of the relational model, the ease with which a database structure can be changed. Finally, in section 4.7, we will look at a very important question: How can we tell whether a system is really relational? With so many systems claiming in their advertising literature to be "relational," this question is especially significant.

In the discussion that follows, we will use SQL as a mechanism for illustrating the concepts. It should be emphasized, however, that many systems which do not support the SQL language provide the features we will be discussing. Although the manner in which this is accomplished varies slightly from one system to another, the basic concepts are the same, and it should be easy for you to transfer the knowledge you gain in this chapter to a non-SQL system.

4.2 VIEWS

Most relational mainframe DBMSs and some of the microcomputer DBMSs support the concept of a view. A **view** is basically an individual user's picture of the database. In many cases, a user can interact with the database via a view. Since a view is usually much less involved than the full database, its use can represent a great simplification. Views also provide a measure of security, since omitting sensitive tables or columns from a view will render them unavailable to anyone who is accessing the database via the view.

To illustrate the idea of a view, let's suppose that Bill is interested in the part number, description, units on hand, and unit price of those parts which are in item class HW. He is not interested in any of the other columns in the *PART* table. Nor is he interested in any of the rows that correspond to parts in other item classes. Life would certainly be simpler for Bill if the other rows and columns were not even present. While we cannot change the structure of the *PART* table and omit some of its rows just for Bill, we can do the next best thing. We can provide him a view that consists of precisely the rows and columns he is interested in. Using SQL, we do this as follows:

```
CREATE VIEW HOUSEWARES AS
      SELECT PARTNUMB, PARTDESC, UNONHAND,
            UNITPRCE
            FROM PART
            WHERE ITEMCLSS = 'HW'
```

The SELECT command, which is called the **defining query**, indicates precisely what is to be included in the view. Notice that it is exactly what Bill wants. Conceptually, given the current data in the *PREMIERE PRODUCTS* database, this view will contain the data shown in Figure 4.1. The data does not really exist in this form, however, nor will it *ever* exist in this form. It is tempting to think that when this view is used, the query will be executed and will produce some sort of temporary table, called *HOUSEWARES*, which Bill could then access. This is *not* what happens.

Figure 4.1	**HOUSEWARES**
HOUSEWARES view	

PARTNUMB	PARTDESC	UNONHAND	UNITPRCE
AX12	IRON	104	17.95
BH22	TOASTER	95	34.95
CA14	SKILLET	2	19.95
CX11	MIXER	112	57.95

Instead, the query acts as a sort of window into the database (see Figure 4.2). As far as Bill is concerned, the whole database is just the portion shown in dark blue. Any change that affects the dark portion of the *PART* table will be seen by Bill. He will be totally unaware, however, of a change that affects any other part of the database.

Figure 4.2	**PART**
PREMIERE PRODUCTS sample data	

PARTNUMB	PARTDESC	UNONHAND	ITEMCLSS	WREHSENM	UNITPRCE
AX12	IRON	104	HW	3	17.95
AZ52	SKATES	20	SG	2	24.95
BA74	BASEBALL	40	SG	1	4.95
BH22	TOASTER	95	HW	3	34.95
BT04	STOVE	11	AP	2	402.99
BZ66	WASHER	52	AP	3	311.95
CA14	SKILLET	2	HW	3	19.95
CB03	BIKE	44	SG	1	187.50
CX11	MIXER	112	HW	3	57.95
CZ81	WEIGHTS	208	SG	2	108.99

The way in which this is implemented is really clever. Suppose, for example, that Bill were to type the following query:

```
SELECT *
      FROM HOUSEWARES
      WHERE UNONHAND > 100
```

The query would *not* be executed in this form. Instead, it would be merged with the query that defines the view to form the query that is actually executed. In this case, the merging of the two would form:

```
SELECT PARTNUMB, PARTDESC, UNONHAND,
      UNITPRCE
      FROM PART
      WHERE ITEMCLSS = 'HW'
           AND UNONHAND > 100
```

Notice the following three things: the selection is from the *PART* table rather than the *HOUSEWARES* view; the * is replaced by those columns which are in the *HOUSE-WARES* view; and the condition involves the condition in the query entered by Bill together with the condition stated in the view definition. This new query is the one that is actually executed.

Bill, however, is unaware that this kind of activity is taking place. It seems to him that there really is a table called *HOUSEWARES* that is being accessed. One advantage of this approach is that since *HOUSEWARES* never exists in its own right, any update to the *PART* table is *immediately* available to someone accessing the database through the view. If *HOUSEWARES* were an actual stored table, this would not be the case.

What if Bill wanted different names for the columns? This could be accomplished by including the desired names in the CREATE VIEW statement. For example, if Bill wanted the names of the part number, description, units on hand, and price columns to be *PNUM, DESC, ON_HAND,* and *PRICE,* respectively, the CREATE VIEW statement would be:

```
CREATE VIEW HOUSEWARES (PNUM, DESC, ON-HAND, PRICE) AS
      SELECT PARTNUMB, PARTDESC, UNONHAND,
          UNITPRCE
          FROM PART
          WHERE ITEMCLSS = 'HW'
```

In this case, when Bill accessed the *HOUSEWARES* view, he would refer to *PNUM, DESC, ON_HAND,* and *PRICE* rather than *PARTNUMB, PARTDESC, UNONHAND,* and *UNITPRCE.*

The *HOUSEWARES* view is an example of a row-and-column subset view; that is, it consists of a subset of the rows and columns in some individual table, in this case the *PART* table. Since the query can be any SQL query, a view could also involve the join of two or more tables.

Suppose, for example, that Joan needed to know the number and name of each sales rep, along with the number and name of the customers represented by each sales rep. It would be much simpler for her if this information were in a single table instead of two tables that had to be joined together. She would really like a single table that contained a sales rep number, sales rep name, customer number, and customer name. Suppose she would also like these columns to be named *SNUM, SNAME, CNUM,* and *CNAME,* respectively. This could be accomplished by using a join in the CREATE VIEW statement, as follows:

```
CREATE VIEW SLSCUST (SNUMB, SNAME, CNUMB, CNAME) AS
      SELECT SLSREP.SLSRNUMB, SLSREP.SLSRNAME,
          CUSTOMER.CUSTNUMB, CUSTOMER.NAME
          FROM SLSREP, CUSTOMER
          WHERE SLSREP.SLSRNUMB =
              CUSTOMER.SLSRNUMB
```

Given the current data in the *PREMIERE PRODUCTS* database, this view is conceptually the table shown in Figure 4.3.

Figure 4.3

SLSCUST view

SLSCUST

SNUMB	SNAME	CNUMB	CNAME
3	MARY JONES	124	SALLY ADAMS
3	MARY JONES	412	SALLY ADAMS
3	MARY JONES	622	DAN MARTIN
6	WILLIAM SMITH	256	ANN SAMUELS
6	WILLIAM SMITH	315	TOM DANIELS
6	WILLIAM SMITH	567	JOE BAKER
6	WILLIAM SMITH	587	JUDY ROBERTS
12	SAM BROWN	311	DON CHARLES
12	SAM BROWN	405	AL WILLIAMS
12	SAM BROWN	522	MARY NELSON

As far as Joan is concerned, this is a real table; she does not need to know what goes on behind the scenes in order to use it. She could find the number and name of the sales rep who represents customer 256, for example, merely by entering:

```
SELECT SNUMB, SNAME
     FROM SLSCUST
        WHERE CNUMB = 256
```

She will be completely unaware that, behind the scenes, her query is actually converted to:

```
SELECT SLSREP.SLSRNUMB, SLSREP.SLSRNAME
     FROM SLSREP, CUSTOMER
   WHERE SLSREP.SLSRNUMB =
           CUSTOMER.SLSRNUMB
           AND CUSTNUMB = 256
```

As you can see, the use of views can greatly simplify the process of querying a database. In general, the more complicated the defining query, the more simplification the user is provided by the view. Not everything here is rosy, however. Problems arise when one tries to update a database through a view, especially a view whose defining query involves a join. This problem and others like it are beyond the scope of this text; for information about them, see [9] and [14]. Nevertheless, views are a helpful feature from which substantial benefits can be achieved.

4.3   INDEXES

If you wanted to find a discussion of a given topic in a book, you could scan the entire book from start to finish, looking for references to the topic you had in mind. More than likely, however, you wouldn't have to resort to this technique. If the book had a good index, you could use it to rapidly locate the pages on which your topic was discussed.

Within relational model systems on both mainframes and microcomputers, the main mechanism for increasing the efficiency with which data is retrieved from the database is the use of **indexes**. Conceptually, these indexes are very much like the index in a book. Consider Figure 4.4, for example, which shows the *CUSTOMER* table for *PREMIERE PRODUCTS* together with one extra column, *REC*. This extra column gives the number of each record within the file. (Customer 124 is on record one; customer 256 is on record two; and so on.) These record numbers are used by the DBMS, not by the users, and that is why we do not normally show them. Here, however, we are dealing with the manner in which the DBMS works, so we do need to be aware of them.

Figure 4.4

CUSTOMER table
with record
numbers

CUSTOMER

REC	CUSTNUMB	CUSTNAME	ADDRESS	BALANCE	CREDLIM	SLSRNUMB
1	124	SALLY ADAMS	481 OAK,LANSING,MI	418.75	500	3
2	256	ANN SAMUELS	215 PETE,GRANT,MI	10.75	800	6
3	311	DON CHARLES	48 COLLEGE,IRA,MI	200.10	300	12
4	315	TOM DANIELS	914 CHERRY,KENT,MI	320.75	300	6
5	405	AL WILLIAMS	519 WATSON,GRANT,MI	201.75	800	12
6	412	SALLY ADAMS	16 ELM,LANSING,MI	908.75	1000	3
7	522	MARY NELSON	108 PINE,ADA,MI	49.50	800	12
8	567	JOE BAKER	808 RIDGE,HARPER,MI	201.20	300	6
9	587	JUDY ROBERTS	512 PINE,ADA,MI	57.75	500	6
10	622	DAN MARTIN	419 CHIP,GRANT,MI	575.50	500	3

Figure 4.5

Index for
CUSTOMER table
on *CUSTNUMB*
column

In order to rapidly access a customer on the basis of his or her number, we might choose to create and use an index as shown in Figure 4.5. The index has two columns.

CUSTNUMB INDEX

CUSTNUMB	REC
124	1
256	2
311	3
315	4
405	5
412	6
522	7
567	8
587	9
622	10

CUSTOMER

REC	CUSTNUMB	CUSTNAME	ADDRESS	BALANCE	CREDLIM	SLSRNUMB
1	124	SALLY ADAMS	481 OAK,LANSING,MI	418.75	500	3
2	256	ANN SAMUELS	215 PETE,GRANT,MI	10.75	800	6
3	311	DON CHARLES	48 COLLEGE,IRA,MI	200.10	300	12
4	315	TOM DANIELS	914 CHERRY,KENT,MI	320.75	300	6
5	405	AL WILLIAMS	519 WATSON,GRANT,MI	201.75	800	12
6	412	SALLY ADAMS	16 ELM,LANSING,MI	908.75	1000	3
7	522	MARY NELSON	108 PINE,ADA,MI	49.50	800	12
8	567	JOE BAKER	808 RIDGE,HARPER,MI	201.20	300	6
9	587	JUDY ROBERTS	512 PINE,ADA,MI	57.75	500	6
10	622	DAN MARTIN	419 CHIP,GRANT,MI	575.50	500	3

The first column contains a customer number, and the second column contains the number of the record on which the customer is found. Since customer numbers are unique, there will be only a single record number. This need not always be the case, however. Suppose, for example, that we wanted to rapidly access all customers who had a given credit limit or all customers who were represented by a given sales rep. We might choose to create and use an index on credit limit as well as an index on sales rep number. These two indexes, along with the index on the customer number are shown in Figure 4.6. In the index on credit limit, the first column contains a credit limit, and the second column contains the numbers of *all* the records on which that credit limit is found. The index on sales rep number is similar. (Actually, the structure used for an index is usually a little more complicated than these examples show. They are perfectly acceptable for our purposes, however. For more information about the structure of these indexes, see [14].)

Figure 4.6

Three indexes for
CUSTOMER table

CUSTNUMB INDEX

CUSTNUMB	REC
124	1
256	2
311	3
315	4
405	5
412	6
522	7
567	8
587	9
622	10

CREDLIM INDEX

CREDLIM	RECs
300	3, 4, 8
500	1, 9, 10
800	2, 5, 7
1000	6

SLSRNUMB INDEX

SLSRNUMB	RECs
3	1, 6, 10
6	2, 4, 8, 9
12	3, 5, 7

CUSTOMER

REC	CUSTNUMB	CUSTNAME	ADDRESS	BALANCE	CREDLIM	SLSRNUMB
1	124	SALLY ADAMS	481 OAK,LANSING,MI	418.75	500	3
2	256	ANN SAMUELS	215 PETE,GRANT,MI	10.75	800	6
3	311	DON CHARLES	48 COLLEGE,IRA,MI	200.10	300	12
4	315	TOM DANIELS	914 CHERRY,KENT,MI	320.75	300	6
5	405	AL WILLIAMS	519 WATSON,GRANT,MI	201.75	800	12
6	412	SALLY ADAMS	16 ELM,LANSING,MI	908.75	1000	3
7	522	MARY NELSON	108 PINE,ADA,MI	49.50	800	12
8	567	JOE BAKER	808 RIDGE,HARPER,MI	201.20	300	6
9	587	JUDY ROBERTS	512 PINE,ADA,MI	57.75	500	6
10	622	DAN MARTIN	419 CHIP,GRANT,MI	575.50	500	3

Typically, an index can be created and maintained for any column or combination of columns in any table. Once an index has been created, it can be used to facilitate retrieval. In powerful mainframe relational systems, the decision concerning which index or indexes to use (if any) during a particular type of retrieval is one function of a part of the DBMS called an **optimizer**. (No reference is made to any index by the user; rather, the system makes the decision behind the scenes.) In less powerful systems and, in particular, in many of the microcomputer systems, the user may have to specifically indicate in some fashion that a given index should be used.

As you would expect, the use of any index is not purely advantageous or disadvantageous. The advantage was already mentioned: an index makes certain types of retrieval more efficient. There are two disadvantages. First, an index occupies space that could be used for something else. Any retrieval that can be made using an index can also be made without the index. The process may be less efficient, but it is still possible. So an index, while it occupies space, is technically not necessary. The other disadvantage is that the index must be updated whenever corresponding data in the database is updated. Without the index, these updates would not have to be performed. The main question that we must ask when considering whether or not to create a given index is: Do the benefits derived during retrieval outweigh the additional storage required and the extra processing involved in update operations?

Indexes can be added and dropped at will. The final decision concerning the columns or combination of columns on which indexes should be built does not have to be made at the time the database is first implemented. If the pattern of access to the database later indicates that overall performance would benefit from the creation of a new index, it can easily be added. Likewise, if it appears that an existing index is unnecessary, it can easily be dropped.

For further information concerning indexes, see [9] and [14].

4.4   THE CATALOG

Information about tables that are known to the system is kept in the system **catalog**. This section will describe the types of things kept in a catalog and the way the catalog can be queried to determine information about the database structure. (This description happens to represent the way things are done in DB2, IBM's mainframe relational DBMS.) Although catalogs in individual relational DBMSs will vary from what is shown here, the general ideas apply to most relational systems.

The catalog we will look at contains three tables; *SYSTABLES* (information about the tables known to SQL), *SYSCOLUMNS* (information about the columns within these tables), and *SYSINDEXES* (information about the indexes that are defined on these tables). While these tables would have many columns, only a few are of concern to us here.

SYSTABLES contains columns *NAME, CREATOR,* and *COLCOUNT*. The *NAME* column identifies the name of a table. The *CREATOR* column contains an identification of the person or group that created the table. The *COLCOUNT* column contains the number of columns within the table that is being described. If, for example, the user whose ID is SALESX01 created the sales rep table and the sales rep table had five columns, there would be a row in the *SYSTABLES* table in which *NAME* was SLSREP, *CREATOR* was SALESX01, and *COLCOUNT* was 5. Similar rows would exist for all tables known to the system.

SYSCOLUMNS contains columns *NAME, TBNAME,* and *COLTYPE*. The *NAME* column identifies the name of a column in one of the tables. The table in which the column is found is stored in *TBNAME*, and the data type for the column is found in *COLTYPE*. There will be a row in *SYSCOLUMNS* for each column in the *SLSREP* table, for example. On each of these rows, *TBNAME* will be SLSREP. On one of these rows, *NAME* will be SLSRNUMB and *COLTYPE* will be DECIMAL(2). On another row, *NAME* will be SLSRNAME and *COLTYPE* will be CHAR(15).

SYSINDEXES contains columns *NAME, TBNAME,* and *CREATOR*. The name of the index is found in the *NAME* column. The name of the table on which the index was built is found in the *TBNAME* column. The ID of the person or group that created the index is found in the *CREATOR* column.

The system catalog is a relational database of its own. Consequently, in general, the same types of queries that are used to retrieve information from relational databases can be used to retrieve information from the system catalog. The following queries illustrate this process.

1. List the name and creator of all tables known to the system.

```
SELECT NAME, CREATOR
    FROM SYSTABLES
```

2. List all of the columns in the *CUSTOMER* table as well as their associated data types.

```
SELECT NAME, COLTYPE
    FROM SYSCOLUMNS
    WHERE TBNAME = 'CUSTOMER'
```

3. List all tables that contain a column called *SLSRNUMB*.

```
SELECT TBNAME
    FROM SYSCOLUMNS
    WHERE NAME = 'SLSRNUMB'
```

Thus, information about the tables that are in place in our relational database, the columns they contain, and the indexes built on them can be obtained from the catalog by using the same SQL syntax that is used to query any other relational database. We don't need to worry about updating these tables; the system will do it for us automatically every time a change is made in the database structure.

4.5 INTEGRITY RULES

There are two very special rules that should be enforced by a relational DBMS. They were defined by Codd in [6], and they relate to two special types of keys: **primary keys** and **foreign keys**. The two integrity rules are called **entity integrity** and **referential integrity**.

Entity Integrity

In some DBMSs, when we describe a database, we can indicate that certain columns can accept a special value called **null**. Essentially, setting the value in a given column to null is like not filling it in at all. It is used when a value is unknown or inapplicable. It is *not* the same as blank or zero, which are actual values. For example, the value of zero in *BALANCE* indicates that the customer has a zero balance. A value of null, on the other hand, indicates that, for whatever reason, the customer's balance is unknown.

If we indicate that the column *BALANCE* can be null, we are saying that this situation (a customer with an unknown balance) is something we want to allow. If we don't want to allow it, we indicate that *BALANCE* cannot be null.

The decision as to whether to allow nulls is generally made on a column-by-column basis. There is one type of column for which we should *never* allow nulls, however, and that is the **primary key**. After all, the primary key is supposed to uniquely identify a given row, and this could not happen if nulls were allowed. How, for example, could we tell two customers apart if both had a null customer number? The restriction that the primary key cannot allow null values is called entity integrity.

> *Definition:* **Entity integrity** is the rule that no column that participates in the primary key may accept null values.

This property guarantees that each entity will indeed have its own identity. In other words, preventing the primary key from accepting null values ensures that one entity can be distinguished from another.

Referential Integrity

In the relational model as we've been discussing it up until now, relationships are not explicit. They are accomplished by having common columns in two or more tables. The one-to-many relationship between sales reps and customers, for example, is accomplished by including *SLSRNUMB*, the primary key of the *SLSREP* table, as a column in the *CUSTOMER* table.

This approach has its problems. First of all, relationships are not very obvious. If we were not already familiar with the relationships within the *PREMIERE PRODUCTS* database, we would have to note the matching columns in separate tables in order to be aware of a relationship. Even then, we couldn't be sure. Two columns having the same name could be just a coincidence. These columns might have nothing to do with each other. Second, what if the key to the *SLSREP* table were *SLSRNUMB* but the corresponding column within the *CUSTOMER* table happened to be called *SLSRNUMB*? Unless we were aware that these two columns were really the same, the relationship between customers and sales reps would not be clear. In a database having as few tables and columns as the *PREMIERE PRODUCTS* database, these problems might not be major ones. But picture a database that has twenty tables, each one containing an average of thirty columns. As the number of tables and columns increases, so do the problems.

There is also another problem. Nothing about the model itself would prevent a user from storing a customer whose sales rep number did not correspond to any sales rep already in the database. Clearly this is not a desirable situation.

Fortunately, a solution has been found for these two problems, and involves the use of foreign keys. A **foreign key** is a column (or collection of columns) in one table whose value is required to match the value of the primary key for some row in another table. The *SLSRNUMB* in the *CUSTOMER* table is a foreign key that must match the primary key of the *SLSREP* table. In practice, this simply means that the sales rep number for any customer must be the same as the number of some sales rep who is already in the database.

There is one possible exception to this. Some organizations do not require a customer to have a sales rep. This situation could be indicated in the *CUSTOMER* table by setting such a customer's sales rep number to null. Technically, however, a null sales rep number would violate the restrictions that we have indicated for a foreign key. So if we were to use a null sales rep number, we would have to modify the definition of foreign keys to include the possibility of nulls. We would insist, though, that if the foreign key contained a value *other than null*, it would have to match the value of the primary key in some row in the other table. (In our example, for instance, a customer's sales rep number could be null, but if it were not, then it would have to be the number of an actual sales rep.) The general property we have just described is called referential integrity.

> *Definition:* **Referential integrity** is the rule that if table A contains a foreign key that matches the primary key of table B then values of this foreign key either must match the value of the primary key for some row in table B or must be null.

The problems just mentioned are solved through the use of foreign keys. Indicating that the *SLSRNUMB* in the *CUSTOMER* table is a foreign key that must match the *SLSREP* table makes the relationship between customers and sales reps explicit. We do not need to look for common columns in several tables. Further, with foreign keys, matching columns that have different names no longer pose a problem. For example, it would not matter if the name of the foreign key in the *CUSTOMER* table happened to be *SLSRNUMB* while the primary key in the *SLSREP* table happened to be *SLSRNUMB*, the only thing that would matter is that this column was a foreign key that matched the *SLSREP* table. Finally, through referential integrity, it is possible for a customer not to have a sales rep number, but it is not possible for a customer to have *an invalid sales rep number*; that is, a customer's sales rep number *must* either be null or the number of a sales rep who is already in the database.

For other perspectives on integrity in the relational model, see [9] and [14].

4.6   CHANGING THE STRUCTURE OF A RELATIONAL DATABASE

One of the best things about relational DBMSs is the ease with which the database structure can be changed. New tables can be added and old ones can be removed. Columns can be added or deleted. The characteristics of columns can be changed. New indexes can be created and old ones can be dropped. Though the exact manner in which these changes are accomplished varies from one system to another, many systems allow all of these changes to be made quickly and easily. Since SQL is so widely used, we will use it as a vehicle to illustrate the manner in which these changes may be accomplished.

Alter

Changing the structure of a table in SQL is accomplished through the ALTER table command. Virtually every implementation of SQL allows new columns to be added to the end of an existing table. For example, let's suppose that we now wish to maintain a customer type for each customer in the *PREMIERE PRODUCTS* database. We can decide to call regular customers type R, distributors type D, and special customers type S. To implement this change, we need a new column in the customer table. This can be added as follows:

```
ALTER TABLE CUSTOMER
      ADD CUSTTYPE          CHAR(1)
```

At this point, the *CUSTOMER* table contains an extra column, *CUSTTYPE*.

For rows added from this point on, the value of *CUSTTYPE* will be assigned as the row is added. For existing rows, some value of *CUSTTYPE* must be assigned. The simplest approach (from the point of view of the DBMS, *not* the user) is to assign the value NULL as a *CUSTTYPE* on all existing rows. (This requires that *CUSTTYPE* accept null values, and some systems do require this. This means that any column added to a table definition *will* accept nulls; the user has no choice in the matter.) A more flexible approach, and one that is supported by some systems, is to allow the user to specify an initial value. In our example, if most customers were type R, we might set all of the customer types for existing customers to R and later change those customers of type D or type S to the appropriate value. To change the structure and set the value of *CUSTTYPE* to R for all existing records, we would type:

```
ALTER TABLE CUSTOMER
      ADD CUSTTYPE          CHAR(1)    INIT = 'R'
```

Some systems allow existing columns to be deleted. The syntax for deleting the *WREHSENM* column from the *PART* table would typically be something like this:

```
ALTER TABLE PART
      DELETE WREHSENM
```

Finally, some systems allow changes in the data types of given columns. A typical use of such a provision would be to increase the length of a character field that was found to be inadequate. Assuming that the *NAME* column in the *CUSTOMER* table needed to be increased to 30 characters, the ALTER statement would be something like this:

```
ALTER TABLE CUSTOMER
      CHANGE COLUMN NAME TO CHAR(30)
```

Drop

A table that is no longer needed can be deleted with the DROP command. If the *SLSREP* table were no longer needed in the *PREMIERE PRODUCTS* database, the command would be:

```
DROP TABLE SLSREP
```

The table would be erased, as would all indexes and views defined on the table. References to the table would be removed from the system catalog.

4.7 WHAT DOES IT TAKE TO BE RELATIONAL?

This chapter will conclude with a discussion of this interesting question. If you look at ads for both microcomputer and mainframe DBMSs, you will rarely find one that doesn't claim the DBMS is "relational." In order to make some sense out of all these claims, we need a yardstick to measure them with. In other words, we need to know what it really means for a system to be relational. For the answer to this question, let's first turn to the person who initially proposed the relational model, Dr. E. F. Codd.

In [7], Codd defines a relational system as one in which at least the following two properties hold:

1. Users perceive databases as collections of tables and are not aware of the presence of any additional structures.
2. The operations of SELECT, PROJECT, and JOIN from the relational algebra are supported. This support is independent of any predefined access paths; that is, it makes no difference which indexes do or do not exist. If, for example, a join can be performed only when indexes exist for the columns on which the join is to take place, the system should not be considered relational.

Codd's definition states that for a system to be relational, it must support the SELECT, PROJECT, and JOIN operations of the **relational algebra**, but the definition does *not* indicate that the system must use this terminology. SQL supports these three operations through the SELECT statement, which is considerably more powerful than the relational algebra SELECT indicated in Codd's definition. The following are examples of the manner in which SQL supports the SELECT, PROJECT, and JOIN operations.

SELECT (choose certain rows from a table):

```
SELECT *
    FROM CUSTOMER
    WHERE CREDLIM = 500
```

PROJECT (choose certain columns from a table):

```
SELECT CUSTNUMB, NAME, ADDRESS
    FROM CUSTOMER
```

JOIN (combine tables based on matching columns):

```
SELECT SLSRNUMB, SLSRNAME, SLSRADDR,
    TOTCOMM, COMMRATE, CUSTNUMB,
    NAME, ADDRESS, BALANCE, CREDLIM
    FROM SLSREP, CUSTOMER
    WHERE SLSREP.SLSRNUMB =
        CUSTOMER.SLSRNUMB
```

Let's turn now to C. J. Date, another individual who has been heavily involved in the development of the relational model. In [9], Date discusses a classification scheme for systems that support at least the relational structure, that is, systems in which the only structure that the user perceives is the table. Such systems fall into one of the following four categories:

1. **Tabular system.** In a **tabular system**, the only structure perceived by the user is the table, but the system does not support the SELECT, PROJECT, and JOIN operations in the unrestricted fashion indicated by Codd. (In this case, either the system does not support SELECT, PROJECT, and JOIN, or, if it does, the support

relies on predefined access paths.) The microcomputer file-management systems and some of the database management systems are in this category.

2. **Minimally relational.** Actually, the operations SELECT, PROJECT, and JOIN are not the only operations in the relational algebra. There are eight operations altogether. A **minimally relational** system is one that supports the tabular structure together with the SELECT, PROJECT, and JOIN operations but that does not support the complete set of operations from the relational algebra. The vast majority of microcomputer DBMSs fall into this category.

3. **Relationally complete.** Any system that supports the tabular structure and all the operations of the relational algebra (without requiring appropriate indexes) is said to be **relationally complete.** Many relational mainframe DBMSs and some microcomputer DBMSs fall into this category. In particular, any system that supports a *full* implementation of SQL is relationally complete.

4. **Fully relational.** A system that supports the tabular structure, all the operations of the relational algebra, and the two integrity rules (entity and referential integrity) described earlier in this section is said to be **fully relational.** This is the goal for which systems are (or should be) striving. Currently, the principal failing on the part of many systems is lack of support for referential integrity. Much progress is occurring in this area, and soon we shall see a number of fully relational systems.

SUMMARY

1. Views are used to give each user his or her own picture of the database.
   a. A view is defined in SQL through the use of a defining query.
   b. When a query is entered which references a view, it is merged with the defining query to produce the query that is actually executed.
   c. Retrieving data through a view presents no problem, but updating the database through a view is often prohibited.
2. Indexes are often used to facilitate retrieval. Indexes may be created on any column or combination of columns.
3. The catalog is a feature of many relational model DBMSs which stores information about the structure of a database. The system updates the catalog automatically. Users can retrieve data from the catalog in the same manner in which they retrieve data from the database.
4. There are two special integrity rules for relational databases:
   a. Entity integrity is the property that no column that is part of the primary key can accept null values.
   b. Referential integrity is the property that the value in any foreign key either must be null or must match an actual value of the primary key of another table.
5. Relational DBMSs provide facilities that allow users to easily change the structure of a database. Two examples of such faciiities are as follows:
   a. ALTER allows columns to be added to a table, columns to be deleted, or characteristics of columns to be changed.
   b. DROP allows a table to be deleted from a database.
6. According to Codd, a DBMS cannot be considered to be relational unless the following two conditions pertain:
   a. Users perceive a database as simply a collection of tables.
   b. The DBMS supports at least the SELECT, PROJECT, and JOIN operations of the relational algebra.

7. According to Date, DBMSs in which users perceive databases as collections of tables can be classified as:
   a. Tabular, if data is viewed as tables and if the system does not support SELECT, PROJECT, and JOIN independently of any predefined access paths.
   b. Minimally relational, if SELECT, PROJECT, and JOIN are supported but the full set of operations in the relational algebra is not.
   c. Relationally complete, if the DBMS supports all operations of the relational algebra.
   d. Fully relational, if the DBMS supports all operations of the relational algebra and both integrity rules (entity and referential integrity).

KEY TERMS

catalog	minimally relational
defining query	null
entity integrity	optimizer
foreign key	referential integrity
fully relational	relationally complete
index	tabular
integrity rules	view

EXERCISES

1. What is a view? How is it defined? Does the data described in a view definition ever exist in that form? What happens when a user accesses a database through a view?
2. Define a view called *SMALLCUST*. It consists of the customer number, name, address, balance, and credit limit for all customers whose credit limit is $500 or less.
   a. Write the view definition for *SMALLCUST*.
   b. Write an SQL query to retrieve the number and name of all customers in *SMALLCUST* whose balance is over their credit limit.
   c. Convert the query from (b) to the query that will actually be executed.
3. Define a view called *CUSTORD*. It consists of the customer number, name, balance, order number, and order date for all orders currently on file.
   a. Write the view definition for *CUSTORD*.
   b. Write an SQL query to retrieve the customer number, name, order number, and order date for all orders in *CUSTORD* for customers whose balance is more than $100.
   c. Convert the query from (b) to the query that will actually be executed.
4. What are the advantages of using indexes? The disadvantages?
5. On relational mainframe DBMSs, who or what is responsible for the decision to use a particular index? What about on microcomputer DBMSs?
6. What is the catalog? What are three items about which the catalog maintains information?
7. Why is it a good idea for the DBMS to update the catalog automatically when a change is made in the database structure? Could users cause problems by updating the catalog themselves?
8. State the two integrity rules. Indicate the reasons for enforcing each rule.
9. How can the structure of a table be changed in SQL? What general types of changes are possible? Which commands are used to implement these changes?
10. List the two properties specified by Codd which a system must satisfy in order to be considered relational.
11. List the four categories of systems proposed by Date and describe the characteristics of systems in each category.

5 *Database Design I: Normalization*

OBJECTIVES

1. Present the idea of functional dependence.
2. Define the term primary key.
3. Define first normal form (1NF), second normal form (2NF), and third normal form (3NF).
4. Describe the problems associated with relations (tables) that are not in 1NF, 2NF, or 3NF, along with the mechanism for converting to all three.
5. Discuss the problems associated with incorrect conversions to 3NF.

5.1 INTRODUCTION

We have discussed the basic relational model, its structure, and the various ways of manipulating data within a relational database. In this chapter, we discuss the **normalization** process and its underlying concepts and features. Normalization enables us to analyze the design of a relational database to see whether it is bad; that is, normalization gives us a method for identifying the existence of potential problems, called **update anomalies**, in the design. The normalization process also supplies methods for correcting these problems.

The process involves various types of **normal forms. First normal form** (1NF), **second normal form** (2NF), and **third normal form** (3NF) are three of these types. It is these three that will be of the greatest use to us during database design. They form a progression in which a table that is in 1NF is better than a table that is not in 1NF; a table that is in 2NF is better yet; and so on. The goal of this process is to allow us to start with a table or collection of tables and produce a new collection of tables that is equivalent to the original collection (i.e., that represents the same information) but is free of problems. For practical purposes, this means that tables in the new collection will be in 3NF.

These normal forms were initially defined by Codd in 1972 (see [4]). Subsequently, it was discovered that the definition of third normal form was inadequate for certain situations. A revised and stronger definition was provided by Boyce and Codd in 1974 (see [5]). It is this more recent definition of third normal form (sometimes called **Boyce-Codd normal form**) that we shall examine later in this section.

We begin by discussing two crucial concepts that are fundamental to the understanding of the normalization process: functional dependence and keys. We then discuss first, second, and third normal forms. Finally, we look at the application of normalization to database design. We will then be ready to begin our study of the database design process in the next chapter.

Many of the examples in this chapter use data from the *PREMIERE PRODUCTS* example (see Figure 5.1).

Figure 5.1

*PREMIERE
PRODUCTS*
sample data

SLSREP

SLSRNUMB	SLSRNAME	SLSRADDR	TOTCOMM	COMMRATE
3	MARY JONES	123 MAIN,GRANT,MI	2150.00	.05
6	WILLIAM SMITH	102 RAYMOND,ADA,MI	4912.50	.07
12	SAM BROWN	419 HARPER,LANSING,MI	2150.00	.05

CUSTOMER

CUSTNUMB	CUSTNAME	ADDRESS	BALANCE	CREDLIM	SLSRNUMB
124	SALLY ADAMS	481 OAK,LANSING,MI	418.75	500	3
256	ANN SAMUELS	215 PETE,GRANT,MI	10.75	800	6
311	DON CHARLES	48 COLLEGE,IRA,MI	200.10	300	12
315	TOM DANIELS	914 CHERRY,KENT,MI	320.75	300	6
405	AL WILLIAMS	519 WATSON,GRANT,MI	201.75	800	12
412	SALLY ADAMS	16 ELM,LANSING,MI	908.75	1000	3
522	MARY NELSON	108 PINE,ADA,MI	49.50	800	12
567	JOE BAKER	808 RIDGE,HARPER,MI	201.20	300	6
587	JUDY ROBERTS	512 PINE,ADA,MI	57.75	500	6
622	DAN MARTIN	419 CHIP,GRANT,MI	575.50	500	3

ORDERS

ORDNUMB	ORDDTE	CUSTNUMB
12489	90294	124
12491	90294	311
12494	90494	315
12495	90494	256
12498	90594	522
12500	90594	124
12504	90594	522

ORDLNE

ORDNUMB	PARTNUMB	NUMBORD	QUOTPRCE
12489	AX12	11	14.95
12491	BT04	1	402.99
12491	BZ66	1	311.95
12494	CB03	4	175.00
12495	CX11	2	57.95
12498	AZ52	2	22.95
12498	BA74	4	4.95
12500	BT04	1	402.99
12504	CZ81	2	108.99

PART

PARTNUMB	PARTDESC	UNONHAND	ITEMCLSS	WREHSENM	UNITPRCE
AX12	IRON	104	HW	3	17.95
AZ52	SKATES	20	SG	2	24.95
BA74	BASEBALL	40	SG	1	4.95
BH22	TOASTER	95	HW	3	34.95
BT04	STOVE	11	AP	2	402.99
BZ66	WASHER	52	AP	3	311.95
CA14	SKILLET	2	HW	3	19.95
CB03	BIKE	44	SG	1	187.50
CX11	MIXER	112	HW	3	57.95
CZ81	WEIGHTS	208	SG	2	108.99

5.2 FUNCTIONAL DEPENDENCE

The concept of functional dependence is crucial to the material in the rest of this chapter. Functional dependence is a fancy name for what is basically a simple idea. To illustrate it, suppose that the *SLSREP* table for *PREMIERE PRODUCTS* is as shown in Figure 5.2.

The only difference between this *SLSREP* table and the one we have been looking at previously is the addition of an extra column, *PAYCLASS* (pay class).

Figure 5.2

SLSREP table with additional column, *PAYCLASS*

SLSREP

SLSRNUMB	SLSRNAME	SLSRADDR	TOTCOMM	PAYCLASS	COMMRATE
3	MARY JONES	123 MAIN,GRANT,MI	2150.00	1	.05
6	WILLIAM SMITH	102 RAYMOND,ADA,MI	4912.50	2	.07
12	SAM BROWN	419 HARPER,LANSING,MI	2150.00	1	.05

Let's suppose further that one of the policies at Premiere Products is that all sales reps in any given pay class get the same commission rate. If you were asked to describe this policy in another way, you might say something like, "A sales rep's pay class *determines* his or her commission rate." Or you might say, "A sales rep's commission rate *depends on* his or her pay class." If you said either of these things, you would be using the word *determines* or the words *depends on* in exactly the fashion that we will be using them. If we wanted to be formal, we would precede either expression with the word *functionally*. Thus we might say, "A sales rep's pay class *functionally determines* his or her commission rate," or "A sales rep's commission rate *functionally depends on* his or her pay class." The formal definition of functional dependence is as follows:

Definition: An attribute, B, is **functionally dependent** on another attribute, A (or possibly a collection of attributes), if a value for A determines a single value for B at any one time.

We can think of this as follows. If we are given a value for A, do we know that we will be able to find a single value for B? If so, B is functionally dependent on A (often written as A --> B). If B is functionally dependent on A, we also say that A **functionally determines** B.

For example, in the *CUSTOMER* table, is the *CUSTNAME* functionally dependent on *CUSTNUMB*? The answer is yes. If we are given customer number 124, for example, we would find a *single* name, Sally Adams, associated with it.

In the same *CUSTOMER* table, is *ADDRESS* functionally dependent on *CUSTNAME*? Here the answer is no since, given the name Sally Adams, we would not be able to find a single address.

In the *ORDLNE* table, is the *NUMBORD* functionally dependent on *ORDNUMB*? No. *ORDNUMB* does not give enough information. Is it functionally dependent on *PARTNUMB*? No. Again, not enough information is given. In reality, *NUMBORD* is functionally dependent on the **concatenation** (combination) of *ORDNUMB* and *PARTNUMB*.

At this point, a question naturally arises: How do we determine functional dependencies? Can we determine them by looking at sample data? The answer is no.

Consider Figure 5.3, in which customer names happen to be unique. It is very tempting to say that *CUSTNAME* functionally determines *ADDRESS* (or equivalently that *ADDRESS* is functionally dependent on *CUSTNAME*). After all, given the name of a customer, we can find the single address. But what happens when customer 412, whose name also happens to be Sally Adams, is added to the database? We then have the situation exhibited in Figure 5.4.

Figure 5.3

CUSTOMER table

CUSTOMER

CUSTNUMB	CUSTNAME	ADDRESS	BALANCE	CREDLIM	SLSRNUMB
124	SALLY ADAMS	481 OAK,LANSING,MI	418.75	500	3
256	ANN SAMUELS	215 PETE,GRANT,MI	10.75	800	6
311	DON CHARLES	48 COLLEGE,IRA,MI	200.10	300	12
315	TOM DANIELS	914 CHERRY,KENT,MI	320.75	300	6
405	AL WILLIAMS	519 WATSON,GRANT,MI	201.75	800	12
522	MARY NELSON	108 PINE,ADA,MI	49.50	800	12
567	JOE BAKER	808 RIDGE,HARPER,MI	201.20	300	6
587	JUDY ROBERTS	512 PINE,ADA,MI	57.75	500	6
622	DAN MARTIN	419 CHIP,GRANT,MI	575.50	500	3

Figure 5.4

CUSTOMER table
with second Sally
Adams

CUSTOMER

CUSTNUMB	CUSTNAME	ADDRESS	BALANCE	CREDLIM	SLSRNUMB
124	SALLY ADAMS	481 OAK,LANSING,MI	418.75	500	3
256	ANN SAMUELS	215 PETE,GRANT,MI	10.75	800	6
311	DON CHARLES	48 COLLEGE,IRA,MI	200.10	300	12
315	TOM DANIELS	914 CHERRY,KENT,MI	320.75	300	6
405	AL WILLIAMS	519 WATSON,GRANT,MI	201.75	800	12
412	SALLY ADAMS	16 ELM,LANSING,MI	908.75	1000	3
522	MARY NELSON	108 PINE,ADA,MI	49.50	800	12
567	JOE BAKER	808 RIDGE,HARPER,MI	201.20	300	6
587	JUDY ROBERTS	512 PINE,ADA,MI	57.75	500	6
622	DAN MARTIN	419 CHIP,GRANT,MI	575.50	500	3

If the name we are given is Sally Adams, we can no longer find a single address. Thus we were misled by our original sample data. The only way to really determine the functional dependencies that exist is to examine the user's policies.

5.3   KEYS

A second underlying concept of the normalization process is that of the primary key. It builds on functional dependence, and it completes the background required for an understanding of the normal forms.

> *Definition:* Attribute A (or a collection of attributes) is the **primary key** for a relation (table), R, if
>
> 1. *All* attributes in R are functionally dependent on A.
> 2. No subcollection of the attributes in A (assuming A is a collection of attributes and not just a single attribute) also has property 1.

For example, is *CUSTNAME* the primary key for the *CUSTOMER* table? No, since the other attributes are not functionally dependent on name. (Note that the answer would be different in an organization that had a policy enforcing uniqueness of customer names.)

Is *CUSTNUMB* the primary key for the *CUSTOMER* table? Yes, since all attributes in the *CUSTOMER* table are functionally dependent on *CUSTNUMB*.

Is *ORDNUMB* the primary key for the *ORDLNE* table? No, since it does not uniquely determine *NUMBORD* or *QUOTPRCE*.

Is the combination of the *ORDNUMB* and the *PARTNUMB* the primary key for the *ORDLNE* table? Yes, since all attributes can be determined by this combination, and nothing less will do.

Is the combination of the *PARTNUMB* and the *PARTDESC* the primary key for the *PART* table? No. Though it is true that all attributes of the *PART* table can be determined by this combination, something less, namely, the *PARTNUMB* alone, also has this property.

Occasionally (but not often) there might be more than one possibility for the primary key. For example, in an *EMPLOYEE* table either the *EMPNUMB* or the *SSNUMB* (social security number) could serve as the key. In this case one of these is designated as the primary key. The other is referred to as a **candidate key**. A candidate key is a collection of attributes that has the same properties presented in the definition of the primary key. (Technically, the definition given for primary key really defines candidate key. From all of the candidate keys one is chosen to be the primary key. The candidate keys that are not chosen to be the primary key are often referred to as **alternate keys**.)

Note: The primary key is frequently called simply the *key* in other studies on database management and the relational model. We will continue to use the term primary key in order to clearly distinguish among the several different concepts of a key that we will encounter.

5.4   *FIRST NORMAL FORM*

A relation (table) that contains a repeating group is called an **unnormalized relation**. (Technically, it is not a relation at all.) Removal of repeating groups is the starting point in our quest for relations that are as free of problems as possible. Relations without repeating groups are said to be in first normal form.

> *Definition:* A relation (table) is in **first normal form** (1NF) if it does not contain repeating groups.

As an example, consider the following *ORDERS* table, in which there is a repeating group consisting of *PARTNUMB* and *NUMBORD*. As the example shows, there is one row per order with *PARTNUMB, NUMBORD* repeated as many times as is necessary.

```
          PK                          RG
ORDERS (ORDNUMB, ORDDTE, PARTNUMB, NUMBORD)
```

(This notation indicates a table called *ORDERS*, consisting of a primary key, *ORDNUMB*, an attribute, *ORDDTE*, and a repeating group containing two attributes, *PARTNUMB* and *NUMBORD*.) Figure 5.5 shows a sample of this table.

Figure 5.5

Sample unnormalized table

ORDERS

ORDNUMB	ORDDTE	PARTNUMB	NUMBORD
12489	90294	AX12	11
12491	90294	BT04	1
		BZ66	1
12494	90494	CB03	4
12495	90494	CX11	2
12498	90594	AZ52	2
		BA74	4
12500	90594	BT04	1
12504	90594	CZ81	2

To convert this table to 1NF, the repeating group is removed, giving the following:

```
ORDERS(ORDNUMB, ORDDTE, PARTNUMB, NUMBORD)
```

The corresponding example of the new table is shown in Figure 5.6.

Figure 5.6

Result of normalization (conversion to 1NF)

ORDERS

ORDNUMB	ORDDTE	PARTNUMB	NUMBORD
12489	90294	AX12	11
12491	90294	BT04	1
12491	90294	BZ66	1
12494	90494	CB03	4
12495	90494	CX11	2
12498	90594	AZ52	2
12498	90594	BA74	4
12500	90594	BT04	1
12504	90594	CZ81	2

Note that the second row of the unnormalized table indicates that part BZ66 and part BT04 are both present for order 12491. In the normalized table, this information is represented by *two* rows, the second and third. The primary key to the unnormalized *ORDERS* table was the *ORDNUMB* alone. The primary key to the normalized table is now the combination of *ORDNUMB* and *PARTNUMB*. In general it will be true that the primary key will expand in converting a non-1NF table to 1NF. It will typically include the original primary key concatenated with the key to the repeating group; i.e., the attribute that distinguishes one occurrence of the repeating group from another within a given row in the table. In this case, *PARTNUMB* is the key to the repeating group and thus becomes part of the primary key of the 1NF table.

5.5 SECOND NORMAL FORM

Even though the following table is in 1NF, problems exist within the table that will cause us to want to restructure it. Consider the table:

```
ORDERS(ORDNUMB, ORDDTE, PARTNUMB, PARTDESC, NUMBORD, QUOTPRCE)
```

with the functional dependencies:

```
ORDNUMB --> ORDDTE
PARTNUMB --> PARTDESC
ORDNUMB, PARTNUMB --> NUMBORD, QUOTPRCE
```

Thus *ORDNUMB* determines *ORDDTE*, *PARTNUMB* determines *PARTDESC*, and the concatenation of *ORDNUMB* and *PARTNUMB* determines *NUMBORD* and *QUOTPRCE*. Consider the sample of this table shown in Figure 5.7.

Figure 5.7

Sample *ORDERS* table

ORDERS

ORDNUMB	ORDDTE	PARTNUMB	PARTDESC	NUMBORD	QUOTPRCE
12489	90294	AX12	IRON	11	14.95
12491	90294	BT04	STOVE	1	402.99
12491	90294	BZ66	WASHER	1	311.95
12494	90494	CB03	BIKE	4	175.00
12495	90494	CX11	MIXER	2	57.95
12498	90594	AZ52	SKATES	2	22.95
12498	90594	BA74	BASEBALL	4	4.95
12500	90594	BT04	STOVE	1	402.99
12504	90594	CZ81	WEIGHTS	2	108.99

As you can see in the example, the description of a specific part, BT04 for example, occurs several times in the table. This redundancy causes several problems. It is certainly wasteful of space, but that in itself is not nearly as serious as some of the other problems. These other problems are called **update anomalies** and they fall into four categories:

1. **Update.** A change to the description of part BT04 requires not one change but several—we have to change each row in which BT04 appears. This certainly makes the update process much more cumbersome; it is more complicated logically and takes more time to update.
2. **Inconsistent data.** There is nothing about the design that would prohibit part BT04 from having two different descriptions in the database. In fact, if it occurs in twenty rows, it could conceivably have twenty *different* descriptions in the database!!!
3. **Additions.** We have a real problem when we try to add a new part and its description to the database. Since the primary key for the table consists of both *ORDNUMB* and *PARTNUMB*, we need values for both of these in order to add a new row. If we have a part to add but there are as yet no orders for it, what do we use for an *ORDNUMB*? Our only solution would be to make up a dummy order number and then replace it with a real *ORDNUMB* once an order for this part had actually been received. Certainly this is not an acceptable solution!
4. **Deletions.** In the example above, if we delete order 12489 from the database, we also *lose* the fact that part AX12 is called IRON.

The problems just described occur because we have an attribute, *PARTDESC*, that is dependent on only a portion of the primary key, *PARTNUMB*, and *not* on the complete primary key. This leads to the definition of second normal form. Second normal form represents an improvement over first normal form since it eliminates these update anomalies in these situations. First, we need to define nonkey attribute.

Definition: An attribute is a **nonkey attribute** if it is not a part of the primary key.

We can now provide a definition for second normal form.

Definition: A relation (table) is in **second normal form** (2NF) if it is in first normal form and no nonkey attribute is dependent on only a portion of the primary key.

For another perspective on 2NF, consider Figure 5.8. This type of diagram, sometimes called a **dependency diagram**, indicates all of the functional dependencies present in the *ORDERS* table through arrows. The arrows above the boxes indicate the normal dependencies that should be present; i.e., the primary key functionally determines all other attributes. (In this case, the concatenation of *ORDNUMB* and *PARTNUMB* determines all other attributes.) It is the arrows below the boxes that prevent the table from being in 2NF. These arrows represent what is often termed **partial dependencies**, which are dependencies on something less than the key. In fact, an alternative definition for 2NF is that a table is in 2NF if it is in 1NF but contains no partial dependencies.

Figure 5.8

Dependencies in *ORDERS* table (blue arrows represent partial dependencies)

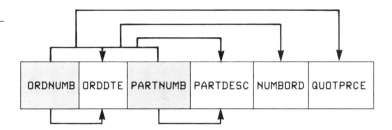

Either way we view 2NF, we can now name the fundamental problem with the *ORDERS* table: it is *not* in 2NF. While it may be pleasing to have a name for the problem, what we really need, of course, is a method to *correct* it. Such a method follows.

First, for each subset of the set of attributes that make up the primary key, begin a table with this subset as its primary key. For the *ORDERS* table, this would give:

```
(ORDNUMB,
(PARTNUMB,
(ORDNUMB, PARTNUMB,
```

Next, place each of the other attributes with the appropriate primary key; that is, place each one with the minimal collection on which it depends. For the *ORDERS* table this would yield:

```
(ORDNUMB, ORDDTE)
(PARTNUMB, PARTDESC)
(ORDNUMB, PARTNUMB, NUMBORD, QUOTPRCE)
```

Each of these tables can now be given a name that is descriptive of the meaning of the table, such as *ORDERS, PART,* or *ORDLNE,* for example. Figure 5.9 shows samples of the tables involved.

Figure 5.9

Conversion to 2NF

ORDERS

ORDNUMB	ORDDTE	PARTNUMB	PARTDESC	NUMBORD	QUOTPRCE
12489	90294	AX12	IRON	11	14.95
12491	90294	BT04	STOVE	1	402.99
12491	90294	BZ66	WASHER	1	311.95
12494	90494	CB03	BIKE	4	175.00
12495	90494	CX11	MIXER	2	57.95
12498	90594	AZ52	SKATES	2	22.95
12498	90594	BA74	BASEBALL	4	4.95
12500	90594	BT04	STOVE	1	402.99
12504	90594	CZ81	WEIGHTS	2	108.99

is replaced by

ORDERS

ORDNUMB	ORDDTE
12489	90294
12491	90294
12494	90494
12495	90494
12498	90594
12500	90594
12504	90594

PART

PARTNUMB	PARTDESC
AX12	IRON
AZ52	SKATES
BA74	BASEBALL
BH22	TOASTER
BT04	STOVE
BZ66	WASHER
CA14	SKILLET
CB03	BIKE
CX11	MIXER
CZ81	WEIGHTS

ORDLNE

ORDNUMB	PARTNUMB	NUMBORD	QUOTPRCE
12489	AX12	11	14.95
12491	BT04	1	402.99
12491	BZ66	1	311.95
12494	CB03	4	175.00
12495	CX11	2	57.95
12498	AZ52	2	22.95
12498	BA74	4	4.95
12500	BT04	1	402.99
12504	CZ81	2	108.99

Note that the update anomalies have been eliminated. A description appears only once, so we do not have the redundancy that we did in the earlier design. Changing the description of part BT04 to OVEN is now a simple process involving a single change. Since the description for a part occurs in one single place, it is not possible to have multiple descriptions for a single part in the database at the same time. To add a new part and its description, we create a new row in the *PART* table and thus there is no need to have an order exist for that part. Also, deleting order 12489 does not cause part number AX12 to be deleted from the *PART* table, and thus we still have its description (IRON) in the database. Finally, we have not lost any information in the process. The data in the original design can be reconstructed from the data in the new design.

5.6   THIRD NORMAL FORM

Problems can still exist with tables that are in 2NF. Consider the following *CUSTOMER* table:

    CUSTOMER(CUSTNUMB, CUSTNAME, ADDRESS, SLSRNUMB, SLSRNAME)

with the functional dependencies:

    CUSTNUMB --> CUSTNAME, ADDRESS, SLSRNUMB, SLSRNAME
    SLSRNUMB --> SLSRNAME

(*CUSTNUMB* determines all of the other attributes. In addition *SLSRNUMB* determines *SLSRNAME*.)

If the primary key of a table is a single column, the table is automatically in second normal form. (If the table were not in 2NF, some column would be dependent on only a *portion* of the primary key, which is impossible when the primary key is just one column.) Thus, the *CUSTOMER* table is in second normal form.

As the sample of this table, shown in Figure 5.10, demonstrates, this table possesses problems similar to those encountered earlier, even though it is in second normal form. In this case it is the name of a sales rep that can occur many times in the table; see sales rep 12 (Sam Brown), for example. This redundancy results in the same exact set of problems that was described in the previous *ORDERS* table.

Figure 5.10

Sample *CUSTOMER* table

CUSTOMER

CUSTNUMB	CUSTNAME	ADDRESS	SLSRNUMB	SLSRNAME
124	SALLY ADAMS	481 OAK,LANSING,MI	3	MARY JONES
256	ANN SAMUELS	215 PETE,GRANT,MI	6	WILLIAM SMITH
311	DON CHARLES	48 COLLEGE,IRA,MI	12	SAM BROWN
315	TOM DANIELS	914 CHERRY,KENT,MI	6	WILLIAM SMITH
405	AL WILLIAMS	519 WATSON,GRANT,MI	12	SAM BROWN
412	SALLY ADAMS	16 ELM,LANSING,MI	3	MARY JONES
522	MARY NELSON	108 PINE,ADA,MI	12	SAM BROWN
567	JOE BAKER	808 RIDGE,HARPER,MI	6	WILLIAM SMITH
587	JUDY ROBERTS	512 PINE,ADA,MI	6	WILLIAM SMITH
622	DAN MARTIN	419 CHIP,GRANT,MI	3	MARY JONES

In addition to the problem of wasted space, we have similar update anomalies, as follows:

1. **Update.** A change to the name of a sales rep requires not one change but several. Again the update process becomes very cumbersome.
2. **Inconsistent data.** There is nothing about the design that would prohibit a sales rep from having two different names in the database. In fact, if the same sales rep

represents twenty different customers (and thus would be found on twenty different rows), he or she could have twenty different names in the database.

3. **Additions.** In order to add sales rep 47, whose name is Mary Daniels, to the database, we must have at least one customer whom she represents. If she has not yet been assigned any customers, then either we cannot record the fact that her name is Mary Daniels or we have to create a fictitious customer for her to represent. Again, this is not a very desirable solution to the problem.

4. **Deletions.** If we were to delete all of the customers of sales rep 6 from the database, then we would also lose the name of sales rep 6.

These update anomalies are due to the fact that *SLSRNUMB* determines *SLSRNAME* but *SLSRNUMB* is not the primary key. As a result, the same *SLSRNUMB* and consequently the same *SLSRNAME* can appear on many different rows.

We've seen that 2NF is an improvement over 1NF, but in order to eliminate 2NF problems, we need an even better strategy for creating tables in our database. Third normal form gives us that strategy. Before we look at third normal form, however, we need to become familiar with the special name that is given to any column that determines another column (like *SLSRNUMB* in the *CUSTOMER* table).

Definition: Any attribute (or collection of attributes) that determines another attribute is called a **determinant**.

Certainly the **primary key** in a table will be a determinant. In fact, by definition, any **candidate key** will be a determinant. (Remember that a candidate key is an attribute or collection of attributes which could have functioned as the primary key.) In this case, *SLSRNUMB* is a determinant but it is certainly not a candidate key, and that is the problem.

Definition: A relation (table) is in **third normal form** (3NF) if it is in second normal form and if the only determinants it contains are candidate keys.

Again, for an additional perspective, we will consider a dependency diagram, as shown in Figure 5.11. As before, the arrows above the boxes represent the normal dependencies of all attributes on the primary key. It is the arrow below the boxes that causes the problem. The presence of this arrow makes *SLSRNUMB* a determinant. If there were arrows from *SLSRNUMB* to all of the attributes, *SLSRNUMB* would be a candidate key and we would not have a problem. The absence of these arrows indicates that this table possesses a determinant that is not a candidate key. Thus, the table is not in 3NF.

Figure 5.11

Dependencies in *CUSTOMER* table (*SLSRNUMB* is a determinant since it functionally determines *SLSRNAME*)

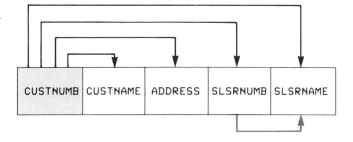

We have now named the problem with the *CUSTOMER* table: it is not in 3NF. What we need is a scheme to correct the deficiency in the *CUSTOMER* table and in all tables having similar deficiencies. Such a method follows.

First, for each determinant that is not a candidate key, remove from the table the attributes that depend on this determinant. Next, create a new table containing all the attributes from the original table that depend on this determinant. Finally, make the determinant the primary key of this new table.

In the *CUSTOMER* table, for example, *SLSRNAME* is removed since it depends on the determinant *SLSRNUMB*, which is not a candidate key. A new table is formed, consisting of *SLSRNUMB* as the primary key and *SLSRNAME*. Specifically

```
CUSTOMER(CUSTNUMB, CUSTNAME, ADDRESS, SLSRNUMB, SLSRNAME)
```

is replaced by

```
CUSTOMER(CUSTNUMB, CUSTNAME, ADDRESS, SLSRNUMB)
```

and

```
SLSREP(SLSRNUMB, SLSRNAME)
```

Figure 5.12 shows samples of the tables involved.

Figure 5.12

Conversion to 3NF

CUSTOMER

CUSTNUMB	CUSTNAME	ADDRESS	SLSRNUMB	SLSRNAME
124	SALLY ADAMS	481 OAK,LANSING,MI	3	MARY JONES
256	ANN SAMUELS	215 PETE,GRANT,MI	6	WILLIAM SMITH
311	DON CHARLES	48 COLLEGE,IRA,MI	12	SAM BROWN
315	TOM DANIELS	914 CHERRY,KENT,MI	6	WILLIAM SMITH
405	AL WILLIAMS	519 WATSON,GRANT,MI	12	SAM BROWN
412	SALLY ADAMS	16 ELM,LANSING,MI	3	MARY JONES
522	MARY NELSON	108 PINE,ADA,MI	12	SAM BROWN
567	JOE BAKER	808 RIDGE,HARPER,MI	6	WILLIAM SMITH
587	JUDY ROBERTS	512 PINE,ADA,MI	6	WILLIAM SMITH
622	DAN MARTIN	419 CHIP,GRANT,MI	3	MARY JONES

is replaced by

CUSTOMER

CUSTNUMB	CUSTNAME	ADDRESS	SLSRNUMB
124	SALLY ADAMS	481 OAK,LANSING,MI	3
256	ANN SAMUELS	215 PETE,GRANT,MI	6
311	DON CHARLES	48 COLLEGE,IRA,MI	12
315	TOM DANIELS	914 CHERRY,KENT,MI	6
405	AL WILLIAMS	519 WATSON,GRANT,MI	12
412	SALLY ADAMS	16 ELM,LANSING,MI	3
522	MARY NELSON	108 PINE,ADA,MI	12
567	JOE BAKER	808 RIDGE,HARPER,MI	6
587	JUDY ROBERTS	512 PINE,ADA,MI	6
622	DAN MARTIN	419 CHIP,GRANT,MI	3

SLSREP

SLSRNUMB	SLSRNAME
3	MARY JONES
6	WILLIAM SMITH
12	SAM BROWN

Have we now corrected all previously identified problems? A sales rep's name appears only once, thus avoiding redundancy and making the process of changing a sales rep's name a very simple one. It is not possible with this design for the same sales rep to have two different names in the database. To add a new sales rep to the database, we add a row in the *SLSREP* table so that it is not necessary to have a customer whom the sales rep represents. Finally, deleting all of the customers of a given sales rep will not remove the sales rep's record from the *SLSREP* table, so we do retain the sales rep's name; all of the data in the original table can be reconstructed from the data in the new collection of tables. All previously mentioned problems have indeed been solved.

5.7　INCORRECT DECOMPOSITIONS

It is important to note that the decomposition of a table into two or more 3NF tables *must* be accomplished by the method indicated even though there are other possibilities that might seem at first glance to be legitimate. Let us examine two other decompositions of the *CUSTOMER* table into 3NF tables in order to understand the difficulties they pose.

What if, in the decomposition process,

CUSTOMER(<u>CUSTNUMB</u>, CUSTNAME, ADDRESS, SLSRNUMB, SLSRNAME)

is replaced by

CUSTOMER(<u>CUSTNUMB</u>, CUSTNAME, ADDRESS, SLSRNUMB)

and

SLSREP(<u>CUSTNUMB</u>, SLSRNAME)

Samples of these tables are shown in Figure 5.13.

Figure 5.13

Incorrect
decomposition

CUSTOMER

CUSTNUMB	CUSTNAME	ADDRESS	SLSRNUMB	SLSRNAME
124	SALLY ADAMS	481 OAK,LANSING,MI	3	MARY JONES
256	ANN SAMUELS	215 PETE,GRANT,MI	6	WILLIAM SMITH
311	DON CHARLES	48 COLLEGE,IRA,MI	12	SAM BROWN
315	TOM DANIELS	914 CHERRY,KENT,MI	6	WILLIAM SMITH
405	AL WILLIAMS	519 WATSON,GRANT,MI	12	SAM BROWN
412	SALLY ADAMS	16 ELM,LANSING,MI	3	MARY JONES
522	MARY NELSON	108 PINE,ADA,MI	12	SAM BROWN
567	JOE BAKER	808 RIDGE,HARPER,MI	6	WILLIAM SMITH
587	JUDY ROBERTS	512 PINE,ADA,MI	6	WILLIAM SMITH
622	DAN MARTIN	419 CHIP,GRANT,MI	3	MARY JONES

is replaced by

CUSTOMER

CUSTNUMB	CUSTNAME	ADDRESS	SLSRNUMB
124	SALLY ADAMS	481 OAK,LANSING,MI	3
256	ANN SAMUELS	215 PETE,GRANT,MI	6
311	DON CHARLES	48 COLLEGE,IRA,MI	12
315	TOM DANIELS	914 CHERRY,KENT,MI	6
405	AL WILLIAMS	519 WATSON,GRANT,MI	12
412	SALLY ADAMS	16 ELM,LANSING,MI	3
522	MARY NELSON	108 PINE,ADA,MI	12
567	JOE BAKER	808 RIDGE,HARPER,MI	6
587	JUDY ROBERTS	512 PINE,ADA,MI	6
622	DAN MARTIN	419 CHIP,GRANT,MI	3

SLSREP

CUSTNUMB	SLSRNAME
124	MARY JONES
256	WILLIAM SMITH
311	SAM BROWN
315	WILLIAM SMITH
405	SAM BROWN
412	MARY JONES
522	SAM BROWN
567	WILLIAM SMITH
587	WILLIAM SMITH
622	MARY JONES

Both of the new tables are in 3NF. In addition, by joining these two tables together on *CUSTNUMB* we can reconstruct the original *CUSTOMER* table. The result, however, still suffers from some of the same kinds of problems that the original *CUSTOMER* table did. Consider, for example, the redundancy in the storage of sales reps' names, the problem encountered in changing the name of sales rep 12, and the difficulty of adding a new sales rep for whom there are as yet no customers. In addition, since the sales rep number is in one table and the sales rep name is in another, we have actually *split a functional dependence across two different tables*. Thus, this decomposition, while it may appear to be valid, is definitely not a desirable way to create 3NF tables.

There is another decomposition that we might choose, and that is to replace

CUSTOMER(<u>CUSTNUMB</u>, CUSTNAME, ADDRESS, SLSRNUMB, SLSRNAME)

by

CUSTOMER(<u>CUSTNUMB</u>, CUSTNAME, ADDRESS, SLSRNAME)

and

SLSREP(<u>SLSRNUMB</u>, SLSRNAME)

Samples of these tables are shown in Figure 5.14.

Figure 5.14

Second incorrect
decomposition

CUSTOMER

CUSTNUMB	CUSTNAME	ADDRESS	SLSRNUMB	SLSRNAME
124	SALLY ADAMS	481 OAK,LANSING,MI	3	MARY JONES
256	ANN SAMUELS	215 PETE,GRANT,MI	6	WILLIAM SMITH
311	DON CHARLES	48 COLLEGE,IRA,MI	12	SAM BROWN
315	TOM DANIELS	914 CHERRY,KENT,MI	6	WILLIAM SMITH
405	AL WILLIAMS	519 WATSON,GRANT,MI	12	SAM BROWN
412	SALLY ADAMS	16 ELM,LANSING,MI	3	MARY JONES
522	MARY NELSON	108 PINE,ADA,MI	12	SAM BROWN
567	JOE BAKER	808 RIDGE,HARPER,MI	6	WILLIAM SMITH
587	JUDY ROBERTS	512 PINE,ADA,MI	6	WILLIAM SMITH
622	DAN MARTIN	419 CHIP,GRANT,MI	3	MARY JONES

is replaced by

CUSTOMER

CUSTNUMB	CUSTNAME	ADDRESS	SLSRNAME
124	SALLY ADAMS	481 OAK,LANSING,MI	MARY JONES
256	ANN SAMUELS	215 PETE,GRANT,MI	WILLIAM SMITH
311	DON CHARLES	48 COLLEGE,IRA,MI	SAM BROWN
315	TOM DANIELS	914 CHERRY,KENT,MI	WILLIAM SMITH
405	AL WILLIAMS	519 WATSON,GRANT,MI	SAM BROWN
412	SALLY ADAMS	16 ELM,LANSING,MI	MARY JONES
522	MARY NELSON	108 PINE,ADA,MI	SAM BROWN
567	JOE BAKER	808 RIDGE,HARPER,MI	WILLIAM SMITH
587	JUDY ROBERTS	512 PINE,ADA,MI	WILLIAM SMITH
622	DAN MARTIN	419 CHIP,GRANT,MI	MARY JONES

SLSREP

SLSRNUMB	SLSRNAME
3	MARY JONES
6	WILLIAM SMITH
12	SAM BROWN

This seems to be a possibility. Not only are both tables in 3NF, but joining them together based on *SLSRNAME* seems to reconstruct the data in the original table. Or does it? Suppose that the name of sales rep 6 is also Mary Jones. In that case, when we join the two new tables together, we will get a row in which customer 124 (Sally Adams) is associated with sales rep 3 (Mary Jones) and *another* row in which customer 124 is associated with sales rep 6 (William Smith). Since we obviously want decompositions that preserve the original information, this scheme is not appropriate.

Q & A

Question:

Using the types of entities found in a college environment (faculty, students, departments, courses, etc.), create an example of a table that is in 1NF but not in 2NF and an example of a table that is in 2NF but not in 3NF. In each case justify the answers and show how to convert to the higher forms.

Answer: There are many possible solutions. If your solution differs from the one we will look at, this does not mean that it is an unsatisfactory solution.

To create a 1NF table that is not in 2NF, we need a table that (a) has no repeating groups and (b) has at least one attribute that is dependent on only a portion of the primary key. For an attribute to be dependent on a portion of the primary key, the key must contain at least two attributes. Following is a picture of what we need:

(___1___, ___2___, 3 , 4)

This table contains four attributes, numbered 1, 2, 3, and 4, in which attributes 1 and 2 functionally determine both attributes 3 and 4. In addition, neither attribute 1 nor attribute 2 can determine *all* other attributes, otherwise the key would contain only this one attribute. Finally, we want part of the key, say, attribute 2, to determine another attribute, say, attribute 4. Now that we have the pattern we need, we would like to find attributes from within the college environment to fit it. One example would be:

(STUNUMB, CRSENUMB, GRADE, CRSEDESC)

In this example, the concatenation of *STUNUMB* (student number) and *CRSENUMB* (course number) determines both *GRADE* and *CRSEDESC* (course description). Both of these are required to determine *GRADE*, and thus the primary key consists of their concatenation (nothing less will do). The *CRSEDESC*, however, is only dependent on the *CRSENUMB*. This violates second normal form. To convert this table to 2NF we would replace it by the two tables

(STUNUMB, CRSENUMB, GRADE)

and

(CRSENUMB, CRSEDESC)

We would of course now give these tables appropriate names.

To create a table that is in 2NF but not in 3NF, we need a 2NF table in which there is a determinant that is *not* a candidate key. If we choose a table that has a single attribute as the primary key, it is automatically in 2NF, so the real problem is the determinant. We need a table like the following:

(___1___, 2 , 3)

This table contains three attributes, numbered 1, 2, and 3, in which attribute 1 determines each of the others and is thus the primary key. If, in addition, attribute 2 determines attribute 3, it is a determinant. If it does not also determine attribute 1, then it is not a candidate key. One example that fits this pattern would be:

(STUNUMB, ADVNUMB, ADVNAME)

Here the student number (*STUNUMB*) determines both the student's advisor's number (*ADVNUMB*) and the advisor's name (*ADVNAME*). *ADVNUMB* determines *ADVNAME* but *ADVNUMB* does not determine *STUNUMB*, since one advisor can have many advisees. This table is in 2NF but not in 3NF. To convert it to 3NF, we replace it by

(STUNUMB, ADVNUMB)

and

(ADVNUMB, ADVNAME)

Q & A

Question: Convert the following table to 3NF:

                   STUDENT(<u>STUNUMB</u>, STUNAME, NUMBCRED,
                   ADVNUMB, ADVNAME,
                              <u>CRSENUMB</u>, CRSEDESC, GRADE)

                   In this table, *STUNUMB* determines *STUNAME, NUMBCRED, ADVNUMB,* and *ADVNAME. ADVNUMB* determines *ADVNAME. CRSENUMB* determines *CRSEDESC.* The combination of a *STUNUMB* and a *CRSENUMB* determines a *GRADE.*

Answer: **Step 1.** Remove the repeating group to convert to 1NF. This yields:

                   STUDENT(<u>STUNUMB</u>, STUNAME, NUMBCRED,
                              ADVNUMB, ADVNAME,
                              <u>CRSENUMB</u>,
                              CRSEDESC, GRADE)

                   This table is now in 1NF, since it has no repeating groups. It is not, however, in 2NF, since *STUNAME* is dependent only on *STUNUMB,* which is only a portion of the primary key.

                   Step 2. Convert the 1NF table to 2NF. First, for each subset of the primary key, start a table with that subset as its key yielding:

                   (<u>STUNUMB</u>,
                   (<u>CRSENUMB</u>,
                   (<u>STUNUMB</u>, <u>CRSENUMB</u>,

                   Next, place the rest of the attributes with the minimal collection on which they depend, giving:

                   (<u>STUNUMB</u>, STUNAME, NUMBCRED, ADVNUMB,
                              ADVNAME)
                   (<u>CRSENUMB</u>, CRSEDESC)
                   (<u>STUNUMB</u>, <u>CRSENUMB</u>, GRADE)

                   Finally, we assign names to each of the newly created tables:

                   STUDENT(<u>STUNUMB</u>, STUNAME, NUMBCRED,
                              ADVNUMB, ADVNAME)
                   COURSE(<u>CRSENUMB</u>, CRSEDESC)
                   GRADE(<u>STUNUMB</u>, <u>CRSENUMB</u>, GRADE)

                   While these tables are all in 2NF, both *COURSE* and *GRADE* are also in 3NF. The *STUDENT* table is not, however, since it contains a determinant, *ADVNUMB,* that is not a candidate key.

                   Step 3. Convert the 2NF *STUDENT* table to 3NF by removing the attribute that depends on the determinant *ADVNUMB* and placing it in a separate table:

                   (<u>STUNUMB</u>, STUNAME, NUMBCRED, ADVNUMB)
                   (<u>ADVNUMB</u>, ADVNAME)

Step 4: Name these tables and put the entire collection together, giving:

```
        STUDENT(STUNUMB, STUNAME, NUMBCRED,
ADVNUMB)
        ADVISOR(ADVNUMB, ADVNAME)
        COURSE(CRSENUMB, CRSEDESC)
        GRADE(STUNUMB, CRSENUMB, GRADE)
```

SUMMARY

1. Column B is functionally dependent on column (or collection of columns) A if a value of A uniquely determines a value of B at any point in time.
2. The primary key is a column (or collection of columns), A, such that all other columns are functionally dependent on A and no subcollection of the columns in A also has this property.
3. If there is more than one possible choice for the primary key, one of the possibilities is chosen to be *the* primary key. The others are referred to as candidate keys.
4. A relation (table) is in first normal form (1NF) if it does not contain repeating groups.
5. A relation (table) is in second normal form (2NF) if it is in 1NF and if no column that is not a part of the primary key is dependent on only a portion of the primary key.
6. A determinant is any column that functionally determines another column.
7. A relation (table) is in third normal form (3NF) if it is in 2NF and if the only determinants it contains are candidate keys.
8. A collection of relations (tables) that is not in 3NF possesses inherent problems (called update anomalies). Replacing this collection by an equivalent collection of relations (tables) that is in 3NF removes these anomalies. This replacement must be done carefully, following a method like the one proposed in this text. If not, other problems, such as those discussed in this chapter, may very well be introduced.

KEY TERMS

alternate key
Boyce-Codd normal form
  (BCNF)
candidate key
database design
dependency diagram
determinant
first normal form (1NF)

functional dependence
normalization
partial dependency
primary key
repeating group
second normal form (2NF)
third normal form (3NF)
unnormalized relation

EXERCISES

1. Define functional dependence.
2. Give an example of an attribute, A, and another attribute, B, such that B is functionally dependent on A. Give an example of an attribute, C, and an attribute, D, such that D is not functionally dependent on C.
3. Define primary key.
4. Define candidate key.
5. Define first normal form.

6. Define second normal form. What types of problems are encountered in tables that are not in second normal form?

7. Define third normal form. What types of problems are encountered in tables that are not in third normal form?

8. Consider a student table containing student number; student name; student's major department; student's advisor's number; student's advisor's name; student's advisor's office number; student's advisor's phone number; student's number of credits; and student's class standing (freshman, sophomore, and so on). List the functional dependencies that exist, along with the assumptions that would support these dependencies.

9. Using the types of entities found in Henry's system (books, authors, publishers, and so on), create an example of a table that is in 1NF but not in 2NF and an example of a table that is in 2NF but not in 3NF. In each case, justify the answers and show how to convert to the higher forms.

10. Convert the following table to an equivalent collection of tables that is in 3NF.

```
PATIENT(HHNUMB, HHNAME, HHADDR, HHBAL, PATNUMB, PATNAME,
    SERVCODE, SERVDESC, SERVFEE, SERVDATE)
```

This is a table concerning information about patients of a dentist. Each patient belongs to a household. The head of the household is designated as HH in the table. The following dependencies exist in *PATIENT*:

```
PATNUMB --> HHNUMB, HHNAME, HHADDR, HHBAL, PATNAME
HHNUMB --> HHNAME, HHADDR, HHBAL
SERVCODE --> SERVDESC, SERVFEE
PATNUMB, SERVCODE --> SERVDATE
```

11. List the functional dependencies in the following table, subject to the specified conditions. Convert this table to an equivalent collection of tables that are in 3NF.

```
INVOICE(INVNUMB, CUSTNUMB, CUSTNAME, ADDRESS, INVDATE
    PARTNUMB, PARTDESC, UNITPRCE, NUMBSHIP)
```

This table concerns invoice information. For a given invoice (identified by the invoice number) there will be a single customer. The customer's number, name, and address appear on the invoice as well as the invoice date. Also, there may be several different parts appearing on the invoice. For each part that appears, the part number, description, price, and number shipped will be displayed. The price is from the current master price list.

12. Using your knowledge of a college environment, determine the functional dependencies that exist in the following table. After these have been determined, convert this table to an equivalent collection of tables that are in 3NF.

```
STUDENT(STUNUMB, STUNAME, NUMBCRED, ADVNUMB, ADVNAME, DEPTNUMB,
    DEPTNAME, CRSENUMB, CRSEDESC, CRSETERM, GRADE)
```

Database Design II: Design Methodology

OBJECTIVES

1. Discuss the general process and goals of database design.
2. Define user views and explain their function.
3. Present a methodology for database design at the information level as well as examples illustrating the use of this methodology.
4. Explain how to produce a pictorial representation of a database design.
5. Explain the process of mapping an information-level design to a design that is appropriate for a relational model system.

6.1 INTRODUCTION

Now that we have learned how to identify and correct bad designs, we will turn our attention to the design process itself; that is, the process of determining the tables and columns that will make up the database and determining the relationships between the various tables.

Database design is often approached as a two-step process. In the first step, a database is designed which satisfies the requirements as cleanly as possible. This step is called **information-level design**, and it is taken *independently* of any particular DBMS that will ultimately be used. In the second step, which is called the **physical-level design**, the information-level design is transformed into a design for the specific DBMS that will be used. Naturally, the characteristics of that DBMS must come into play during this step.

In this text, we will focus on the information-level design process and discuss that portion of the physical-level design process which is geared toward producing a legitimate design for a typical microcomputer DBMS. This approach represents a subset of the design methodology given in [14]. That methodology, which encompasses both the information and physical levels of design, is intended to be used for the design of complex databases that may be implemented on a variety of DBMSs, on mainframes or microcomputers, and where performance can be a very important concern. For the majority of microcomputer applications, the database design process as presented in this text is more than sufficient. If you are interested in delving more deeply into the process, see Chapters 6, 7, and 12 of [14].

6.2   INFORMATION-LEVEL DESIGN

User Views

No matter which approach is adopted with regard to database design, a complete database design that will satisfy all of the requirements can only rarely be a one-step process. Unless the requirements are exceptionally simple, it is usually necessary to subdivide the overall job of database design into smaller tasks. This is often done through the separate consideration of individual pieces of the design problem. In design problems for large organizations, these pieces are often called user views, and we will use the same terminology here. A **user view** is the view of data that is necessary to support the operations of a particular user. For each user view, a database structure to support the view must be designed and then merged into a cumulative design. Each user view, in general, will be much simpler than the total collection of requirements, so working on these individual tasks will be much more manageable than attempting to turn the design of the entire database into one large task.

The General Database Design Methodology

The database design methodology set forth in this text involves representing individual user views, refining them to eliminate any problems, and then merging them into a cumulative design. A "user" could be a person or a group that will use the system, a report the system must produce, or a type of transaction that the system must support. In the last two instances, you might think of the user as the person who will use the report or enter the transaction. In fact, if the same individual required three separate reports, for example, we would probably be better off to consider each of the reports as a separate user view, even though only one "user" was involved, since the smaller the user view, the easier it is to work with.

We now turn to the methodology itself. For each user view, we need to complete the following four steps:

1. Represent the user view as a collection of tables.
2. Normalize these tables.
3. Represent all keys.
4. Merge the result of the previous steps into the design.

We will now examine each of these steps in detail.

6.3   THE METHODOLOGY

Represent the User View as a Collection of Tables

When given a user view or some sort of stated requirement, we must develop a collection of tables that will support it. In some cases, the collection of tables may be obvious to us. Let's suppose, for example, that a given user view involves departments and employees. Let's assume further that each department can employ many employees but that each employee is assigned to exactly one department (a typical restriction). The design

```
DEPT (DEPTNUMB, DEPTNAME, DEPTLOC)
EMPLOYEE (EMPNUMB, EMPNAME, EMPADDR, WAGERATE, SSNUMB, DEPTNUMB)
```

may have naturally occurred to you and is an appropriate design. You will undoubtedly find that the more designs you have done, the easier it will be for you to develop such a collection without resorting to any special procedure. The real question is, What procedure should be followed if a correct design is not so obvious? In this case, we can take the following four steps:

Step 1. Determine the entities involved and create a separate table for each type of entity. At this point, you do not need to do anything more than give the table a name. For example, if a user view involves departments and employees, we can create a *DEPT* table and an *EMPLOYEE* table. At this point, we will write down something like this:

```
DEPT (
EMPLOYEE (
```

That is, we will write down the name of a table and a left parenthesis, *and that is all*. Later steps will fill in the attributes in these tables.

Step 2. Determine the primary key for each of these tables. This will fill in one or two attributes (depending on how many attributes make up the primary key). Other attributes will not be filled in until a later step. It may seem strange, but even though we have yet to determine the attributes in the table, we can usually determine the primary key. For example, the primary key to an *EMPLOYEE* table will probably be the employee number, and the primary key to a *DEPT* table will probably be the department number.

The primary key is the unique identifier, so the essential question here is, What does it take to uniquely identify an employee or a department? Even if we are in the process of trying to automate a system that was previously manual, some unique identifier can still usually be found in the manual system. If not, it is probably time to assign one. Let's say, for example, that in a particular manual system customers did not have numbers. The customer base was small enough that the organization felt they were not needed. Now is a good time to assign them, however, since the company is computerizing. These numbers would then be the unique identifier we are seeking.

Now let's add these primary keys to what we have written down already. At this point, we will have something like the following:

```
DEPT (DEPTNUMB,
EMPLOYEE (EMPNUMB,
```

That is, we will have the name of the table and the primary key, but that is all. Later steps will fill in the other attributes.

Step 3. Determine the properties for each of these entities. We can look at the user requirements and then determine the other properties of each entity which are required. These properties, along with the key identified in step 2, will become attributes in the appropriate tables. For example, an employee entity may require *EMPNAME*, *EMPADDR*, *WAGERATE*, and *SSNUMB* (social security number). The department entity may require *DEPTNAME* (department name) and *DEPTLOC* (department location). Adding these to what is already in place would produce the following:

```
DEPT (DEPTNUMB, DEPTNAME, DEPTLOC)
EMPLOYEE (EMPNUMB, EMPNAME, EMPADDR, WAGERATE, SSNUMB)
```

Step 4. Determine relationships among the entities. The basic relationships are one-to-many, many-to-many, and one-to-one. We will now see how to handle each of these types of relationships.

One-to-many. A one-to-many relationship is implemented by including the primary key of the "one" table as a foreign key in the "many" table. Let's suppose, for example, that each employee is assigned to a single department but that a department can have many employees. Thus *one* department is related to *many* employees. In this case, we would include the primary key of the *DEPT* table (the "one") as a foreign key in the *EMPLOYEE* table (the "many"). Thus, the tables would now look like this:

```
DEPT (DEPTNUMB, DEPTNAME, DEPTLOC)
EMPLOYEE (EMPNUMB, EMPNAME, EMPADDR, WAGERATE, SSNUMB, DEPTNUMB)
```

Many-to-many. A many-to-many relationship is implemented by creating a new table whose key is the combination of the keys of the original tables. Let's suppose that each employee can be assigned to multiple departments and that each department can have many employees. In this case, we would create a new table whose primary key would be the combination of *EMPNUMB* and *DEPTNUMB*. Since the new table represents the fact that an employee *works in* a department, we might choose to call it *WORKSIN*, in which case the collection of tables is as follows:

```
DEPT (DEPTNUMB, DEPTNAME, DEPTLOC)
EMPLOYEE (EMPNUMB, EMPNAME, EMPADDR, WAGERATE, SSNUMB)
WORKSIN (EMPNUMB, DEPTNUMB)
```

In some situations, no other attributes will be required in the new table. The other attributes in the *WORKSIN* table would be those attributes which depended on both the employee and the department, if such attributes existed. One possibility, for example, would be the date when the employee was first assigned to the department, since it depends on *both* the employee *and* the department.

One-to-one. If each employee is assigned to a single department and each department consists of only one employee, the relationship between employees and departments is one-to-one. The simplest way for us to implement a one-to-one relationship is to treat it as a one-to-many relationship. But which is the "one" part of the relationship and which is the "many" part? Sometimes looking to the future helps. For instance, in the example we are discussing, we might ask, If the relationship changes in the future, is it more likely that one employee will be assigned to many departments or that one department may consist of several employees rather than just one? If we feel, for example, that it is more likely that a department would be allowed to contain more than one employee, we would make *EMPLOYEE* the "many" part of the relationship. If the answer is that both things might very well happen, we might even treat the relationship as many-to-many. If neither change were likely to occur, we could actually resort to flipping a coin in order to choose the "many" part of the relationship.

Normalize These Tables

Normalize each table, with the target being third normal form.

Represent All Keys

Identify all keys. The types of keys we must identify are primary keys, alternate keys, secondary keys, and foreign keys.

1. *Primary:* The primary key has already been determined in the earlier steps.
2. *Alternate:* An **alternate key** is an attribute or collection of attributes that could have been chosen as a primary key but was not. It is not common to have alternate keys; but if they do exist, and if the system is to enforce their uniqueness, they should be so noted.
3. *Secondary:* If there are any **secondary keys** (attributes that are of interest strictly for the purpose of retrieval), they should be represented at this point. If a user were to indicate, for example, that rapidly retrieving an employee on the basis of his or her name was important, we would designate *EMPNAME* as a secondary key.
4. *Foreign:* This is in many ways the most important category, since it is through **foreign keys** that relationships are established and that certain types of integrity constraints are enforced in the database. Remember that a foreign key is an attribute (or collection of attributes) in one table that is required to either match

the value of the primary key for some row in another table or be null. (This is the property called **referential integrity**.) Consider, for example, the following tables:

```
DEPT (DEPTNUMB, DEPTNAME, DEPTLOC)
EMPLOYEE (EMPNUMB, EMPNAME, EMPADDR, WAGERATE, SSNUMB, DEPTNUMB)
```

As before, *DEPTNUMB* in the *EMPLOYEE* table indicates the department to which the employee is assigned. We say that *DEPTNUMB* in the *EMPLOYEE* table is a foreign key that *identifies DEPT*. Thus, the number in this attribute on any row in the *EMPLOYEE* table must either be the number of a department that is already in the database, or be null. (Null would indicate that, for whatever reason, the employee is not assigned to a department.)

Database Design Language (DBDL)

We need a mechanism for representing the tables and keys together with the restrictions previously discussed. The standard mechanism for representing tables is fine but it does not go far enough. There is no routine way to represent alternate, secondary, or foreign keys, nor is there a way of representing foreign key restrictions. There is no way of indicating that a given field or attribute can accept null values. Since the methodology is based on the relational model, however, it is desirable to represent tables with the standard method. We will add additional features capable of representing additional information. The end result is **Database Design Language** (or **DBDL**).

Figure 6.1a shows sample DBDL documentation for the *EMPLOYEE* table. In DBDL, tables and their primary keys are represented in the usual manner. Any field that is allowed to be null, such as the *EMPADDR* attribute in the *EMPLOYEE* table, is followed by an asterisk. Underneath the table, the various types of keys are listed. Each is preceded by an abbreviation indicating the type of key (AK – alternate key, SK – secondary key, FK – foreign key). It is sufficient to list the attribute or collection of attributes that forms an alternate or secondary key. In the case of foreign keys, however, we must also represent the table that is identified by the foreign key; i.e., the table whose primary key the foreign key must match. This is accomplished in DBDL by following the foreign key with an arrow pointing to the table that the foreign key identifies.

Figure 6.1a

DBDL for
EMPLOYEE relation

```
EMPLOYEE (EMPNUMB, EMPNAME, EMPADDR*, SSNUMB, DEPTNUMB,...)
       AK    SSNUMB
       SK    EMPNAME
       FK    DEPTNUMB --> DEPT
```

Figure 6.1b summarizes the details of DBDL. Examples of DBDL will be presented throughout this chapter. The only feature of DBDL not listed is actually more of a tip than a rule. When several tables are listed, a table containing a foreign key should be listed after the table that the foreign key identifies, if possible.

Figure 6.1b

Summary of DBDL

DBDL (Database Design Language)

1. Relations, attributes, and primary keys are represented in the usual way.
2. Attributes that are allowed to be null are followed by an asterisk.
3. Alternate keys are identified by the letters AK followed by the attribute(s) that comprise the alternate key.
4. Secondary keys are identified by the letters SK followed by the attribute(s) that comprise the secondary key.
5. Foreign keys are identified by the letters FK followed by the attribute(s) that comprise the foreign key. Foreign keys are followed by an arrow pointing to the relation identified by the foreign key.

In the example shown in Figure 6.1a, we are saying that there is a table called *EMPLOYEE*, consisting of fields *EMPNUMB, EMPNAME, EMPADDR, SSNUMB* (social security number), *DEPTNUMB,* and so on. The *EMPADDR* field is the only one that can accept null values. The primary key is *EMPNUMB*. Another possible key is *SSNUMB*. We are interested in being able to retrieve information efficiently, based on the employee's name, so we have designated *EMPNAME* as a secondary key. The *DEPTNUMB* is a foreign key identifying the department to which the employee is assigned (it identifies the appropriate department in the *DEPT* table).

A Pictorial Representation of the Database

For many people, a pictorial representation of the structure of the database is quite useful. (As the old saying goes, "A picture is worth a thousand words.") Fortunately, there is an easy procedure for including a diagram representing the database structure in DBDL. The type of diagram we will use is often called a data structure diagram. The procedure for constructing such a diagram from tables represented in DBDL is as follows:

1. Draw a rectangle for each table in the DBDL design. Label the rectangle with the name of the corresponding table.
2. For each foreign key, draw an arrow from the rectangle that corresponds to the table being identified to the rectangle that corresponds to the table containing the foreign key.
3. In the rare event that you have two arrows joining the same two rectangles, label the arrows with names that are indicative of the meaning of the relationships represented by the arrows.
4. If the diagram you have drawn is cluttered or messy, redraw the diagram. If possible, avoid crossing arrows, since this makes the diagram more difficult to understand.

Figure 6.2 shows the DBDL from Figure 6.1a together with a corresponding data structure diagram. Notice that there is a *DEPT* rectangle and an *EMPLOYEE* rectangle. Further, since the *EMPLOYEE* table contains a foreign key identifying the *DEPT* table, there is an arrow from *DEPT* to *EMPLOYEE*. This arrow visually emphasizes the relationship between departments and employees. Such arrows represent one-to-many relationships (*one* department to *many* employees) with the arrow pointing to the "many" part of the relationship.

Figure 6.2

DBDL with data structure diagram

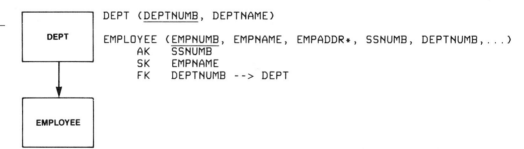

```
DEPT (DEPTNUMB, DEPTNAME)

EMPLOYEE (EMPNUMB, EMPNAME, EMPADDR*, SSNUMB, DEPTNUMB,...)
     AK    SSNUMB
     SK    EMPNAME
     FK    DEPTNUMB --> DEPT
```

Merge the Result into the Design

As soon as we have completed steps 1 through 3 for a given user view, we can merge these results into the overall design. If the view on which we have been working happens to be the first user view, then the cumulative design will be identical to the design for this first user. Otherwise, we add all the tables for this user to those which are currently in the cumulative design. We combine tables that have the same primary key to form a new table. This table has the same primary key as those tables which have been combined.

The new tables also contain all the attributes from both tables. In the case of duplicate attributes, we remove all but one copy of the attribute. For example, if the cumulative collection already contained the following:

EMPLOYEE (<u>EMPNUMB</u>, EMPNAME, WAGERATE, SSNUMB, DEPTNUMB)

and the user view just completed contained the following:

EMPLOYEE (<u>EMPNUMB</u>, EMPNAME, EMPADDR)

then the two tables would be combined, since they would have the same primary key. All of the attributes from both tables would appear in the new table, but without duplicates. Thus, *EMPNAME* would appear only once, even though it is in each of the individual tables. The result would be the following:

EMPLOYEE (<u>EMPNUMB</u>, EMPNAME, WAGERATE, SSNUMB, DEPTNUMB, EMPADDR)

If we wanted to, we could reorder the attributes at this point. We might feel, for example, that placing *EMPADDR* immediately after *EMPNAME* would put it in a more natural position. This would give the following:

EMPLOYEE (<u>EMPNUMB</u>, EMPNAME, EMPADDR, WAGERATE, SSNUMB, DEPTNUMB)

We would then check the new design to ensure that it was still in third normal form. If it was not, we would convert it to 3NF before proceeding.

The process, which is summarized in Figure 6.3, is repeated for each user view until all user views have been examined. At that point, the design is reviewed in order to resolve any problems that may remain and to ensure that the needs of all individual users can indeed be met. Once this has been done, the information-level design is complete.

Figure 6.3

Information-level design methodology

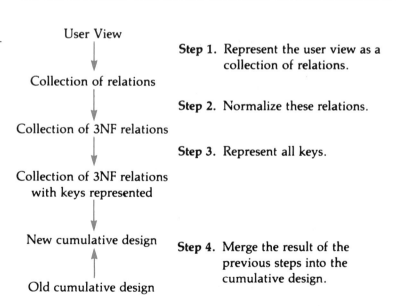

6.4   DATABASE DESIGN EXAMPLES

Let's now look at some examples of database design.

Example 1: For an initial example of the design methodology, let's complete an information-level design for a database that must satisfy the following constraints and requirements.

1. For a sales rep, store the sales rep's number, name, address, total commission and commission rate.
2. For a customer, store the customer's number, name, address, balance, and credit limit. In addition, store the number and name of the sales rep who represents this customer. Upon further checking with the user we determine that a sales rep can represent many customers but a customer must have exactly one sales rep (i.e., a customer *must have* a sales rep and cannot have more than *one*.)
3. For a part, store the part's number, description, units on hand, item class, the number of the warehouse in which the part is located, and the price.
4. For an order, store the order number, order date, the number, name, and address of the customer who placed the order, and the number of the sales rep who represents that customer. In addition, for each line item within the order, store the part number and description, the number of the part that was ordered, and the quoted price. The following information has also been obtained from the user:
   a. Each order must be placed by a customer who is already in the customer file.
   b. There is only one customer per order.
   c. On a given order, there is at most one line item for a given part. For example, part BT04 cannot appear on several lines within the same order.
   d. The quoted price may be the same as the current price in the part master file, but it need not be. This allows the enterprise the flexibility to sell the same parts to different customers for different prices. It also allows you to change the basic price for a part without necessarily affecting other orders that are currently on file.

What are the user views in the preceding example? In particular, how should the design proceed if we are given requirements that are not specifically stated in the form of user views? We might actually be lucky enough to be confronted with a series of well-thought-out user views in a form that can readily be merged into our design. On the other hand, we might only be given a set of requirements like the set we have encountered in this example. Or we might be given a list of reports and updates that a system must support. If we happen to be given the job of interviewing users and documenting their needs as a preliminary to the design process, we can make sure that their views are specified in a form that will be easy to work with when the design process starts. On the other hand, we may just have to take this information as we get it.

If the user views are not spelled out as user views per se, then we should consider each requirement that is specified to be a user view. Thus each report or update transaction that the system must support, as well as any other requirement, such as any of those just stated, can be considered an individual user view. In fact, even if the requirements are presented as user views, we may wish to split up a user view that is particularly complex into smaller pieces and consider each piece a user view for the design process.

Let us now proceed with the example.

User View 1: This requirement, or user view, poses no particular difficulty. Only one table is required to support this view:

```
SLSREP (SLSRNUMB, SLSRNAME, SLSRADDR, TOTCOMM, COMMRATE)
```

This table is in 3NF. Since there are no foreign, alternate, or secondary keys, the DBDL representation of the table is precisely the same as the relational model representation.

Notice that we have assumed that the sales rep's number (*SLSRNUMB*) is the primary key to the table. This is a fairly reasonable assumption. But since this information was not given in the first requirement, we would need to verify its accuracy with the user.

In each of the following requirements, we shall assume that the obvious attribute (customer number, part number, and order number) is the primary key. Since this is the first user view, the "merge" step of the design methodology will produce a cumulative design consisting of this one table (see Figure 6.4).

Figure 6.4

Cumulative design after first user view

```
┌──────────┐
│          │
│  SLSREP  │
│          │
└──────────┘
```

SLSREP (<u>SLSRNUMB</u>, SLSRNAME, SLSRADDR, TOTCOMM, COMMRATE)

User View 2: Because the first user view was relatively simple, we were able to come up with the necessary table without having to go through the steps mentioned in the discussion of the design methodology. The second user view is a little more complicated, however, so let's use the steps suggested earlier to determine the tables. (If you've already spotted what the tables should be, you have a natural feel for the process. If so, please be patient while we work through the process.)

We'll take two different approaches to this requirement so that we can see how they can both lead to the same result. The only difference between the two approaches concerns the entities that we initially identify. In the first approach, suppose we identify two entities, *sales reps* and *customers*. We would then begin with the two following tables:

```
SLSREP (
CUSTOMER (
```

After determining the unique identifiers, we add the primary keys, which would give:

```
SLSREP (SLSRNUMB,
CUSTOMER (CUSTNUMB,
```

Adding attributes for the properties of each of these entities would yield:

```
SLSREP (SLSRNUMB, SLSRNAME
CUSTOMER (CUSTNUMB, CUSTNAME, ADDRESS, BALANCE, CREDLIM
```

Finally, we would deal with the relationship: *one* sales rep is related to *many* customers. To implement this one-to-many relationship, we would include the key of the "one" table in the "many" table as a foreign key. In this case, we would include *SLSRNUMB* in the *CUSTOMER* table. Thus, we would have the following:

```
SLSREP (SLSRNUMB, SLSRNAME)
CUSTOMER (CUSTNUMB, CUSTNAME, ADDRESS, BALANCE, CREDLIM,
          SLSRNUMB)← B/c 1:M relationship
```

Both tables are in 3NF, so we can move on to representing the keys. Before doing that, however, let's investigate another approach that could have been used to determine the tables.

Suppose we didn't realize that there were really two entities and thought there was only a single entity, *customers*. We would thus begin only the single table as follows:

```
CUSTOMER (
```

Adding the unique identifier as the primary key would give this:

```
CUSTOMER (CUSTNUMB,
```

Finally, adding the other properties as additional attributes would yield:

```
CUSTOMER (CUSTNUMB, CUSTNAME, ADDRESS, BALANCE, CREDLIM,
          SLSRNUMB, SLSRNAME)
```

A problem appears, however, when we examine the functional dependencies that exist in *CUSTOMER*. *CUSTNUMB* determines all the other fields, as it should. But *SLSRNUMB* determines *SLSRNAME*, yet *SLSRNUMB* is not a candidate key. This table, which is in 2NF, since no attribute depends on a portion of the key, is not in 3NF. Thus, converting to 3NF would produce the following two tables:

```
CUSTOMER (CUSTNUMB, CUSTNAME, ADDRESS, BALANCE, CREDLIM,
          SLSRNUMB)
SLSREP (SLSRNUMB, SLSRNAME)
```

Note that these are precisely the same tables that we determined with the other approach. It just took us a little longer to get there.

It is these two tables that we merge into the design. Besides the obvious primary keys, *CUSTNUMB* for *CUSTOMER* and *SLSRNUMB* for *SLSREP*, the *CUSTOMER* table now contains a foreign key, *SLSRNUMB*.

There are no alternate keys, nor did the requirements state anything that would lead to a secondary key. If there were a requirement to retrieve the customer based on his or her name, for example, we would probably choose to make *CUSTNAME* a secondary key. (Since names are not unique, *CUSTNAME* is not an alternate key.)

At this point, we could represent the table *SLSREP* in DBDL in preparation for merging this collection of tables into the collection we already have. Looking ahead, however, we see that since this table has the same primary key as the table *SLSREP* from the first user view, the two tables will be merged. A single table will be formed that has the common key *SLSRNUMB* as its primary key and that contains all of the other attributes from both tables without duplication. For this second user view, the only attribute in *SLSREP* besides the primary key is *SLSRNAME*. This attribute is the same as the attribute called *SLSRNAME* already present in *SLSREP* from the first user view. Thus, nothing will be added to the *SLSREP* table that is already in place. The cumulative design now contains the two tables *SLSREP* and *CUSTOMER*, as shown in Figure 6.5.

Figure 6.5

Cumulative design after second user view

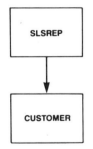

```
SLSREP (SLSRNUMB, SLSRNAME, SLSRADDR, TOTCOMM, COMMRATE)

CUSTOMER (CUSTNUMB, CUSTNAME, ADDRESS, BALANCE, CREDLIM, SLSRNUMB)
     FK   SLSRNUMB --> SLSREP
```

User View 3: Like the first user view, this one poses no special problems. Only one table is required to support it:

```
PART (PARTNUMB, PARTDESC, UNONHAND, ITEMCLSS, WREHSENM, UNITPRCE)
```

Figure 6.6

Cumulative design
after third user
view

This table is in 3NF. The DBDL representation is identical to the relational model representation.

Since *PARTNUMB* is not the primary key of any table we have already encountered, merging this table into the cumulative design produces a design with the three tables *SLSREP, CUSTOMER,* and *PART* (see Figure 6.6).

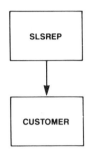

```
SLSREP (SLSRNUMB, SLSRNAME, SLSRADDR, TOTCOMM, COMMRATE)

CUSTOMER (CUSTNUMB, CUSTNAME, ADDRESS, BALANCE, CREDLIM, SLSRNUMB)
     FK    SLSRNUMB --> SLSREP

PART (PARTNUMB, PARTDESC, UNONHAND, ITEMCLSS, WREHSENM, UNITPRCE)
```

User View 4: This user view is a bit more complicated, and we could approach it in several ways. Suppose we felt that only a single entity was being mentioned, namely *orders.* In that case, we would create a single table, as follows:

```
ORDERS (
```

Since orders are uniquely identified by order numbers, we would add *ORDNUMB* as the primary key, giving:

```
ORDERS (ORDNUMB,
```

Examining the various properties of an order, such as the date, the customer number, and so on, as listed in the requirement, we would add appropriate attributes, giving:

```
ORDERS (ORDNUMB, ORDDTE, CUSTNUMB, CUSTNAME, ADDRESS, SLSRNUMB,
```

What about the fact that we are supposed to store the part number, description, number ordered, and quoted price for each order line on this order? One way of doing this would be to include all these attributes within the *ORDERS* table as a repeating group (since there can be many order lines on an order). This would yield:

```
ORDERS (ORDNUMB, ORDDTE, CUSTNUMB, CUSTNAME, ADDRESS, SLSRNUMB,
        PARTNUMB, PARTDESC, NUMBORD, QUOTPRCE)
```

At this point, we have a table that does contain all the necessary attributes. Now we must convert this table to an equivalent collection of tables that are in 3NF. Since this table is not even in 1NF, we would remove the repeating group and expand the key to produce the following:

```
ORDERS (ORDNUMB, ORDDTE, CUSTNUMB, CUSTNAME, ADDRESS, SLSRNUMB,
        PARTNUMB, PARTDESC, NUMBORD, QUOTPRCE)
```

In the new *ORDERS* table, we have the following functional dependencies:

```
ORDNUMB --> ORDDTE, CUSTNUMB, CUSTNAME, ADDRESS, SLSRNUMB
CUSTNUMB --> CUSTNAME, ADDRESS, SLSRNUMB
PARTNUMB --> PARTDESC
ORDNUMB, PARTNUMB --> NUMBORD, QUOTPRCE
```

From the discussion of the quoted price in the statement of the requirement, it should be noted that quoted price does indeed depend on *both* the order number and the part number, not on the part number alone. Since some attributes depend on only a portion of the primary key, the *ORDERS* table is not in 2NF. Converting to 2NF would yield the following:

```
ORDERS (ORDNUMB, ORDDTE, CUSTNUMB, CUSTNAME, ADDRESS, SLSRNUMB)
PART (PARTNUMB, PARTDESC)
ORDLNE (ORDNUMB, PARTNUMB, NUMBORD, QUOTPRCE)
```

The tables *PART* and *ORDLNE* are in 3NF. The *ORDERS* table is not in 3NF, since *CUSTNUMB* determines *NAME, ADDRESS,* and *SLSRNUMB,* but *CUSTNUMB* is not a candidate key. Converting the *ORDERS* table to 3NF and leaving the other tables untouched would produce the following design for this requirement.

```
ORDERS (ORDNUMB, ORDDTE, CUSTNUMB)
CUSTOMER (CUSTNUMB, CUSTNAME, ADDRESS, SLSRNUMB)
PART (PARTNUMB, PARTDESC)
ORDLNE (ORDNUMB, PARTNUMB, NUMBORD, QUOTPRCE)
```

This is the collection of tables that will be represented in DBDL and then merged into the cumulative design. Again, however, we can look ahead and see that *CUSTOMER* will be merged with the existing *CUSTOMER* table, and *PART* will be merged with the existing *PART* table. In neither case will anything new be added to the *CUSTOMER* and *PART* tables already in place, so the *CUSTOMER* and *PART* tables for this user view will not affect the overall design. The representation for this user view in DBDL is shown in Figure 6.7.

Figure 6.7

DBDL for fourth user view

```
CUSTOMER (CUSTNUMB, CUSTNAME, ADDRESS, SLSRNUMB)

PART (PARTNUMB, PARTDESC)

ORDERS (ORDNUMB, ORDDTE, CUSTNUMB)
     FK    CUSTNUMB --> CUSTOMER

ORDLNE (ORDNUMB, PARTNUMB, NUMBORD, QUOTPRCE)
     FK    ORDNUMB --> ORDERS
     FK    PARTNUMB --> PART
```

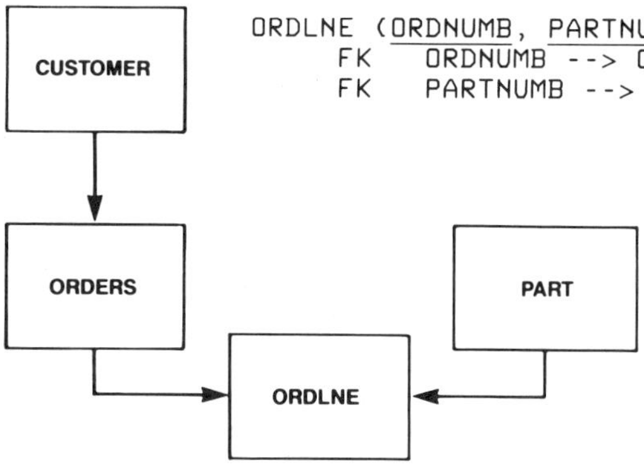

At this point, we have completed the process for each user. We should now review the design to make sure that it will cleanly fulfill all of the requirements. If problems are encountered or new information comes to light, the design must be modified accordingly. Based on the assumption that we do not have to further modify the design here, the final information-level design is shown in Figure 6.8, which is the result of merging the design for the fourth user view into the cumulative design.

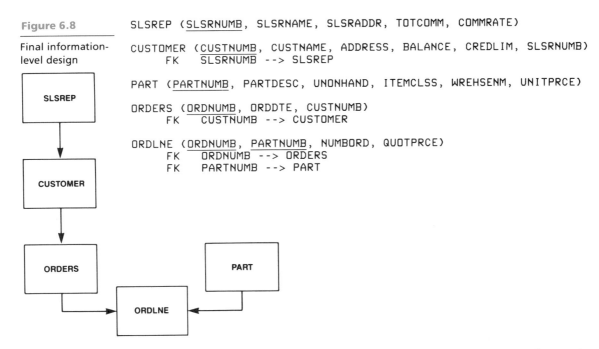

Figure 6.8

Final information-level design

```
SLSREP (SLSRNUMB, SLSRNAME, SLSRADDR, TOTCOMM, COMMRATE)

CUSTOMER (CUSTNUMB, CUSTNAME, ADDRESS, BALANCE, CREDLIM, SLSRNUMB)
    FK    SLSRNUMB --> SLSREP

PART (PARTNUMB, PARTDESC, UNONHAND, ITEMCLSS, WREHSENM, UNITPRCE)

ORDERS (ORDNUMB, ORDDTE, CUSTNUMB)
    FK    CUSTNUMB --> CUSTOMER

ORDLNE (ORDNUMB, PARTNUMB, NUMBORD, QUOTPRCE)
    FK    ORDNUMB  --> ORDERS
    FK    PARTNUMB --> PART
```

Example 2: We will now design a database for Henry. Henry wants to keep information on books, authors, publishers, and branches. The only user is Henry, but we don't want to treat the whole project as a single user view: so let's assume we've asked Henry for all the reports the system is to produce, and we will treat each one as a user view. Suppose that Henry has given us the following requirements:

1. For each publisher, list the publisher code, the name, and the city in which the publisher is located.
2. For each branch, list the number, the name, the location, and the number of employees.
3. For each book, list its code, title, the code and name of the publisher, the price, and whether or not it is paperback.
4. For each book, list its code, title, type, and price. In addition, list the number and name of each of the authors of the book. (**Note:** If there is more than one author, they must be listed in the order in which they are listed on the book. This may or may not be alphabetically.)
5. For each branch, list the number and name. In addition, list the code and title of each book currently in the branch as well as the number of units of the book the branch currently has.
6. For each book, list the code and title. In addition, for each branch currently having the book in stock, list the number and name of each branch along with the number of copies available.

With these six reports as the user views, let's move on to the design of Henry's database.

User View 1: The only entity in this user view is *publshr*. The table to support it is as follows:

```
PUBLSHR (PUBCODE, PUBNAME, PUBCITY)
```

This table is in 3NF. The primary key is *PUBCODE*. There are no alternate or foreign keys. Let's assume Henry wants to be able to access a publisher rapidly on the basis of its name. Then we will make *PUBNAME* a secondary key.

Since this is the first user view, there is no previous cumulative design. So at this point the new cumulative design will consist solely of the design for this user view. It is shown in Figure 6.9.

Figure 6.9

DBDL for *BOOK* database after first requirement

PUBLSHR (PUBCODE, PUBNAME, PUBCITY)
        SK PUBNAME

User View 2: The only entity in this user view is *branch.* The table to support it is as follows:

BRANCH (BRNUMB, BRNAME, BRLOC, NUMEMP)

This table is also in 3NF. The primary key is *BRNUMB,* and there are no alternate or foreign keys. Let's assume Henry wants to be able to access a branch rapidly on the basis of its name. Thus we will make *BRNAME* a secondary key.

Since no table in the cumulative design has *BRNUMB* as its primary key, this table will simply be added to the collection of tables in the cumulative design during the merge step. The result is shown in Figure 6.10.

Figure 6.10

DBDL for *BOOK* database after second requirement

PUBLSHR (PUBCODE, PUBNAME, PUBCITY)
        SK PUBNAME

BRANCH (BRNUMB, BRNAME, BRLOC, NUMEMP)
       SK BRNAME

User View 3: There are two entities here, publishers and books, and a one-to-many relationship between them. This leads to the following:

PUBLSHR (PUBCODE, PUBNAME)
BOOK (BKCODE, BKTITLE, PUBCODE, BKPRICE, PB)

where *PUBCODE* in *BOOK* is a foreign key identifying the publisher. Merging these tables with those which are already in place does not add any new attributes to the *PUBLSHR* table but adds the *BOOK* table to the cumulative design. The result of the merge is shown in Figure 6.11.

Figure 6.11

DBDL for *BOOK* database after third requirement

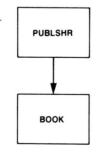

PUBLSHR (PUBCODE, PUBNAME, PUBCITY)
        SK PUBNAME

BRANCH (BRNUMB, BRNAME, BRLOC, NUMEMP)
       SK BRNAME

BOOK (BKCODE, BKTITLE, PUBCODE, BKPRICE, PB)
     FK PUBCODE --> PUBLSHR

User View 4: There are two entities in this user view, books and authors. The relationship between them is many-to-many (an author can write many books and a book can have many authors). Creating tables for each entity and the relationship gives:

```
AUTHOR (AUTHNUMB, AUTHNAME)
BOOK (BKCODE, BKTITLE, BKTYPE, BKPRICE)
WROTE (BKCODE, AUTHNUMB
```

(Since the last table represents the fact that an author *wrote* a particular book, we will call the table *WROTE*).

In this user view, we need to be able to list the authors for a book in the appropriate order. To accomplish this, we will add a sequence number column to the last table. This completes the tables for this user view, which are:

```
AUTHOR (AUTHNUMB, AUTHNAME)
BOOK (BKCODE, BKTITLE, BKTYPE, BKPRICE)
WROTE (BKCODE, AUTHNUMB, SEQNO)
```

The *AUTHOR* table is new. Merging the *BOOK* table adds one additional column, *BKTYPE*. The *WROTE* table is new. The result of the merge step is shown in Figure 6.12.

Figure 6.12

DBDL for *BOOK* database after fourth requirement

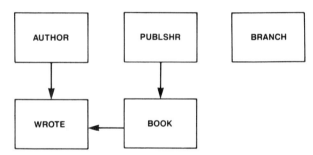

```
PUBLSHR (PUBCODE, PUBNAME, PUBCITY)
    SK PUBNAME

BRANCH (BRNUMB, BRNAME, BRLOC, NUMEMP)
    SK BRNAME

BOOK (BKCODE, BKTITLE, PUBCODE, BKPRICE, PB, BKTYPE)
    FK  PUBCODE --> PUBLSHR

AUTHOR (AUTHNUMB, AUTHNAME)

WROTE (BKCODE, AUTHNUMB, SEQNO)
    FK  BKCODE --> BOOK
    FK  AUTHNUMB --> AUTHOR
```

User View 5: Suppose we were to decide that the only entity mentioned in this requirement was *branch*. We would then create this table:

```
BRANCH (
```

We would then add the primary key, *BRNUMB*, producing the following:

```
BRANCH (BRNUMB,
```

The other properties include the branch name as well as the book code, book title, and number of units on hand. Since a branch will have several books, the last three columns will form a repeating group. We thus have the following:

        BRANCH (BRNUMB, BRNAME, BKCODE, BKTITLE, OH)

We convert this table to 1NF by removing the repeating group and expanding the key. This gives:

        BRANCH (BRNUMB, BRNAME, BKCODE, BKTITLE, OH)

In this table, we have the following functional dependencies:

        BRNUMB --> BRNAME
        BKCODE --> BKTITLE
        BRNUMB, BKCODE --> OH

The table is not in 2NF, since some attributes depend on just a portion of the key. Converting to 2NF gives:

        BRANCH (BRNUMB, BRNAME)
        BOOK (BKCODE, BKTITLE)
        INVENT (BRNUMB, BKCODE, OH)

The primary keys are indicated. We call the final table *INVENT*, since it effectively represents each branch's inventory. In the *INVENT* table, *BRNUMB* is a foreign key that identifies *BOOK*, and *BKCODE* is a foreign key that identifies *BOOK*. In other words, in order for a row to exist in the *INVENT* table, *both* the branch number *and* the book code must already be in the database.

Figure 6.13

DBDL for *BOOK* database after fifth requirement

The *BRANCH* table will merge with the existing *BRANCH* table without adding anything new. Similarly, the *BOOK* table will not add anything new to the existing *BOOK* table. The *INVENT* table is totally new and will appear as part of the new cumulative design, which is given in Figure 6.13.

```
PUBLSHR (PUBCODE, PUBNAME, PUBCITY)
    SK PUBNAME

BRANCH (BRNUMB, BRNAME, BRLOC, NUMEMP)
    SK BRNAME

BOOK (BKCODE, BKTITLE, PUBCODE, BKPRICE, PB, BKTYPE)
    FK  PUBCODE --> PUBLSHR

AUTHOR (AUTHNUMB, AUTHNAME)

WROTE (BKCODE, AUTHNUMB, SEQNO)
    FK  BKCODE --> BOOK
    FK  AUTHNUMB --> AUTHOR

INVENT (BRNUMB, BKCODE, OH)
    FK  BRNUMB --> BRANCH
    FK  BKCODE --> BOOK
```

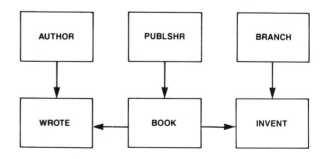

Q & A

Question:	How would the design for this user view have turned out if we had started out with two entities, *branch* and *book*, instead of just the single entity *branch*?
Answer:	In the first step, we would have these two tables:

BRANCH (
BOOK (

Adding the primary keys would give:

BRANCH (BRNUMB,
BOOK (BKCODE,

Filling in the other attributes would give:

BRANCH (BRNUMB, BRNAME)
BOOK (BKCODE, BKTITLE)

Finally, we have to implement the relationship between *BRANCH* and *BOOK*. Since a branch can have many books and a book can be in stock at many branches, the relationship is many-to-many. To implement a many-to-many relationship, we add a new table whose primary key is the combination of the primary keys of the other tables. Doing this, we produce the following:

BRANCH (BRNUMB, BRNAME)
BOOK (BKCODE, BKTITLE)
INVENT (BRNUMB, BKCODE

Finally, we add any column that depends on both *BRNUMB* and *BKCODE* to the *INVENT* table, giving

BRANCH (BRNUMB, BRNAME)
BOOK (BKCODE, BKTITLE)
INVENT (BRNUMB, BKCODE, OH)

Thus we end up with exactly the same collection of tables, which illustrates a point made earlier: there's more than one way of arriving at a correct result.

User View 6: This user view leads to precisely the same set of tables that were created for User View 5.

We have now reached the end of the requirements, and the design shown in Figure 6.13 represents the complete information-level design. You should take a moment to review each of the requirements to make sure they can all be satisfied.

(**Note:** If you compare this design with the *BOOK* database we have seen earlier in the text, you will see a slight difference in the order of some of the columns. In theory it doesn't make any difference. In practice, however, we sometimes rearrange the columns when we are done for convenience. If, for example, you execute a SQL SELECT using the star (*), you will see the columns *in the order in which they occur in the table*. Thus, if we have a particular order we prefer, we will often be sure the columns occur in that order.)

6.5 PHYSICAL-LEVEL DESIGN

Once the information-level design is complete, we are ready to begin producing the specific design that will be implemented with some typical microcomputer DBMS. This is part of the overall process called physical-level design. For further information on the full physical-level design process, see [14].

Since most microcomputer DBMSs are relational (at least they claim to be), and since our final information-level design is presented in a relational format, the basic job of producing the design for the chosen DBMS is not difficult. We simply use the same tables and columns. (At this point, we do need to supply format details, of course, like the fact that *CUSTNUMB* is a three-digit number, but again, this is not difficult.)

If the DBMS happened to support primary, candidate, secondary, and foreign keys, we would be all set. We would simply use these features to implement the various types of keys that are listed in the final DBDL version of the information-level design. Unfortunately, very few systems support all these types of keys. In fact, surprising as it may seem, many of the well-known and widely respected microcomputer DBMSs *don't support any of them!* So we need to devise a scheme for handling these keys in the event that the DBMS cannot do so.

Basically, such a scheme must ensure the uniqueness of primary and candidate keys. It must ensure that values in the foreign keys are legitimate; in other words, that they match the value of the primary key on some row in another table. As far as secondary keys are concerned, we merely need to ensure the efficiency of access to rows on the basis of a value of the secondary key.

For instance, suppose we are implementing the *EMPLOYEE* table shown in Figure 6.1a, in which *EMPNUMB* is the primary key, *SSNUMB* is a candidate key, *EMPNAME* is a secondary key, and *DEPTNUMB* is a foreign key that matches the *DEPT* table. We will have to ensure that the following conditions hold true:

1. Employee numbers are unique.
2. Social security numbers are unique.
3. Access to an employee on the basis of his or her name is rapid. (This restriction differs in that it merely states that a certain type of activity must be efficient, but it is an important restriction nonetheless.)
4. Department numbers are valid; that is, they match the number of a department currently in the database.

The next question is, Who should enforce these restrictions? Two choices are possible, provided the DBMS can't do it. The users of the system could enforce them, or programmers could. In the case of users, they would have to be careful when entering data not to enter two employees with the same employee number, not to enter an employee whose department number was invalid, and so on. Clearly, this would put a tremendous burden on the user.

Provided the DBMS can't enforce the restrictions, the appropriate place for the enforcement to take place is in programs. Thus, the responsibility burden for this enforcement should fall on the programmers who write the programs that users will run to update the database. Incidentally, users *must* update the data through these programs and *not* through the built-in features of the DBMS in such circumstances; otherwise, they would be able to bypass all the controls that we are attempting to program into the system.

Thus, it is the responsibility of programmers to include logic in their programs to enforce all the constraints. With respect to the DBDL shown in Figure 6.1a, this means the following:

1. Before an employee is added, the program should determine three things:
   a. whether an employee with the same employee number is already in the database; if so, the update should be rejected.

    b. whether an employee with the same social security number is already in the database; if so, the update should be rejected; and

    c. whether the department number that was entered matches the number of a department that is already in the database; if it doesn't, the update should be rejected.

2. When an employee is changed, if the department number is one of the values that is changed, the program should check to make sure that the new number also matches the number of a department that is already in the database. If it doesn't, the update should be rejected.

3. When a department is deleted, the program should check to make sure that the database contains no employees for this department. If the department does contain employees and it is allowed to be deleted, these employees will have department numbers that are no longer valid. In that case, the update should be rejected.

These actions must be performed efficiently, and in most systems this means creating and using indexes on all key columns. Thus, an index will be created for each column (or combination of columns) that is a primary key, a candidate key, a secondary key, or a foreign key.

SUMMARY

1. Database design is the process of determining an appropriate database structure to satisfy a given set of requirements. It is a two-part process:

    a. The information-level design, wherein a clean DBMS-independent design is created to satisfy the requirements.

    b. The physical-level design, wherein the final information-level design is converted into an appropriate design for the particular DBMS that will be used.

2. A user view is the view of data necessary to support the operations of a particular user. In order to simplify the design process, the overall set of requirements is split into user views.

3. The information-level design methodology involves applying the following steps to each user view:

    a. Represent the user view as a collection of tables.

    b. Normalize these tables; that is, convert this collection into an equivalent collection that is in 3NF.

    c. Represent all keys: primary, alternate, secondary, and foreign.

    d. Merge the results of the previous step into the cumulative design.

4. The design is represented in a language called DBDL (Database Design Language).

5. To obtain a pictorial representation of a design, apply the following steps to the DBDL design:

    a. Create a rectangle for each table in the DBDL design.

    b. For each foreign key, create an arrow that (1) begins with the rectangle that corresponds to the table identified by the foreign key and (2) terminates at the rectangle that corresponds to the table containing the foreign key. The foreign key may then be removed, although it is not essential to do so.

6. Assuming that a relational or relational-like microcomputer DBMS is going to be used, the physical-level design process consists of creating a table for each table in the DBDL design. Any constraints (primary key, alternate key, or foreign key) that the DBMS cannot enforce must be enforced by the programs in the system, so this fact must be documented for the programmers.

KEY TERMS

alternate key   physical-level design
database design   primary key
Database Design Language   referential integrity
  (DBDL)   relationship
foreign key   secondary key
information-level design   user view
integrity rules

EXERCISES

1. Define the term *user view* as it applies to database design.
2. What is the purpose of breaking down the overall design problem into a consideration of individual user views?
3. Under what circumstances would you not have to break down the overall design into a consideration of individual user views?
4. The information-level design methodology presented in this section contains a number of steps that are to be repeated for each user view. List the steps and briefly describe the kinds of activities that must take place at each step.
5. Describe the function of each of the following types of keys: primary, alternate, secondary, and foreign.
6. Describe the process of mapping an information-level design to a design for a relational model system.
7. Suppose that a given user view contains information about employees and projects. Suppose further that each employee has a unique employee number and that each project has a unique project number. Explain how you would implement the relationship between employees and projects in each of the following scenarios:
   a. Many employees can work on a given project, but each employee can work on only a *single* project.
   b. An employee can work on many projects but each project has a *unique* employee assigned to it.
   c. An employee can work on many projects, and a project can be worked on by many employees.
8. A database at a college is required to support the following requirements:
   a. For a department, store its number and name.
   b. For an advisor, store his or her number and name and the number of the department to which he or she is assigned.
   c. For a course, store its code and description (for example, MTH110, ALGEBRA).
   d. For a student, store his or her number and name. For each course the student has taken, store the course code, the course description, and the grade received. Also, store the number and name of the student's advisor. Assume that an advisor may advise any number of students but that each student has just one advisor.
   Complete the information-level design for this set of requirements. Use your own experience to determine any constraints you need that are not stated in the problem. Represent the answer in DBDL.
9. List the changes that would need to be made in your answer to Exercise 8 if a student could have more than one advisor.

10. Suppose that in addition to the requirements specified in Exercise 8, we must store the number of the department in which the student is majoring. Indicate the changes this would cause in the design in these two situations:
    a. The student must be assigned an advisor who is in the department in which the student is majoring.
    b. The student's advisor does not necessarily have to be in the department in which the student is majoring.
11. Illustrate mapping to the relational model by means of the design shown in Figure 6.13. List the relations. Identify the keys. List the special restrictions that programs will have to enforce.
12. In Example 2 of section 6.4, the claim was made that User View 6 led to the same set of tables that had been created for User View 5. Show that this is true.

Functions of a Database Management System

OBJECTIVES

1. Discuss the following eight functions, or services, that should be provided by a DBMS:
   a. data storage, retrieval, and update
   b. a user-accessible catalog
   c. support for shared update
   d. backup and recovery services
   e. security services
   f. integrity services
   g. services to promote data independence
   h. utility services
2. Discuss the manner in which these services are typically provided.

7.1   INTRODUCTION

A good DBMS should furnish a number of capabilities. As you might expect, the list of features that a full-scale mainframe DBMS could provide would be more extensive than a comparable list for a microcomputer system. Moreover, mainframe systems often furnish these features in a more sophisticated fashion. If you are interested in the features of mainframe systems, see Chapter 2 of [14]. In this chapter, we will focus on microcomputer systems. The list of features that a microcomputer DBMS should furnish includes the following:

1. **Data storage**, **retrieval**, and **update**: the ability to store, retrieve, and update the data that is in the database.
2. A user-accessible **catalog** in which descriptions of data items are stored and which is accessible to users.
3. Support for **shared update**: a mechanism to ensure accuracy when several users are updating the database at the same time.
4. **Backup** and **recovery** services: a mechanism for recovering the database in the event that the database is damaged in any way.
5. **Security** services: a mechanism to ensure that only authorized users can access the database.
6. **Integrity** services: a mechanism to ensure that certain rules are followed with regard to data in the database and any changes that are made in the data.
7. Services to promote **data independence**: facilities to support the independence of programs from the structure of the database.
8. **Utility** services: DBMS-provided services that assist in the general maintenance of the database.

The preceding list is summarized in Figure 7.1.

Figure 7.1	**Functions of a DBMS**
Functions of a DBMS	1. Data storage, retrieval, and update
	2. A user-accessible catalog
	3. Support for shared update
	4. Backup and recovery services
	5. Security services
	6. Integrity services
	7. Services to promote data independence
	8. Utility services

7.2　STORAGE AND RETRIEVAL

A DBMS must furnish users with the ability to store, retrieve, and update the data that is in the database.

This statement about storage and retrieval almost goes without saying. It defines the fundamental capability of a DBMS. Unless a DBMS provides this facility, further discussion of what a DBMS can do is irrelevant. In storing, updating, and retrieving data, it should not be incumbent upon the user to be aware of the system's internal structures or the procedures used to manipulate these structures. This manipulation is strictly the responsibility of the DBMS.

7.3　CATALOG

A DBMS must furnish a **catalog** in which descriptions of data items are stored and which is accessible to users.

This catalog contains crucial information for those who are in charge of a database or who are going to write programs to access a database. Such persons must be able to easily determine what the database "looks like." Specifically, they need to be able to get quick answers to questions like the following:

1. What tables and columns are included in the current structure? What are their names?
2. What are the characteristics of these columns? For example, is the *CUSTNAME* column within the *CUSTOMER* row 20 characters long or 30? Is the *CUSTNUMB* column a numeric field or is it a character field? How many decimal places are in the *UNITPRCE* column?
3. What are the possible values for the various columns? Are there any restrictions on the possibilities for *CREDLIM*, for example?
4. What is the meaning of the various columns? For example, what exactly is *ITEMCLSS*, and what does the item class HW mean?
5. What relationships are present? What is the meaning of each relationship? Must the relationship always exist? For example, must a customer always have a sales rep?
6. Which programs within the system access which data within the database? How do they access it? Do they merely retrieve the data, or do they update it? What kinds of updates do they do? Can a certain program add a new customer, for example, or can it merely make changes regarding information about customers whose names are already in the database? When it makes a change with regard to a customer, can it change all the columns or only the address?

Mainframe DBMSs are often accompanied by a separate entity called a **data dictionary**, which contains answers to all of the previous questions and more. The data dictionary forms a sort of super catalog. Microcomputer DBMSs are not typically accompanied by such a comprehensive tool, but they often have built-in capabilities that furnish answers to at least some of these questions. At a minimum, the capabilities they furnish would allow us to obtain the answers to questions 1 and 2 in the preceding list. Some microcomputer DBMSs provide features that allow us to ask questions along the lines of 3, 4, and 5.

7.4   SHARED UPDATE

A DBMS must furnish a mechanism to ensure accuracy when several users are updating the database at the same time.

Microcomputer databases are often used by just one person at one machine. Sometimes several people may be allowed to update a database but only one person at a time. For example, several people might take turns with one microcomputer to access the database. The advent of microcomputer networks and microcomputer DBMSs that were capable of running on these networks and of allowing several users to access the same database gave rise to a problem that had been a headache to mainframe database management for years: shared update.

By **shared update**, we mean that two or more users are involved in making updates to the database at the same time. On the surface, it might seem that shared update wouldn't present any problem. Why couldn't two or three (or fifty, for that matter) users update the database simultaneously without incurring a problem?

The Problem   To illustrate the problems involved in shared update, let's assume that we have two users, Fred and Sue, who both work for Premiere Products. Fred is currently accessing the database to process orders and, among other things, to increase customers' balances by the amount of the orders. Let's say that Fred is going to increase the balance of customer 124 (Sally Adams) by $100.00. Sue, on the other hand, is accessing the database to post payments and, among other things, to decrease customers' balances by the amount of the payments. As it happens, customer 124 has just made a $100.00 payment, so Sue will decrease her balance by $100.00. The balance of customer 124 was $418.75 prior to the start of Fred and Sue's activity and, since the amount of the increase exactly matches the amount of the decrease, the balance should still be $418.75 after the activity has been completed. But will it? That depends.

How exactly does Fred make the required update? First, the data concerning customer 124 is read from the database into Fred's work area. Second, any changes are made in the data in his work area; in this case, $100.00 is added to the current balance of $418.75, bringing the balance to $518.75. This change has *not* yet taken place in the database, *only* in Fred's work area. Finally, the information is written to the database and the change is now made in the database itself (see Figure 7.2).

Figure 7.2

Fred updates the database

Fred's Work Area

124_..._418.75

Database on Disk

Sue's Work Area

Prior to Step 1 — Before updates

(continued)

Figure 7.2

(continued)

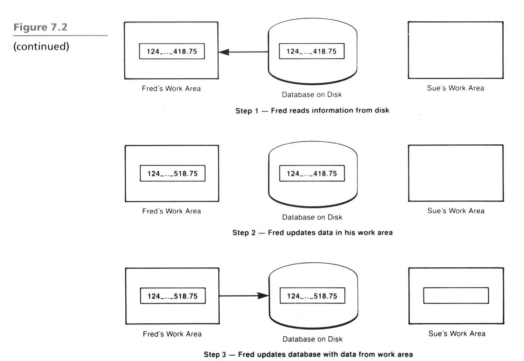

Step 1 — Fred reads information from disk

Step 2 — Fred updates data in his work area

Step 3 — Fred updates database with data from work area

Suppose that Sue begins her update at this point. The data for customer 124 will be read from the database, including the new balance of $518.75. The amount of the payment, $100.00, will then be subtracted from the balance, thus giving a balance of $418.75 *in Sue's work area*. Finally, this new information is written to the database, and the balance for customer 124 is what it should be (see Figure 7.3).

Figure 7.3

Sue updates the database

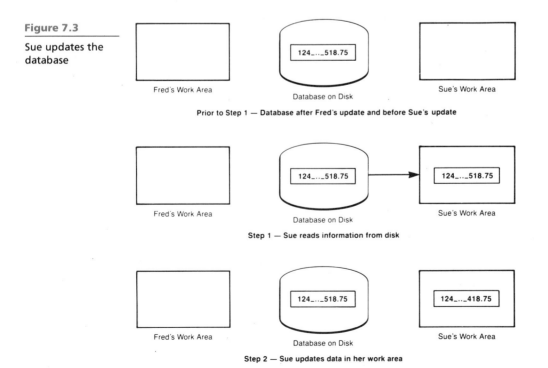

Prior to Step 1 — Database after Fred's update and before Sue's update

Step 1 — Sue reads information from disk

Step 2 — Sue updates data in her work area

Figure 7.3

(continued)

Step 3 — Sue updates database with data from work area

In the preceding scenario, things worked out the right way. But they don't necessarily have to. Do you see how things could happen in a way that would lead to an incorrect result? What if the scenario shown in Figure 7.4 had occurred instead? Here, Fred reads the data from the database into his work area, and at about the same time, Sue reads the data from the database into her work area. At this point, both Fred and Sue have the correct data for customer 124, including a balance of $418.75. Fred adds $100.00 to the balance in his work area, and Sue subtracts $100.00 from the balance in her work area. At this point, in Fred's work area the balance reads $518.75, while in Sue's work area it reads $318.75. Fred now writes to the database. At this moment, customer 124 has a balance of $518.75 in the database. Then Sue writes to the database. Now the balance for customer 124 in the database is *$318.75!* (This is a very good deal for Sally Adams, but not such a good deal for Premiere Products.) Had the updates taken place in the reverse order, the final balance would have been $518.75. In either case, we now have incorrect data in our database (one of the updates has been *lost*). This cannot be permitted to happen.

Figure 7.4

Fred and Sue update the database in a manner that leads to inconsistent data

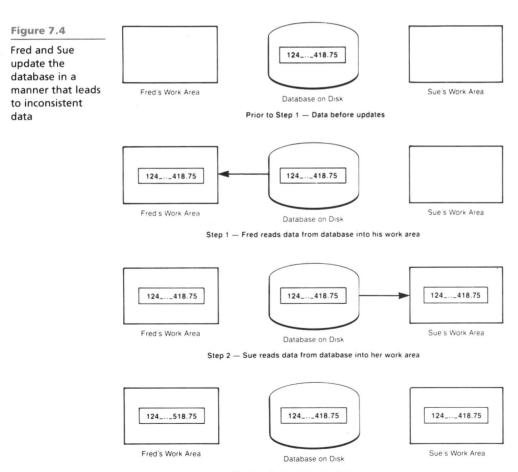

Prior to Step 1 — Data before updates

Step 1 — Fred reads data from database into his work area

Step 2 — Sue reads data from database into her work area

Step 3 — Fred updates data in his work area

(continued)

Figure 7.4

(continued)

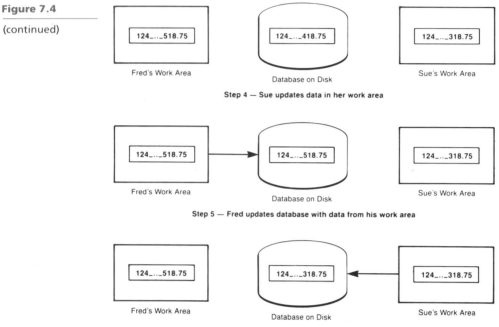

Step 4 — Sue updates data in her work area

Step 5 — Fred updates database with data from his work area

Step 6 — Sue updates the database with data from her work area

Fred's update is lost!

Avoiding the Problem

One way to prevent this situation from occurring is to prohibit shared update. This may seem a little drastic, but it is not really so farfetched. We could permit several users to access the database at the same time, but for *retrieval* only; that is, they would be able to read information *from* the database but they would not be able to write anything *to* the database. When these users entered some kind of transaction to update the database (like posting a payment), the database itself would not be updated at all. Instead, a record would be placed in a separate file of transactions. A record in this file might indicate, for example, that $100.00 had been received from customer 124 on a certain date. Periodically, a single update program would read the records in this transaction file and perform the appropriate updates to the database. Since this program would be the only one to update the database, we would eliminate the problems associated with shared update.

While this approach would avoid one set of problems, it would create another. From the time users started updating, that is, placing records in the update files, until the time the update program actually ran, the data that was in the database would be out of date. Where a customer's balance in the database was $49.50, it would actually be $649.50 if a transaction had been entered that increased it by $600.00. If the customer in question had an $800 credit limit, he or she should be prohibited from charging, say, a $200 item. But according to the data currently in the database, this would not be so. On the contrary, the data in the database would indicate that this customer still had $750.50 of available credit ($800 – $49.50). In a situation that requires the data in the database to be current, this scheme for avoiding the problems of shared update will not work.

Locking   Assuming that we cannot solve the shared update problem by avoiding it, we need a mechanism for dealing with the problem. We need to be able to keep Sue from even beginning the update on customer 124 until Fred has completed his update (or vice versa). This can be accomplished by some kind of **locking** scheme. Suppose that once Fred had read the row for customer 124, it became locked (no other user could access it) and remained locked until Fred had completed the update. For the duration of the lock, any attempt by Sue to read the row would be rejected, and she would be notified that the row was locked. If she chose to do so, she could keep attempting to read the row until it was no longer locked, at which time she could begin her update. This scenario is demonstrated in Figure 7.5. In at least this simple case, the problem of a "lost update" seems to have been solved.

Figure 7.5

Fred and Sue update the database. Locking prevents inconsistent data

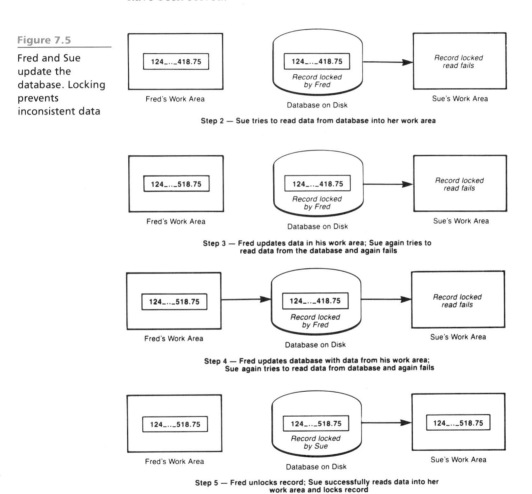

Step 2 — Sue tries to read data from database into her work area

Step 3 — Fred updates data in his work area; Sue again tries to read data from the database and again fails

Step 4 — Fred updates database with data from his work area; Sue again tries to read data from database and again fails

Step 5 — Fred unlocks record; Sue successfully reads data into her work area and locks record

(continued)

Figure 7.5

(continued)

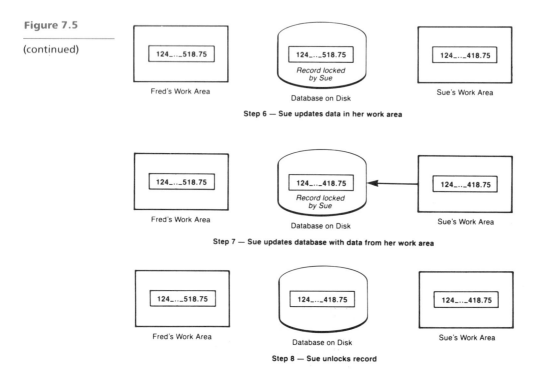

Step 6 — Sue updates data in her work area

Step 7 — Sue updates database with data from her work area

Step 8 — Sue unlocks record

Duration of Locks

How long should a lock be held? If the update involves just changing some values in a single row in a single table (like changing the name of a customer), the lock is no longer necessary once this row has been updated. Sometimes, however, the situation is more involved.

Consider, for example, the process of filling an order. To a user sitting at a terminal, this may seem to involve a single action. A user may merely indicate that an order which is currently on file now needs to be filled. Or he or she may also be required to enter certain data on the order. In either case, however, the process still feels like a single action to the user. Behind the scenes, though, lots of activity may be taking place. We might have to update the *UNONHAND* column in the *PART* table for each part that is on the order in order to reflect the number of units of that part that were shipped and that are consequently no longer on hand. We might also have to update the *BALANCE* column in the *CUSTOMER* table for the customer who placed the order, increasing it by the total amount of the order. Or we might have to update the *TOTCOMM* column in the *SLSREP* table for the sales rep who represents this customer, increasing it by the amount of commission generated by the order.

In circumstances like these, where a single action on the part of a user necessitates several updates in the database, what do we do about locks? How long do we hold each one? For safety's sake, locks should be held until all the required updates have been completed. Serious problems can result from not doing this. (A discussion of such problems is beyond the scope of this text; if you are interested in some of the specifics, see [14].)

Deadlock

Users can hold more than one lock at a time and this may give rise to another potential problem. Let's suppose that Mary has locked the row for customer 124 and is attempting to lock the row for part BT04. Let's also suppose that Tom already has part BT04 locked, so Mary must wait for him to unlock it. Before Tom unlocks part BT04, though, he needs to update, and thus lock, the row for customer 124, which is currently locked by — you guessed it — Mary. Mary is waiting for Tom to act (release the lock for part BT04), while Tom, on the other hand, is waiting for Mary to act (release the lock for customer 124). Without the aid of some outside intervention, this dilemma could be prolonged indefinitely. The term used to describe such situations is **deadlock**, and obviously, some strategy is necessary to either prevent or handle it.

Locking on Microcomputer DBMSs

Mainframe DBMSs typically offer sophisticated schemes for locking as well as for detecting and handling deadlocks (see [14]). Microcomputer DBMSs provide facilities for the same purposes, but they are usually much more limited than the facilities that are provided by mainframe DBMSs. These limitations, in turn, put an additional burden on the programmers who write the programs that allow several users to update the same database simultaneously.

Although the exact features for handling these problems vary from one microcomputer DBMS to another, the following list is fairly typical of the types of facilities provided:

1. Programs can lock a whole table or an individual row within a table (but only one). As long as one program has a row or table locked, no other program may access it.
2. Programs can release any or all of the locks they currently hold.
3. Programs can inquire whether a given row or table is locked.

This list, though it is short, comprises the *complete* set of facilities provided by many systems. Consequently, the following guidelines have been devised for writing programs for a shared-update environment:

1. If more than one row in the same table must be locked during an update, the *whole table* must be locked.
2. When a program attempts to read a row that is locked, it may wait a short period of time and then try to read the row again. This process could continue until the row becomes unlocked. It is usually preferable, however, to impose a limit on the number of times a program may attempt to read the row. In this case, reading is done in a loop, which proceeds until either (a) the read is successful, or (b) the maximum number of times that the program can repeat the operation is reached. Programs vary in terms of what action is taken should the loop be terminated without the read being successful. One possibility is to notify the user of the problem and let him or her decide whether to try the same update again or move on to something else.
3. Since there is no facility to *detect and handle* deadlocks, we must try to *prevent* them. A common approach to this problem is for every program in the system to attempt to lock all of the rows and/or tables it needs before beginning an update. Assuming it is successful in this attempt, each program can then perform the required updates. If any row or table that the program needs is already locked, it should immediately release *all* the locks that it currently holds, wait some specified period of time, and then try the entire process again. (In some cases, it may be better to notify the user of the problem and see whether the user wants to try again.) Effectively, this means that any program that encounters a problem will immediately get out of the way of all the other programs, rather than be involved in a deadlock situation.

4. Since locks prevent other users from accessing a portion of the database, it is important that no user keep rows or tables locked any longer than necessary. This is especially significant for on-line update programs. Suppose, for example, that a user is employing some on-line update program to update customers. Suppose further that once the user enters the number of the customer to be updated, the customer row is locked and remains locked until the user has entered all the new data and the update has taken place. What if the user is interrupted by a phone call before he or she has completed filling in the new data? What if the user goes to lunch? The row might remain locked for an extended period of time. If the update involves several rows, all of which must be locked, the problem becomes that much worse. In fact, in many microcomputer DBMSs, if more than one row from the same table must be locked, the whole table must be locked, which means that whole tables may be locked for extended periods of time. Clearly, this cannot be permitted to occur.

The trick here is for programs to read the information they need at the beginning of the update and then immediately *release all locks*. After the user has entered all the new data, the update takes place as described earlier (that is, attempt to lock all required rows; proceed with the update if successful; release all locks if not successful). This does pose a problem, however. Suppose my program read the data for customer 124 and then released its lock on this customer while the user of my program was filling in new data on the screen. What if a user of your program updated customer 124 in the meantime and the update was completed before my user finished filling in the new data? If my user then were to finish filling in the new data and my program blindly went ahead to update the row for customer 124 with this new data, your user's update would be *lost* (that is, over-written with my user's data). So, my program should take a further precautionary step. Before blindly updating the database with my user's data, my program should make sure that nobody else has updated the data in the meantime. If someone has, my program cannot update the database with my user's data; instead, my user must be informed of the situation and permitted to decide whether he or she wants to redo the update or move on to something else.

How will my program know whether or not some other user has updated the row for customer 124? Several methods can be used to provide the answer. One is to include an additional column in each row, perhaps a three-digit number called *UPDCOUNT*. Every time any program updates a row in any way, it should also update the value in this column by adding 1 to it. (If the previous value was 999, the new value produced by adding one to the old value would be too big for the column. In such a case, the program updating the row would have to set the update count back to zero.) Assuming that every program in the system were to adhere to this approach, we could utilize the following logic:

1. Read all the data from the row for customer 124, including the value of *UPDCOUNT*. (Let's assume for the purposes of this example that the value is 478.) Store this value in some variable for future reference. Unlock the row.
2. Get all the new data from the user.
3. When it is time to do the update, lock the row for customer 124, read the current data, and examine the value of *UPDCOUNT*. If it is still the same (in our case, 478), the row has not been updated and we can finish our update. If it is different (479 or 480, for example), we know that at least one other program has updated the data in the meantime and we cannot complete our update.
4. If we have to lock multiple rows, the same procedure is followed for each one; that is, for each of the rows involved, store its update count in some variable. When it is time to do the update, lock all the rows, read each row's update count, and compare it with the count we have stored. If the counts *all* agree, we can perform the update. If they don't agree, the update cannot take place.

Two crucial points arise from the preceding discussion. First, the logic to support shared update certainly adds a fair amount of complexity to each of the programs in the system. Second, cooperation among programs is *essential*. Every program must do its job. If one program doesn't update the *UPDCOUNT* column, for example, another program may assume that a row has not been updated when, in fact, it has been. If a program doesn't release all its locks when it encounters a row or table it needs that is locked by some other program, the possibility of deadlock arises. If a program does not release its locks while its user is entering data on the screen, the performance of the whole system may suffer.

One might naturally ask at this point, Is it worth it? Is the ability to have several users updating the database simultaneously worth the complexity that it adds to every program in the system? In some cases, the answer will be no. Shared update may be far from a necessity. In other cases, however, shared update will be necessary to the productivity of the users of the system. In these cases, implementation either of the ideas we have been discussing or of some similar scheme is essential to the proper performance of the system.

7.5   RECOVERY

A DBMS must furnish a mechanism for recovering the database in the event that the database is damaged in any way.

A database can be damaged or destroyed in a number of ways. Users can enter data that is incorrect; programs that are updating the database can abort during an update; a hardware problem can occur; and so on. After such an event has occurred, the database may contain invalid data. It may even have been totally destroyed.

Obviously, a situation in which data has been damaged or destroyed cannot be allowed to go uncorrected. The database must be returned to a correct state. This process is called **recovery**; we say that we **recover** the database.

The simplest approach to recovery involves periodically making a copy of the database (called a **backup**, or a **save**). If a problem occurs, the database is recovered by copying this backup copy over it. In effect, the damage is undone by returning the database to the state it was in when the last backup was made.

Unfortunately, other activity besides that which caused the destruction is also undone. Suppose the database is backed up at 10:00 at night and users begin updating it at 8:00 the next morning. Suppose further that at 11:30 that morning, something happens that destroys the database. If the previous night's backup is used to recover the database, the *entire* database is returned to the state it was in at 10:00 the previous night. *All* updates made in the morning are lost, not just the update or updates that were in progress at the time the problem occurred. This would mean that during the final part of the recovery process, users would have to redo all the work they had done since 8:00 in the morning.

As you might expect, mainframe DBMSs provide sophisticated facilities to avoid the costly and time-consuming process of having users redo their work. These facilities maintain a record (called a **journal**) of all updates to the database. (If you are interested in the manner in which these features work, see Chapter 2 of [14].)

Such features are not generally available at this time on microcomputer DBMSs. Most of them furnish users with a simple way to make backup copies and to recover the database later by copying the backup over the database; but this is all they furnish in this regard.

Given this state of affairs, how should we handle backup and recovery in any application system we develop with a microcomputer DBMS? We could simply use the features of the DBMS to periodically make backup copies, and use the most recent backup if a recovery were necessary. The more crucial it was to avoid redoing work, the more often we would make backup copies. (If a backup were made every eight hours, for example, we might have to redo up to eight hours of work. If one were made every two hours, on the other hand, at most two hours of work would have to be redone.)

In many situations, this approach, although not particularly desirable, is acceptable. For some systems, however, it is not. In such cases, the necessary recovery features that are not supplied by the DBMS must be included in the application programs. Each of the programs that update the database could, for example, also write a record to a separate file, the journal, indicating the update that had taken place. A separate program could be written that would look at this file and recreate all of the updates indicated by the records in the file. The recovery process would then consist of (1) copying the backup over the actual database and (2) running this special program.

While this approach does simplify the recovery process for the users of the system, it also causes some problems. First, each of the programs in the system becomes more complicated because of the extra logic involved in adding records to the special file. Second, a separate program to update the database with the information in this file must be written. Finally, every time a user completes an update, the system now has extra work to do, and this additional processing may slow down the system to an unacceptable pace. Thus, in any application, we must determine whether the ease of recovery furnished by this approach is worth the price we may have to pay for it. The answer will vary from one system to another.

7.6 SECURITY

A DBMS must furnish a mechanism that restricts access to the database to authorized users. The term **security** refers to the protection of the database against unauthorized (or even illegal) access, either intentional or accidental. The most common features used by microcomputer DBMSs to provide for security are passwords and encryption.

Passwords

Many microcomputer DBMSs furnish sophisticated schemes whereby system administrators can assign passwords. Each password may be associated with a list of actions that the user who furnishes it is permitted to take. A user who furnished the password XY1JE, for example, might be allowed to view and alter any customer data, whereas another user who furnished the password GS36Y might be permitted to view and alter a customer's name or address, view but not alter a customer's credit limit, and not even view a customer's balance.

Encryption

Encryption refers to the storing of the data in the database in an encrypted format. Any time a user stores or modifies data in the database, the DBMS will encrypt the data before actually updating the database. Before a legitimate user retrieves the data via the DBMS, the data will be decrypted. The whole encryption process is transparent to a legitimate user; that is, he or she is not even aware it is happening. However, if an unauthorized user attempts to bypass all the controls of the DBMS and get to the database directly, he or she will be able to see only the encrypted version of the data.

Views If a DBMS provides a facility that allows various users to have their own **views** of a database, this can be used for security purposes. Tables or columns to which the user does not have access in his or her view effectively do not exist for that user.

For further discussion of security, including security features found on mainframe systems, see [14].

7.7 INTEGRITY

A DBMS must furnish a mechanism to ensure that both the data in the database and changes in the data follow certain rules.

In any database, there will be conditions, called **integrity constraints**, that must be satisfied by the data within the database. The types of constraints that may be present fall into the following four categories:

1. **Data type.** The data entered for any column should be consistent with the data type for that column. For a numeric column, only numbers should be allowed to be entered. If the column is a date, only a legitimate date (in the form MMDDYY or MM/DD/YY) should be permitted. For instance, 13/07/92 would be an illegitimate date that should be rejected by the DBMS.

2. **Legal values.** It may be that for certain columns, not every possible value that is of the right type is legitimate. For example, even though *CREDLIM* is a numeric column, only the values 300, 500, 800, and 1,000 may be valid. It may be that only numbers between 2.00 and 800.00 are legal values for *UNITPRCE*.

3. **Format.** It may be that certain columns have a very special format that must be followed. Even though the column *PARTNUMB* is a character field, for example, only specially formatted strings of characters may be acceptable. Legitimate part numbers may have to consist of two letters followed by a hyphen, followed by a three-digit number. This is an example of a format constraint.

4. **Key constraints.** There are two types of key constraints: primary key constraints and foreign key constraints. Primary key constraints enforce the uniqueness of the primary key. For example, forbidding the addition of a sales rep whose number matched the number of a sales rep already in the database would be a primary key constraint. Foreign key constraints enforce the fact that a value for a foreign key must match the value of the primary key for some row in another table. Forbidding the addition of a customer whose sales rep *was not already in the database* would be an example of a foreign key constraint.

An integrity constraint can be treated in one of four ways:

1. The constraint can be ignored, in which case no attempt is made to enforce the constraint.
2. The burden of enforcing the constraint can be placed on the users of the system. This means that users must be careful that any changes they make in the database do not violate the constraint.
3. The burden can be placed on programmers. Logic to enforce the constraint is then built into programs. Users must update the database only by means of these programs and not through any of the built-in entry facilities provided by the DBMS, since these would allow violation of the constraint. The programs are designed to reject any attempt on the part of the user to update the database in such a way that the constraint is violated.
4. The burden can be placed on the DBMS. The constraint is specified to the DBMS, which then rejects any attempt to update the database in such a way that the constraint is violated.

Q & A

Question: Which of these approaches is best?

Answer:

The fourth approach. Here is why.

The first approach is undesirable, since it can lead to invalid data in the database (two customers with the same number, part numbers with an invalid format, illegal credit limits, and so on).

The second approach is a little better, since at least an attempt is made to enforce the constraints. Yet it puts the burden of enforcement on the user. Not only does this mean extra work for the user, but any mistake on the part of a single user, no matter how innocent, can lead to invalid data in the database.

The third approach removes the burden of enforcement from the user and places it on the programmers. This is better still, since it means that users will be unable to violate the constraints. The disadvantage is that all of the update programs in the system are made more complex. This complexity makes the programmers less productive and makes the programs more difficult to create and to modify. It also makes changing an integrity constraint more difficult, since this may mean changing all of the programs that update the database. Further, any program in which the logic that is used to enforce the constraints is faulty could permit some constraint to be violated *without our even being aware that this had happened* until some problem that occurred at a later date brought it to our attention. Finally, we would have to carefully guard against a user bypassing the programs in the system in order to enter data directly into the database (for example, by using some built-in facility of the DBMS). If this should happen, all of the controls we had so diligently placed into our programs would be helpless to prevent a violation of the constraints.

The best approach is the one in which we put the burden on the DBMS. We would specify any constraints to the DBMS and the DBMS would ensure that they are never violated.

Unfortunately, most microcomputer DBMSs don't have all the necessary capabilities to enforce the various types of integrity constraints (neither do many mainframe DBMSs). Usually, the approach that is taken is a combination of the (3) and (4) in the foregoing list. We let the DBMS enforce any of the constraints that it is capable of enforcing; other constraints are enforced by application programs. We might also create a special program whose sole purpose would be to examine the data in the database to determine whether any constraints had been violated; this program would be run periodically. Corrective action could be taken to remedy any violations that were discovered by means of this program.

Current Microcomputer DBMSs and Integrity

We conclude this section with a discussion of the constraints that current microcomputer DBMSs are able to enforce.

- Virtually all microcomputer DBMSs do an excellent job of enforcing data-type constraints. At a minimum, they typically allow data types of numeric, character, and date. Users are prevented from entering nonnumeric data into numeric columns or invalid dates into date columns.

- Most microcomputer DBMSs do not provide direct support for enforcing constraints that involve legal values. In cases where they do provide such support, it is often for a range of numbers (unit price must be between 2.00 and 8.00), but not for selected numbers (credit limit must be 300, 500, 800, or 1,000). Of the systems that don't provide direct support, some will supply support if users update the data through custom-generated forms; in other words, the constraints can be specified during the description of a form and the constraints will be enforced for any user who employs that form to update the data in the database. However, if the database is updated in some other way, these constraints will not be enforced.

- The way in which many current microcomputer DBMSs treat format constraints is similar to the way in which they treat legal values. The constraints are not implemented directly, but some fairly sophisticated format-type constraints can be built into custom-created forms. In the case of some DBMSs, however, a substantial number of format constraints can be communicated to the DBMS directly, and these constraints will be enforced no matter how the data is entered into the database.

- Most microcomputer DBMSs (and many mainframe DBMSs) are weakest in the area of key constraints. Many DBMSs do not allow a primary key to be specified and thus will certainly not enforce any uniqueness. Most systems allow users to build an index on the column or columns that constitute the primary key, but many systems do not allow users to specify the uniqueness that is essential for the primary key. Thus, the burden of enforcing the uniqueness will typically be placed on the programs. True support for foreign key constraints is almost totally lacking and the burden of enforcing these constraints (like the requirement that the sales rep for a given customer already be in the database) will also be placed on programs.

With time, the better DBMSs will improve in this area. And as each improvement takes place, some of the burden that was formerly placed on programs will be placed where it belongs — on the DBMS.

For a further discussion of integrity, including the integrity features found on mainframe systems, see [14].

7.8 DATA INDEPENDENCE

A DBMS must include facilities that provide programs with independence in terms of their relationship to the structure of the database.

Some of you may have written or worked with application systems that accessed a collection of files. Were any changes ever required in the types of data stored in the files? Did users ever propose any further requirements that necessitated the addition of columns? What about changing the characteristics of a column (for example, expanding the number of characters in the *NAME* column from 25 to 30)? What about additional processing requirements (for example, a new requirement to rapidly access a customer on the basis of his or her name)?

If any of these things have happened to you, you know that even the simplest of them can be very painful. Adding a new column or changing the characteristics of an existing column, for example, usually entails writing a program that will read each record from the existing file and will write a corresponding record with the new layout to a new file. In addition, each of the programs in the existing system must be changed to reflect the new layout, and these changes must be tested.

One of the advantages of working with a DBMS is **data independence**, that is, the property that changes can be made in the layout of a database without application programs necessarily being affected. Let's examine how the various types of changes that can be made in the structure of the database would affect programs that access the database, we are using a good microcomputer DBMS.

1. **Addition of a column.** No program *should* need to be changed except, of course, those programs which will utilize the new column. Some programs *may* need to be changed, however. If, for example, a program used something like the SQL "SELECT * FROM ..." to select all of the columns from a given table, the user would suddenly be presented with an extra column. To prevent this from happening, the output of the program would have to be restricted to only the desired columns. To avoid the imposition of this extra work, it's a good idea to list specific columns in a SQL SELECT command instead of using the *.

2. **Changing the length of a column.** In general, programs should not have to change because the length of a column has been changed. For the most part, the DBMS will handle all the details concerning this change in length. If, however, a program is designed to set aside a certain portion of the screen or a report for the column and the length of the column has increased to the point where the previously allocated space is inadequate, clearly the program will need to be changed.

3. **Creating a new index.** Typically, a simple command is all that is required to create a new index. But in order to *use* the index, reference must be made to it when the database is opened. Thus, in any program that will use this new index, the statement that opens the database must be changed. In most systems, any index that has been referenced in this way is updated automatically when the data is updated, so no changes are required in those portions of the program which are devoted to updating. To ensure the use of the index in a given query, some special action *may* be required, but the necessary changes are usually very simple ones.

4. **Adding or changing a relationship.** This change is the trickiest of all and is best illustrated with an example. Let's suppose that at Premiere Products we now have the following requirements:

a. Customers are assigned to territories.
b. Each territory is assigned to a single sales rep.
c. A sales rep can have more than one territory.
d. A customer is represented by the sales rep who covers the territory to which the customer is assigned.

To implement these changes, we might choose to restructure the database as follows:

```
SLSREP (SLSRNUMB, SLSRNAME, SLSRADDR,
        TOTCOMM, COMMRATE)
TERRITORY (TERRNUMB, TERRDESC,
        SLSRNUMB)
CUSTOMER (CUSTNUMB, NAME, ADDRESS, BALANCE,
        CREDLIM, TERRNUMB)
```

Now let's suppose that a user is accessing the database via the following view, called *SLSCUST*:

```
CREATE VIEW SLSCUST (SNUMB, SNAME, CNUMB, CNAME) AS
    SELECT SLSREP.SLSRNUMB, SLSREP.SLSRNAME,
        CUSTOMER.CUSTNUMB, CUSTOMER.NAME
        FROM SLSREP, CUSTOMER
        WHERE SLSREP.SLSRNUMB =
            CUSTOMER.SLSRNUMB
```

The defining query is no longer legitimate, since there is no *SLSRNUMB* column in the *CUSTOMER* table. A relationship still exists between sales reps and customers, however. The difference is that we now must go through the *TERRITORY* table to relate the two. If users have been accessing the tables directly to form the relationship, their programs will have to change. If they are using the *SLSCUST* view, then only the definition of the view will have to change. The new definition will be as follows on the next page.

```
CREATE VIEW SLSCUST (SNUMB, SNAME, CNUMB, CNAME) AS
    SELECT SLSREP.SLSRNUMB, SLSREP.SLSRNAME,
        CUSTOMER.CUSTNUMB, CUSTOMER.NAME
    FROM SLSREP, TERRITORY, CUSTOMER
    WHERE SLSREP.SLSRNUMB =
        TERRITORY.SLSRNUMB
        AND TERRITORY.TERRNUMB =
        CUSTOMER.TERRNUMB
```

The defining query is now more complicated than it was before, but this will not affect users of the view. They will continue to access the database in exactly the same way they did before, and their programs will not need to change.

We've now seen how the use of views can allow changes to be made in the logical structure of the database without application programs being affected. As helpful as this is, however, all is not quite as positive as it might seem. For one thing, this entire discussion would not even be relevant to the many DBMSs that do not permit the use of views. Second, even those DBMSs which support views often limit the types of update that can be accomplished through a view. In particular, if the view involves a join, often little or no updating is to be allowed to take place. So the benefits that can be derived from the use of views may very well be unavailable to the user who needs to update the database. This problem is the focus of a great deal of current research and should be resolved in the near future.

7.9 UTILITIES

A DBMS should provide a set of utility services.

In addition to the services already discussed, a DBMS can provide a number of utility-type services that assist in the general maintenance of the database. Following is a list of such services that may be provided by a microcomputer DBMS.

1. Services that permit changes to be made in the database structure (adding new tables or columns, deleting existing tables or columns, changing the name or characteristics of a column, and so on).
2. Services that permit the addition of new indexes and the deletion of indexes that are no longer wanted.
3. Access to DOS services from within the DBMS.
4. Services that provide export to and import from other microcomputer software products. For example, these services allow data to be transferred in a relatively easy fashion between the DBMS and a spreadsheet, word processing, or graphics program.
5. Several of the services that form a part of the fourth-generation environment (see Chapter 9) are also furnished by some of the better microcomputer DBMSs. These include such things as easy-to-use edit and query capabilities, screen generators, report generators, and so on.
6. Access to both procedural and nonprocedural languages. (With a procedural language, the computer must be told precisely how a given task is to be accomplished; BASIC, Pascal, and COBOL are examples of procedural languages. With a nonprocedural language, the task is merely described to the computer, which then determines how to accomplish it. SQL is an example of a nonprocedural language.)
7. An easy-to-use menu-driven interface that allows users to tap the power of the DBMS without having to resort to a complicated set of commands.

SUMMARY

1. A DBMS must furnish users with the ability to store, retrieve, and update data that is in the database.
2. A DBMS must furnish a catalog in which descriptions of the structure of a database are stored and which can be queried by users.
3. A DBMS must provide support for shared update (more than one user updating the database at the same time).
    a. If care is not taken, incorrect results can be produced in the database.
    b. Locking is one approach that ensures correct results. As long as a portion of the database is locked by one user, other users cannot gain access to it.
    c. Deadlock is the term used to describe the situation wherein two or more users are each waiting on the other to give up a lock before they can proceed. Mainframe DBMSs have sophisticated facilities for detecting and handling deadlock. Most microcomputer DBMSs do not have such facilities, which means that programs that access the database must be written in such a way that deadlocks are avoided.
4. A DBMS must provide facilities for recovering the database in the event that it is damaged or destroyed. Most microcomputer DBMSs provide facilities for periodically making a backup copy of the database. To recover the database when it is damaged or destroyed, the backup is copied over the database.
5. A DBMS must provide security facilities; that is, features that prevent unauthorized access to the database. Such facilities typically include passwords; encryption (the storing of data in an encoded form); and views (which limit users to accessing only the tables and columns included in the view).
6. An integrity constraint is a rule that data in the database must follow. A DBMS should include features that prevent integrity constraints from being violated.
7. A DBMS must include facilities that promote data independence, the property that the database structure can change without application programs necessarily being affected.
8. The DBMS must provide a set of utility services.

KEY TERMS

backup	integrity	save
catalog	journal	security
data dictionary	locking	shared update
data independence	lost update	utility services
deadlock	passwords	view
encryption	recovery	

EXERCISES

1. What do we mean when we say that a DBMS should provide facilities for storage, retrieval, and update?
2. What is the purpose of the catalog? What types of information are usually found in the catalogs that accompany microcomputer DBMSs? What additional types of information are often found in the catalogs that accompany mainframe DBMSs?
3. What is meant by shared update?
4. Describe a situation, other than the one given in the text, in which uncontrolled shared update would produce incorrect results.
5. What is meant by locking?
6. How long should locks be held?

7. What is deadlock? How does it occur?

8. Are most microcomputer systems capable of detecting and breaking deadlocks?

9. Assuming that we are using a microcomputer DBMS that provides the locking facilities described in the text, how should programs be written to (a) avoid deadlock, (b) guarantee correct results, and (c) keep any individual user from tying up portions of the database for extended periods of time?

10. What is meant by recovery? What facilities are typically provided by microcomputer DBMSs to handle backup and recovery? What main feature is lacking in such facilities? What problems can this cause for users?

11. What is meant by security?

12. How are passwords used by microcomputer DBMSs to promote security?

13. What is encryption? How does it relate to security?

14. How do views relate to security?

15. What is meant by integrity? What is an integrity constraint? Describe four different ways of handling integrity constraints. Which approach is the most desirable?

16. What is meant by data independence? What benefit does it provide?

17. Name some utility services that a DBMS should provide.

8

Database Administration

OBJECTIVES

1. Discuss the need for database administration.
2. Explain the role of DBA in formulating and implementing database policies.
3. Discuss the role of DBA with regard to the data dictionary, user training, and selection and support of DBMS.
4. Discuss the role of DBA in the database design process.

8.1 INTRODUCTION

We have already seen that the database approach confers many benefits. On the other hand, it involves potential hazards, especially when the database serves more than one user. Problems are associated with shared update, as with security: Who is allowed to access various parts of the database, and in what way? How do we prevent unauthorized accesses? Just managing the database involves fundamental difficulties. Each user must be made aware of the database structure, or at least that portion of the database which he or she is allowed to access. Any changes that are made in the structure must be communicated to all users, along with information about how the changes will affect them. Backup and recovery must be carefully coordinated, much more so than in a single-user environment, and this presents another complication.

In order to surmount these problems, the services of a person or group commonly referred to as **database administration (DBA)** are essential. DBA (usually a group rather than an individual) is responsible for supervising both the database and the use of the DBMS. (Sometimes the term DBA is used to refer to the database administrator, the individual who is in charge of this group. Usually the context makes clear which meaning is intended.)

In this chapter, we will investigate the role of DBA, which is summarized in Figure 8.1. In section 8.2, we'll discuss DBA's role in formulating and implementing important policies with respect to the database and its use. In section 8.3, we will examine DBA's role in the use of the data dictionary. In section 8.4, we'll see that DBA plays a crucial role in training various users. We'll discuss DBA's role in the selection and support of the DBMS in section 8.5. Finally, in section 8.6, we'll discuss the role DBA plays in the database design process.

Figure 8.1

Responsibilities of DBA

Responsibilities of DBA

1. Policy Formulation and Implementation
   a. Access Privileges
   b. Security
   c. Planning for Disaster
   d. Archives
2. Data Dictionary Management
3. Training
4. DBMS Support
   a. DBMS Evaluation and Selection
   b. Responsibility for DBMS
5. Database Design

In this text, we'll be focusing on the role of DBA in a microcomputer environment. The role of DBA in a mainframe environment is similar; for a detailed discussion, see [14]. For another perspective on DBA in a microcomputer environment, see [8] and [12].

8.2   POLICY FORMULATION AND IMPLEMENTATION

DBA formulates database policies and communicates these policies to users. DBA is also charged with the implementation of these policies.

**Access
Privileges**

Access to every table and column in the database is not a necessity for every user. Sam, for example, is an employee at Premiere Products whose main responsibility is the inventory. While he may very well need access to the entire *PART* table, does he also need access to the *SLSREP* table? He should probably be able to print inventory reports, but should he be able to change the layout of these reports? Betty, whose responsibility is customer mailings, clearly requires access to customers' names and addresses, but what about their balances or credit limits? Should she be able to change an address? While sales rep 3 (Mary Jones) should be able to obtain information about her own customers, should she be able to obtain the same information about other customers? Figure 8.2 illustrates permitted and denied access for these employees.

We don't have enough information about the policies of Premiere Products to answer the foregoing questions. The DBA, however, must answer questions like these and take steps to ensure that users access the database only in ways to which they are entitled. Policies concerning such access should be clearly documented for and communicated to all concerned parties.

Figure 8.2a

Permitted and
denied access for
Sam

SLSREP

SLSRNUMB	SLSRNAME	SLSRADDR	TOTCOMM	COMMRATE
3	MARY JONES	123 MAIN,GRANT,MI	2150.00	.05
6	WILLIAM SMITH	102 RAYMOND,ADA,MI	4912.50	.07
12	SAM BROWN	419 HARPER,LANSING,MI	2150.00	.05

Access Denied

SAM

Access Permitted

PART

PARTNUMB	PARTDESC	UNONHAND	ITEMCLSS	WREHSENM	UNITPRCE
AX12	IRON	104	HW	3	17.95
AZ52	SKATES	20	SG	2	24.95
BA74	BASEBALL	40	SG	1	4.95
BH22	TOASTER	95	HW	3	34.95
BT04	STOVE	11	AP	2	402.99
BZ66	WASHER	52	AP	3	311.95
CA14	SKILLET	2	HW	3	19.95
CB03	BIKE	44	SG	1	187.50
CX11	MIXER	112	HW	3	57.95
CZ81	WEIGHTS	208	SG	2	108.99

Figure 8.2b

Permitted and denied access for Betty

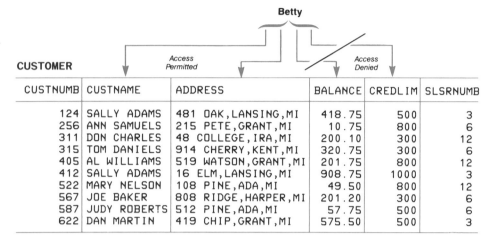

Figure 8.2c

Permitted and denied access for Mary Jones

CUSTOMER

CUSTNUMB	CUSTNAME	BALANCE	CREDLIM	SLSRNUMB	
124	SALLY ADAMS	418.75	500	3	←
256	ANN SAMUELS	10.75	800	6	←
311	DON CHARLES	200.10	300	12	←
315	TOM DANIELS	320.75	300	6	←
405	AL WILLIAMS	201.75	800	12	←
412	SALLY ADAMS	908.75	1000	3	←
522	MARY NELSON	49.50	800	12	←
567	JOE BAKER	201.20	300	6	←
587	JUDY ROBERTS	57.75	500	6	←
622	DAN MARTIN	575.50	500	3	←

Denied · Denied · **Mary Jones** *Permitted Accesses*

Security

As we noted earlier, the term **security** refers to the prevention of unauthorized access to the database. Of course, this includes access by someone who has no right to access the database *at all,* (for example, someone who is not connected with Premiere Products). It can also include users who have legitimate access to some portion of the database but who are attempting to access a portion of it to which they are *not* entitled. Figure 8.3 illustrates both types of security violation.

Figure 8.3a

Attempted security violation: John is not an authorized user

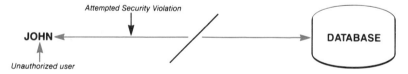

Figure 8.3b

Attempted security violation: although Betty is an authorized user, she is not authorized to access customers' balances

Betty

Attempted Security Violation →

CUSTOMER

CUSTNUMB	CUSTNAME	ADDRESS	BALANCE	CREDLIM	SLSRNUMB
124	SALLY ADAMS	481 OAK,LANSING,MI	418.75	500	3
256	ANN SAMUELS	215 PETE,GRANT,MI	10.75	800	6
311	DON CHARLES	48 COLLEGE,IRA,MI	200.10	300	12
315	TOM DANIELS	914 CHERRY,KENT,MI	320.75	300	6
405	AL WILLIAMS	519 WATSON,GRANT,MI	201.75	800	12
412	SALLY ADAMS	16 ELM,LANSING,MI	908.75	1000	3
522	MARY NELSON	108 PINE,ADA,MI	49.50	800	12
567	JOE BAKER	808 RIDGE,HARPER,MI	201.20	300	6
587	JUDY ROBERTS	512 PINE,ADA,MI	57.75	500	6
622	DAN MARTIN	419 CHIP,GRANT,MI	575.50	500	3

DBA must take steps to ensure that the database is secure. Once access privileges have been specified, DBA should draw up policies to explain them and should then distribute these policies to authorized users.

Whatever facilities are present in the DBMS, such as passwords, encryption, and/or views, should be utilized by DBA to implement these policies. Any necessary features that the DBMS lacks should be supplemented by DBA through the use of special programs. Figure 8.4 shows security features of the DBMS both with and without DBA.

Figure 8.4a

Security features of DBMS as sole security

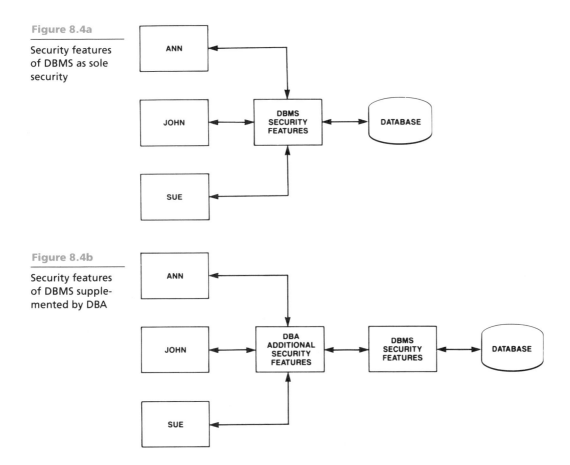

Figure 8.4b

Security features of DBMS supplemented by DBA

One security feature that we have mentioned, passwords, deserves further attention. Sometimes people think that simply establishing a password scheme will ensure security. After all, Tim can't get access to Pam's data if he doesn't know her password (assuming, of course, that Tim doesn't have a password of his own that allows access to the same data).

But what if Tim observes Pam entering her password? What if he guesses her password? You might think this sort of occurrence is so unlikely that there is nothing to worry about. In fact, it is not so unlikely. Many people often choose passwords that they can remember easily. A very common choice, for example, is the name of a family member. So if Pam is a typical user, Tim might very well be able to obtain her password just by trying names of family members. Other users choose unusual passwords (or have such passwords assigned to them), but, in order to remember them, they often have these passwords written down somewhere. Without giving it much thought, such users may be careless about the paper on which a password is written, giving people like Tim still another vehicle for obtaining one. Figure 8.5 illustrates the careless use of passwords.

It is up to DBA to educate users on the use of passwords. The pitfalls we have just discussed should be stressed, as should precautionary measures, including the need for frequent changes of passwords.

Other Threats

The type of security we've been discussing concerns harm done by unauthorized users. A database can be harmed in another way as well, and that is through some physical occurrence such as an aborted program, a disk problem, a power outage, a computer malfunction, and so on. This issue was discussed in Chapter 7 in the material on recovery, but it is listed here as well since it is DBA's responsibility to establish and implement backup and recovery procedures. As in other cases, DBA will use the built-in features of the DBMS where possible and will supplement them where they are lacking.

For example, many microcomputer DBMSs lack facilities to maintain a journal (or log) of changes in the database. Thus, recovery is usually limited to copying the most recent backup over the live database. This means, as we have already seen, that any changes made since this backup have to be redone by the users. If this presents a major problem, DBA may decide to supplement the DBMS facilities. A typical solution would be to have each program that updates the database also make appropriate entries in a journal (see Figure 8.6a) and then make use of this journal in the recovery process (see Figure 8.6b). The database is first recovered by copying the backup version over the live database; and then it is brought up-to-date through a special DBA-created program that updates the database with changes recorded in the journal.

Figure 8.6a

Programs involved in database processing also maintain a journal

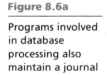
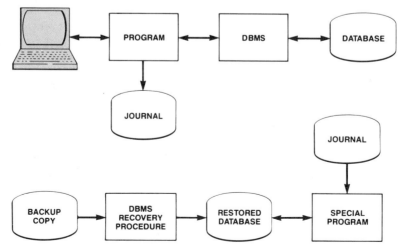

Figure 8.6b

Journal is used in recovery

Archives

Often, data needs to be kept in the database for only a limited time. An order that has been filled, has appeared on some statement, and has been paid is in one sense no longer important. Should the order be left in the database? If data is always left in the database as a matter of policy, the database will continually grow. Along with this, the disk space that is occupied by the database expands, and the performance of programs that access the database may deteriorate. Both of these events can ultimately lead to difficulties. This is a reason for removing an already filled order and all of its associated order lines from the database.

On the other hand, it may be necessary to retain such data for future reference. Possible reasons could include customer inquiries, government regulations, auditing requirements, and so on. Apparently, we may have a conflict on our hands: we may need to remove something, and yet we may not be able to afford to remove it.

The solution is to use what is known as a **data archive**. In ordinary usage an archive (technically archive*s*) is a place where public records and documents are kept. A data archive is similar. It is a place where a record of certain corporate data is kept. In fact, we often refer to a data archive simply as an archive. In the case of the aforementioned order, we would remove it from the database and place it in the archive, thus storing it for future reference (see Figure 8.7).

Typically, the archive will be kept on some magnetic form (for example, a disk, a diskette, or a tape). On mainframes, the most common medium for an archive is tape. On micros, tape might also be used, but another common choice is a collection of diskettes. Still another option would be to keep the data in printed form only. In any case, it is up to DBA to establish and implement procedures for the use and maintenance of the archive.

Figure 8.7

Movement of order 12498 from live database to archive

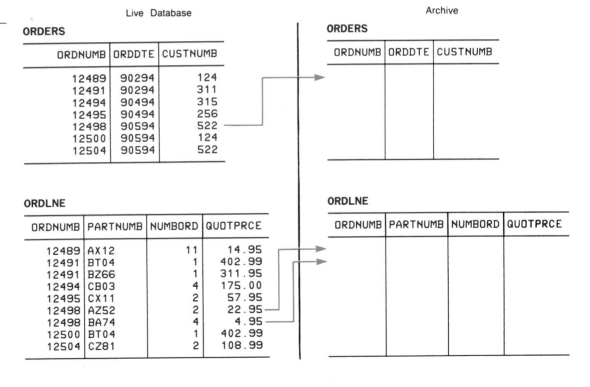

8.3   DATA DICTIONARY MANAGEMENT

In addition to administering the database, DBA also manages the **data dictionary**. The data dictionary is essentially the catalog mentioned in Chapter 7, but it often contains a wider range of information, including, at least, information on tables, columns, indexes, and programs.

DBA establishes naming conventions for tables, columns, indexes, and so on. It creates the data definitions for all tables as well as any data validation rules. It is also charged with the update of the contents of the data dictionary. The creation and distribution of appropriate reports from the data dictionary is another of its responsibilities.

8.4   TRAINING

DBA provides training in the use of the DBMS and in how to access the database. It also coordinates the training of users. In cases where training is provided by the vendor of software the organization has purchased, DBA handles the scheduling in order to make sure that the right users receive the training they require.

8.5   DBMS SUPPORT

DBMS Evaluation and Selection

DBA is responsible for the evaluation and selection of the DBMS. In order to oversee this responsibility, it sets up a checklist like the one shown in Figure 8.8. (This checklist applies specifically to a relational system, since this text deals with microcomputer DBMSs and most of these are, at least in part, relational. If we had not already selected a model for our focus, a category called "Choice of Data Model" would have to be added to the list. A corresponding checklist for mainframe systems would be slightly larger than the list shown in Figure 8.8. This example, however, is quite representative of a checklist for a relational DBMS.) DBA must evaluate each prospective purchase of a DBMS in terms of all the categories shown in the figure. An explanation of the various categories follows.

Figure 8.8

DBMS evaluation checklist

1. Data Definition
   a. Data types
      (1) Numeric
      (2) Character
      (3) Date
      (4) Logical (T/F)
      (5) Memo
      (6) Money
      (7) Other
   b. Support for nulls
   c. Support for primary keys
   d. Support for foreign keys
   e. Unique indexes
   f. Views

2. Data Restructuring
   a. Possible restructuring
      (1) Add new tables
      (2) Delete old tables
      (3) Add new columns
      (4) Change layout of existing columns
      (5) Delete columns
      (6) Add new indexes
      (7) Delete old indexes
   b. Ease of restructuring

3. Nonprocedural Languages
   a. Nonprocedural languages supported
      (1) SQL
      (2) QBE
      (3) Natural language
      (4) Own language. Award points on the basis of ease of use as well as the types of operations (e.g., joining, sorting, grouping, calculating various statistics) which are available in the language. SQL can be used as a standard against which such a language can be judged.
   b. Optimization done by one of the following:
      (1) User, in formulating the query
      (2) DBMS (through built-in optimizer)
      (3) No optimization possible. System will only do sequential searches.

4. Procedural Languages
   a. Procedural languages supported
      (1) Own language. Award points on the basis of the quality of this language both in terms of the types of statements and control structures available and the database manipulation statements included in the language.

Figure 8.8

(continued)

    (2) COBOL
    (3) FORTRAN
    (4) C
    (5) Pascal
    (6) BASIC
    (7) Other
  b. Can nonprocedural language be used in conjunction with the procedural language (e.g., could SQL be embedded in COBOL programs)?

5. Data Dictionary
  a. Types of entities
    (1) Tables
    (2) Columns
    (3) Indexes
    (4) Relationships
    (5) Programs
    (6) Other
  b. Integration of data dictionary with other components of the system

6. Shared Update
  a. Level of locking
    (1) Column
    (2) Row
    (3) Table
  b. Type of locking
    (1) Shared
    (2) Exclusive
    (3) Both
  c. Responsibility for handling deadlock
    (1) Programs
    (2) DBMS (automatic rollback of transaction causing deadlock)

7. Backup and Recovery Services
  a. Backup facilities
  b. Journaling facilities
  c. Recovery facilities
    (1) Recover from backup copy only
    (2) Recover using backup copy and journal
  d. Rollback of individual transactions

8. Security
  a. Passwords
    (1) Access to database only
    (2) Read or write access to any column or combination of columns
  b. Encryption
  c. Views
  d. Difficulty in bypassing security controls

9. Integrity
  a. Support for entity integrity
  b. Support for referential integrity
  c. Support for data-type integrity
  d. Support for other types of integrity constraints

10. Limitations
  a. Number of tables
  b. Number of columns
  c. Length of individual column
  d. Total length of all columns in a table
  e. Number of rows per table
  f. Number of files that can be open at the same time
  g. Types of hardware supported
  h. Types of LANs supported
  i. Other

11. Documentation
  a. Clearly written manuals
  b. Tutorial
    (1) Written
    (2) On-line
  c. On-line help available
    (1) General help
    (2) Context-sensitive help

12. Vendor Support
  a. Type of support available
  b. Quality of support available
  c. Cost of support
  d. Reputation of support

13. Performance
  a. Tests comparing the performance of various DBMSs in such areas as sorting, indexing, reading all rows, changing data values in all rows, and so on, are available from a variety of periodicals.
  b. If you have special requirements, you may want to design your own benchmark tests that could be performed on each DBMS under consideration.

14. Cost
  a. Cost of basic DBMS
  b. Cost of any additional components
  c. Cost of any additional hardware that is required
  d. Cost of network version (if required)
  e. Cost and types of support

15. Future Plans
  a. What does vendor plan for future of system?
  b. What is the history of the vendor in terms of keeping the system up-to-date?
  c. When changes are made in the system, what is involved in converting to the new version?
    (1) How easy is the conversion?
    (2) What will it cost?

16. Other Considerations (Fill in your own special requirements.)
  a. Special Purpose Reports
  b. ?
  c. ?
  d. ?

1. **Data definition.** What types of data are supported? Is support for nulls provided? What about primary and foreign keys? The DBMS will undoubtedly provide indexes, but is it possible to specify that an index is unique and then have the system enforce the uniqueness? Is support for views provided?

2. **Data restructuring.** What type of database restructuring is possible? How easy is it to do this restructuring? Will the system do most of the work or will the DBA have to create special programs for this purpose?

3. **Nonprocedural languages.** What type of nonprocedural language is supported? The possibilities are SQL, QBE, natural language, or a DBMS built-in language. If one of the standard languages is supported, how good a version is provided by the DBMS? If the DBMS furnishes its own language, how good is it? How does its functionality compare to that of SQL?

How does the DBMS achieve optimization of queries? Either the DBMS itself optimizes each query; the user must do so by the manner in which he or she states the query; or no optimization occurs. Most desirable, of course, is the first alternative.

4. **Procedural languages.** What types of procedural languages are supported? Are they common languages, such as COBOL, FORTRAN, BASIC, Pascal, or C, or does the DBMS come with its own language? In the latter case, how complete is the language? Does it contain all the required types of statements and control structures? What facilities are provided for accessing the database? Is it possible to make use of the nonprocedural language while using the procedural language?

5. **Data dictionary.** What kind of data dictionary support is available? Is it a simple catalog, or can it contain more, such as information about programs and the various data items these programs access? How well is the data dictionary integrated with other components of the system (for example, the nonprocedural language)?

6. **Shared update.** Is support provided for shared update? What is the unit that may be locked (column, row, or table)? Are exclusive locks the only ones permitted or are shared locks also allowed? (A shared lock permits other users to read the data; with an exclusive lock, no other user may access the data in any way.) How is deadlock handled? Will the DBMS take care of it, or is it the responsibility of programs to ensure that it is handled correctly?

7. **Backup and recovery services.** What type of backup and recovery facilities are provided? Can the DBMS maintain a journal of changes in the database and use the journal during the recovery process? If a transaction has aborted, is the DBMS capable of rolling it back (that is, undoing the updates of the transaction)?

8. **Security.** What type of security features does the system make available? Are passwords supported? Do passwords simply regulate whether a user may access the database, or is it possible to associate read or write access to a combination of columns with a password? Is encryption supported? Does the system have some type of view mechanism that can be used for security? How difficult is it to bypass the security controls?

9. **Integrity.** What type of integrity constraints are supported? Is there support for entity integrity (the fact that the primary key cannot be null)? What about referential integrity (the property where values in foreign keys must match values already in the database)? Does the DBMS support data-type integrity (the property where values that do not match the data type for the column into which they are being entered are not allowed to occur in the database)? Is there support for any other types of constraints?

10. **Limitations.** What limitations exist with respect to the number of tables, columns, and rows per table? How many files can be open at the same time? (Typically, each table and each index is in a separate file. Thus, a single table with three indexes, all in use at the same time, would account for *four* files. Problems may arise if the number of files that can be open is relatively small and many indexes are in use. On what types of hardware is the DBMS supported? What types of **local area networks** (**LANs**) support the DBMS?

(A local area network is a configuration of several computers all hooked together, thereby allowing users to share a variety of resources. One of these resources is the database. In a local area network, support for shared update is very important, since many users may be updating the database at the same time. The relevant question here, however, is not how well the DBMS supports shared update, but which of the LANs can be used in conjunction with this DBMS?)

11. **Documentation.** How good are the manuals? Are they easy to use? Is there a good index? Is a tutorial, in either printed or on-line form, available to assist users in getting started with the system? Is on-line help available? If so, is it general help or context-sensitive? (Context-sensitive help means that if a user is having trouble and asks for help, the DBMS will provide assistance for that particular problem at the time the user asks for it.)

12. **Vendor support.** What type of support is provided by the vendor, and how good is it? What is the cost? What is the vendor's reputation among current users?

13. **Performance.** How well does the system perform? This is a tough one to answer. One way to determine relative performance is to look into benchmark tests that have been performed on several DBMSs by various periodicals. Beyond this, if an organization has some specialized needs, it may have to set up its own benchmark tests.

14. **Cost.** What is the cost of the DBMS and any components the organization is planning to purchase? Is additional hardware required and, if so, what is the associated cost? If the organization requires a special version of the DBMS for a network, what is the additional cost? What is the cost of vendor support, and what types of support plans are available?

15. **Future plans.** What plans has the vendor made for the future of the system? This information is often difficult to obtain, but we can get an idea by looking at the performance of the vendor with respect to keeping the existing system up-to-date. How easy has it been for users to convert to new versions of the system?

16. **Other considerations.** This is a final, catch-all category that contains any special requirements not covered in the other categories.

Once each DBMS has been examined with respect to all the preceding categories, the results can be compared. Unfortunately, this process can be difficult, owing to the number of categories and their generally subjective nature. To make the process more objective, a numerical ranking can be assigned to each DBMS for its performance in each category (for example, a number between zero and ten, where zero is poor and ten is excellent). Further, the categories can be assigned weights. This allows an organization to signify which categories are more critical to it than others. Then each of the numbers being used in the numerical ranking can be multiplied by the appropriate weight. The results are added up, producing a weighted total. The weighted totals for each DBMS can then be compared, producing the final evaluation.

How does DBA arrive at the numbers to assign each DBMS in the various categories? Several methods are used. It can request feedback from other organizations that are currently using the DBMS in question. It can read journal reviews of the various DBMSs. Sometimes a trial version of the DBMS can be obtained, in which case members of the staff can give it a hands-on test. In practice, all three methods are sometimes combined. Whichever method is used, however, it is crucial that the checklist and weights be carefully thought out; otherwise, the findings may be inadvertently slanted in a particular direction.

Responsibility for the DBMS   Once the DBMS has been selected, DBA continues to be primarily responsible for it. DBA installs the DBMS in a way that is suitable for the organization. And, if the DBMS configuration needs to be changed, it is DBA that will make the changes.

When a new version of the DBMS is released, DBA will review it and determine whether the organization should convert to it. If the decision is made to convert to the new version (or perhaps to a new DBMS), DBA coordinates the conversion. Any fixes to problems in the DBMS which are sent by the vendor are also handled by DBA.

8.6 DATABASE DESIGN

DBA is responsible for carrying out the process of database design. It must ensure that a sound methodology for database design, such as the one discussed in Chapter 6, is established and is followed by all personnel who are involved in the process. It must also ensure that all pertinent information is obtained from the appropriate users.

DBA is responsible for the implementation of the final information-level design; in other words, it is responsible for the physical-level design process. If performance problems surface, it is up to DBA to make the changes that will improve the system's performance. This is called **tuning** the design.

DBA is also responsible for establishing standards for documentation of all the steps in the database design process. It also has to make sure that these standards are followed, that the documentation is kept up-to-date, and that the appropriate personnel have access to the documentation they need.

Requirements don't remain stable over time; they are constantly changing. DBA must review such changes and determine whether a change in the database design is warranted. If so, it must make such changes in the design and in the data in the database. It must also then make sure that all programs affected by the change are modified in any way necessary and that the corresponding documentation is also modified.

SUMMARY

1. Database administration (DBA) is the person or group that is assigned responsibility for supervising the database and the use of the DBMS.
2. DBA formulates and implements policies concerning the following:
    a. Those who can access the database; which portions of the database these persons may access, and in what manner.
    b. Security, that is, the prevention of unauthorized access to the database.
    c. Recovery of the database in the event that it is damaged.
    d. Management of an archive for data that is no longer needed in the database but must be retained for reference purposes.
3. DBA is in charge of maintaining the data dictionary.
4. DBA is in charge of training with respect to the use of the database and the DBMS. Training that is provided by an outside vendor is scheduled by DBA, which ensures that users receive the vendor training they need.
5. DBA is in charge of supporting the DBMS. This has two facets:
    a. The evaluation and selection of a new DBMS; DBA develops a checklist of desirable features for a DBMS and evaluates each prospective purchase of a DBMS against this list.
    b. DBA is responsible for installing and maintaining the DBMS after it has been selected and procured.
6. DBA is in charge of database design, both the information level and the physical level. It is also in charge of evaluating changes in requirements to determine whether a change in the database design is warranted. If so, DBA makes the change and reports it to affected users.

KEY TERMS

access privileges
archive
data archive
data dictionary
database administration
   (DBA)

disaster planning
integrity constraints
integrity support
local area network (LAN)
restructuring
tuning

EXERCISES

1. What is DBA? Why is it necessary?
2. What is DBA's role in regard to access privileges?
3. What is DBA's role in regard to security? What problems can arise in the use of passwords? How should these problems be handled?
4. Suppose a typical microcomputer DBMS is being used by company X. Suppose further that in the event the database is damaged in some way, it is essential that it be recovered *without the users having to redo any work*. What action should DBA take?
5. What are data archives? What purpose do they serve? What is the relationship between databases and data archives?
6. What is DBA's responsibility in regard to the data dictionary?
7. Who trains computer users within an organization? What is DBA's role in this training?
8. Describe the method that should be used to select a new DBMS.
9. For each of the following categories, what kinds of questions would DBA ask in order to evaluate a DBMS?
    a. Data definition
    b. Data restructuring
    c. Nonprocedural languages
    d. Procedural languages
    e. Data dictionary
    f. Shared update
    g. Backup and recovery services
    h. Security
    i. Integrity
    j. Limitations
    k. Documentation
    l. Vendor support
    m. Performance
    n. Cost
    o. Future plans
    p. Other considerations
10. How does DBA obtain the necessary information to award points for the various categories on the checklist?
11. What is DBA's role regarding the DBMS once it has been selected?
12. What is DBA's role in database design?

9

Application Generation

OBJECTIVES

1. Discuss the features offered by the typical application system.
2. Describe the components of an application generator.
3. Discuss the manner in which application generators are used to create an application system.

9.1 INTRODUCTION

When we use the term **application** or **application system**, we mean a software product, typically not just a single program, but a collection of programs, which is designed to satisfy certain *specific* needs of a user. A payroll system is a type of application system, as is a system that provides the functions of maintenance and reporting with regard to sales reps, customers, orders, and so on. The application system used by Lee provided maintenance and reporting with respect to directors, movies, movie stars, and so on. In this particular case, the user, Lee, wrote the application system himself. In other cases, it may be written by employees of the firm that will use it, or by programmers or software firms with which the user has contracted. In still other cases, it may be written by what is called a *software house*, which sells the same application system to a number of users.

The crucial idea is that an application system is written to handle a *specific* task. That task may be very complicated, but it is still something specific. A DBMS, on the other hand, is built to handle *general* types of tasks. It is up to the user of the DBMS to decide the manner in which the DBMS is used. In fact, DBMSs are used to write many, if not most, application systems.

Recently, a new type of software product, called an **application generator**, has emerged. This product, or tool, usually built around and closely integrated with a good DBMS, is designed to greatly improve the ease with which an application system can be developed. Other terms, such as fourth-generation language and fourth-generation environment, have also been applied to this highly valuable tool.

In section 9.2, we will investigate the features that are typically found in today's application systems. The components of application generators are covered in section 9.3. In section 9.4, we will discuss how these components are used to develop an application system. Finally, in section 9.5, we will discuss the other terms that are often used in place of the term application generator.

The focus of this chapter is microcomputer application systems and microcomputer application generators. Much of what will be said here applies equally well to mainframe environments, but they are more complex. For details about mainframes, see Chapter 13 of [14]. For a detailed discussion of how the advent of application generators has affected the whole system development process, see the same reference or [1].

9.2 APPLICATIONS

In what follows, we will attempt to discuss the characteristic **application system**. The features of application systems vary greatly; yet, they have enough in common to warrant a discussion of the "typical" system. You may encounter an application system that lacks one of the features we will mention. Such a system is by no means inferior because of this. It may well be that the feature in question is inappropriate for that particular system or that the system possesses some other feature that addresses the same need.

Menu

The typical application system is **menu-driven**; that is, the user indicates the action that he or she wishes to take by making the appropriate selection from a "menu" of options. A sample menu for a system for Premiere Products is shown in Figure 9.1. In this menu, users indicate their choice by entering the number associated with that option. In another menu style, individual options are highlighted. Through the use of arrow keys or a mouse, users highlight their selection and then press the return key or one of the buttons on the mouse. Figure 9.2 illustrates a menu with a highlighted box.

Figure 9.1

Menu for
*PREMIERE
PRODUCTS*

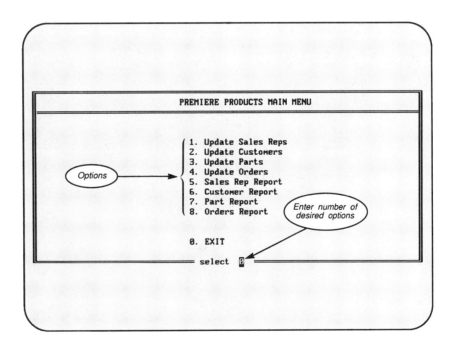

Figure 9.2

Menu for
*PREMIERE
PRODUCTS*

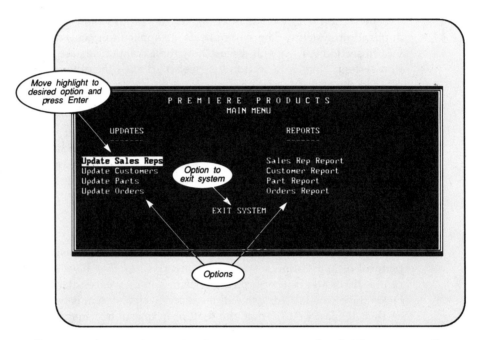

For some choices, the user's selection is now completed. The report will print or the update screen will be displayed, and the user can begin entering data. In other cases, another menu, called a submenu will appear and a second choice will be made in the same fashion as the first. For example, if the first choice was to update customer information, the second choice might be to (1) add, (2) change, or (3) delete a customer. Or, if the first choice was "PRINT REPORTS," the second choice might be to indicate which report was to be printed; whether the report was to go to the screen or the printer; whether to print the report in full detail or only the summary version; or whether all customers or only selected ones were to be printed.

Interactive and Automatic Updates

The systems usually support two types of updates. The most common one is the interactive update, so-called because the user interacts with the system by filling in forms on the screen. The other type is what we might call a "behind the scenes" or automatic update. A customer's balance, for example, might automatically be increased by the amount of an order when the order was shipped. Or fields that contain month-to-date totals would automatically be set back to zero at the end of a month. The user doesn't need to fill in a form on the screen to effect these kinds of updates.

Most updates are made interactively. For example, a **data entry screen** might be used to update customer data at Premiere Products (see Figure 9.3 on the next page). Users would indicate the type of action they wished to take by typing *A* for add, *C* for change, *D* for delete, or *S* for show. When they were through with all their updates, they would type *E* for end and would be returned to the menu.

Figure 9.3

Customer
Maintenance
screen

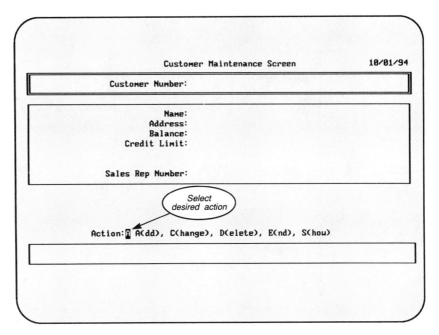

When a user enters the action *A*, the cursor moves to the customer number field and the system asks the user for a customer number. The system will first check to see whether a customer with that number is already filed in the database. If so, the user is given an error message (see Figure 9.4). If not, the user proceeds to fill in the rest of the data. If he or she makes any mistakes (for example, entering a sales rep number that is invalid), the system will give the appropriate error message (see Figure 9.5) and will force the user either to make the necessary correction or to abort the whole transaction.

Figure 9.4

Customer
Maintenance
screen

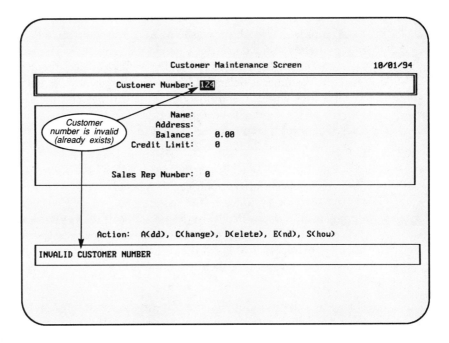

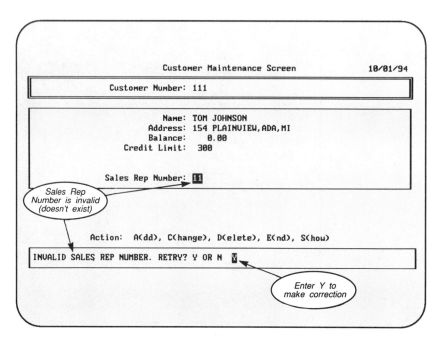

When all the data has been filled in, the system will give the user a chance to either accept the data in its present form or modify it before entering it into the database. Once the user has decided to accept the data as entered on the screen, it is added to the database.

The interaction is similar if the user chooses *C*. The only differences are that (1) the customer number that he or she enters must be the number of a customer who *is* already in the database and (2) current data for this customer is displayed on the screen before the user gets a chance to take further action. The user then steps through the form, entering data only in those fields which he or she wishes to change.

If a user selects *D*, he or she again must enter the customer number of a customer who is already in the database. Current data for this customer is then displayed. Then the user is asked whether he or she really wants to delete the customer. If the response is *Y* (yes), the customer is deleted. Otherwise, the customer is left in the database.

If a user wishes merely to view data on the screen, he or she selects the option *S*. Then he or she enters the customer number of a customer who is already in the database, current data for the customer is displayed, and the user is asked to press any key when he or she has finished looking at the data.

Reports

The typical system offers a variety of reports. Some of them are produced automatically as part of another process, for instance end-of-day or end-of-month routines. Other reports must be specifically requested. Once this has been done, users may be offered choices about how many copies to print, whether to use special paper to print the report, whether the report is to go to the printer or to the screen, whether to produce a detailed or a summary version of the report, and so on.

Queries

The system may support queries. By **query**, we mean a question asked by the user, the answer to which is somewhere in the database. Often, some frequently asked key queries are built into an application system; the user simply selects one from the menu and receives an answer. When we say that a system supports queries, however, we mean that users can ask these questions and get answers *whether or not the question has been pre-programmed*.

Backup and Recovery Services
The typical system provides a convenient way to do backup and recovery. In all probability, it will be the simplified approach discussed in Chapter 7; that is, a backup copy will be made periodically and the database will be recovered by copying the backup version over the live database. As noted earlier, this does entail redoing all work performed since the most recent backup.

Audit Trails
Audit trails are another service provided by the average application system. These are reports that detail whatever activity has taken place. When data is added to the database, the audit trail shows the new data. When it is deleted, the audit trail shows how the data looked *before* the deletion. When data is changed, the audit trail shows how the data looked *before* and *after* the change. The audit trail enables users to ensure the accuracy of updates, and, if a mistake is detected, the data reported in the audit trail aids the user in correcting it.

Utility Services
The typical system also provides utility services. One such service maintains statistics on types of activities users are engaged in and periodically reports on these statistics. Other services periodically check the data in the database for errors (for example, a customer with a sales rep who is not in the database) and print a report describing any errors that have been found.

Administrative Services
The typical application system provides administrative services. These are the services that would be utilized by the person or group (DBA) that is in charge of the system, and they include assigning passwords to new users, changing passwords, specifying whether other security features, such as encryption, are to be used; and varying any of the system's parameters to make the system run more smoothly or to more closely model the workings of the organization. Backup and recovery services, along with some of the utility services, could fall into this category as well.

The features listed in this section are summarized in Figure 9.6.

Figure 9.6

Features of a typical application system

Features of a Typical Application System

1. Menu
2. Interactive and Automatic Updates
3. Reports
4. Queries
5. Backup and Recovery Services
6. Audit Trails
7. Utility Services
8. Administrative Services

9.3 APPLICATION GENERATORS: COMPONENTS

Anyone who has ever been involved in the development of an application system knows that it is a difficult and complicated job, much of which is tedious and repetitive. For example, writing a program in BASIC that employs a simple form on the screen for the interactive updating of customer data is not an easy process; it involves the careful manipulation of many nitty-gritty details. Even putting the form on the screen is tedious. For each piece of the form, the cursor must be moved to the appropriate position on the screen, and then the appropriate characters must be written. For each of the data fields on the screen (the positions where the users must enter data), the cursor must be moved to the correct position and the data must be read and validated. Since the data entered may not be valid, the program must supply a mechanism for handling errors.

Many program statements are required to accomplish these tasks and an application system consists of many programs. Every data entry program in the system must contend with all these details. Creating a program like the one we just discussed is not necessarily difficult, but it's usually time consuming. And until the program has been completed, the user can't really see what the form looks like, that is, how it will appear on the screen. When the user does finally see the form, he or she may or may not like it! If the form is not acceptable to the user, major changes may have to be made to the program.

Clearly, it would be very helpful to have a much simpler way to create data entry programs and all the other types of programs that go into an application system. This is what an application generator is all about. It furnishes an alternate and more productive way to design and develop the features of an application system.

An **application generator** is a many-faceted software product. As you read the following about the various facilities an application generator should offer (see Figure 9.7), keep in mind the goal: the efficient development of application systems.

Figure 9.7	**Features of an Application Generator**
Features of an application generator	1. Programmer's Workbench
	2. Data Dictionary
	3. DBMS, Preferably Relational
	4. Screen Painter (or Screen Generator)
	5. Query Facility
	6. Report Writer
	7. Nonprocedural Language
	8. Procedural Language
	9. Menu Generator
	10. Help Facility
	11. Program Generator

Programmer's Workbench

If you're handier than I am, you probably have a home workbench. It contains the tools you need in order to work efficiently. A **programmer's workbench** is similar, in that it contains tools that simplify various tasks, in this case, those tasks which are necessary to develop an application system. It contains an efficient, easy-to-use editor that provides for the rapid entry and modification of programs. In addition, it contains several utilities for compiling, debugging, and producing reports that are crucial during the application-development process.

Data Dictionary

The **data dictionary** contains descriptions of all the data items in the system. At the very least, it contains the description of all the tables in the system as well as the names and physical characteristics of all the columns that make up these tables. Ideally, it should also contain any indexes that are associated with the tables as well as a description of any relationships between the tables. Information in the dictionary is used by the other components of the application generator.

DBMS, Preferably Relational

To realize all the benefits conferred by the database approach to processing, the application generator should be geared to work with a database rather than with collections of unrelated files. To manage such a database requires a good DBMS, and to achieve the maximum flexibility, the DBMS should be relational. What we are saying here is that a good relational DBMS should be at the heart of the application generator. In essence, the application generator is built around this DBMS.

Screen Painter   A **screen painter**, also called a **screen generator**, is a tool that is used to facilitate the development of screen-oriented data-entry programs. Various screen painters approach the task differently, but they all have one thing in common: to use them, we merely sit down at the keyboard and indicate in some simple way exactly how a form is to appear. As we are doing this, the system displays the form as far as we have constructed it, so we continually get visual feedback on what we are doing. It feels as though we are "painting" the form on the screen and this is how the tool got its name.

The methods of screen painters vary; the one we'll discuss is typical. First, we are presented with a blank screen. To describe a literal part of the screen (that is, a specific string of characters like "Customer Number:"), we use the arrow keys to move the cursor to the correct location and then type the appropriate characters. Figure 9.8 illustrates what the screen might look like at this point. Now we are ready to describe the rest of the **background** (that is, the portion of the form that does not involve data). Figure 9.9 illustrates a completed background.

Figure 9.8

Screen painter

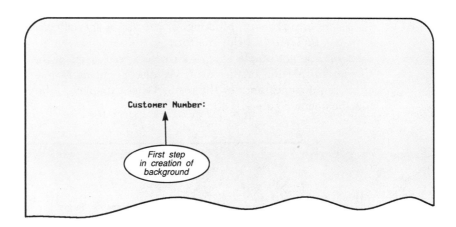

Figure 9.9

Screen painter

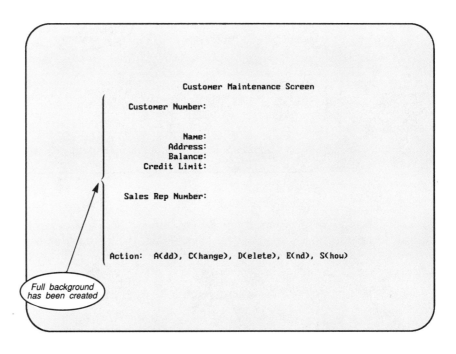

Once the background looks the way we want it to, we can describe the foreground (that is, the data portion of the form). To create the first field in the foreground, we move the cursor to the spot where this field begins. At this point, we are asked to describe the characteristics of the data that is to be entered. Is it a numeric field, a character field, a date, or some other type? How long is it? If it is numeric, how many decimal places does it have? Does it have any special layout (for example, a social security number that must be in the form 999-99-9999)?

There are two commonly used ways of doing this. If the field is a column in the database (like the *CUSTNUMB* column in the *CUSTOMER* table) which has already been described to the data dictionary, all the sought-after information is already in the dictionary, and simply indicating the column's name may be sufficient. If this is not the case, then we may have to supply direct answers to all the questions. Either way, once this has been done, we are finished with the first field in the foreground and can move on to the next.

We may also wish to engage special effects, such as highlighting (making the data in the field brighter), reverse video (dark letters on a light background instead of light letters on a dark background), blinking, or the use of special colors. When this has been done for the *CUSTNUMB* field, the form may look something like the one shown in Figure 9.10. In order to complete the process, we specify each of the data fields that make up the foreground portion of the form. We also might add other features, such as boxes, to enhance the appearance of the form. When completed, the form might look something like the one in Figure 9.11.

Figure 9.10

Screen painter

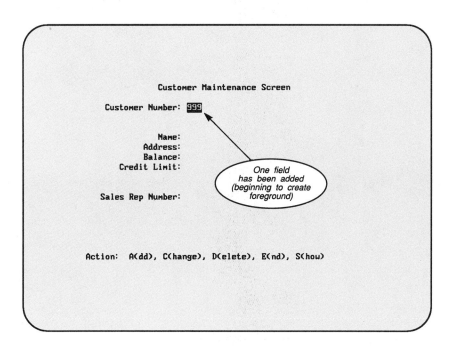

```
                    Customer Maintenance Screen

        Customer Number: 999


                   Name:
                Address:
                Balance:
           Credit Limit:                 One field
                                      has been added
                                     (beginning to create
           Sales Rep Number:              foreground)




        Action:  A(dd), C(hange), D(elete), E(nd), S(how)
```

Figure 9.11

Screen painter

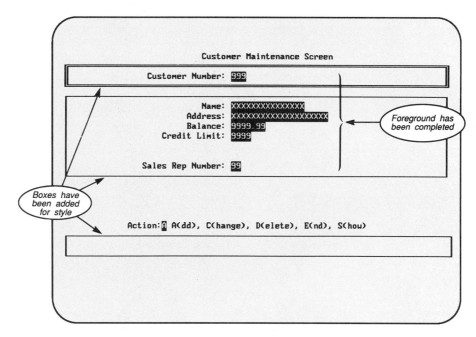

Any one of the background or foreground fields can be moved at any time if we feel it is not in the right position. Some systems accomplish this by having us place the cursor at the first position of the field where it is currently located, hit a special key, move the cursor to the first position of the spot on the screen where we would like to locate the field, and hit another key. The system then completes the work of moving the field.

Query Facility

A **query facility** is a tool that allows queries to the database to be answered in a simple fashion. We may use this facility in developing the application system. For instance, it can be used to build into the application system those queries which we can anticipate in advance. Users are also apt to want to query the database in ways that have not been anticipated (these are called ad hoc queries). Without the use of the query facility, these users would either have to write programs themselves or, more likely, turn in a request to the programming staff to get a program written for them.

Application generators use various types of query facilities. In some, the user types a brief command in a language like **SQL**, which is becoming quite common in such an environment. In others, the user fills in forms by means of a method that is similar to **QBE**. In still others, the user answers a series of questions to indicate his or her queries. **Natural languages** are yet another option. The main thing held in common by all of these approaches is that they provide a simple way for the nontechnical user to specify his or her query.

Report Writer

A **report writer** is similar to a query facility in that it simplifies; specifically, it is intended to facilitate the process of developing reports. It enables the user to describe quickly and easily what a report is to look like; conditions that determine what is to be included in the report; page headings; column headings; totals and subtotals; and so on. We can then use this description in our application. Whenever the user of the application chooses this report, a report will be printed whose layout matches the one that has been described. Further, only the data that satisfies established criteria is included.

Given their similarity, what is the difference between a report writer and a query facility? The line between them is fuzzy at best. When are we submitting a query to which we seek a reply, and when are we requesting a report of all data that meets certain conditions? The answer is not totally clear, and there is a large area of overlap. If we are going to make a distinction, we could base it on three things. First, the answer to a query is usually quite brief; a report is customarily longer. Second, answers to queries are displayed on a screen more often than they are printed, whereas the opposite is true for reports. Finally, fancy formatting is much more of a consideration in a report than in the answer to a query.

Many query facilities can double as report writers. To do so, they include features that produce totals and subtotals and facilities that allow the flexible formatting of output. Provided that they include these features, they can be included in the report writer category.

Nonprocedural Language

If you have studied computer programming at all, the language you studied was probably in the procedural category. With procedural languages, as noted earlier, we must tell the computer precisely *how* to accomplish a desired task. In other words, the computer must be given a *procedure* (also called an *algorithm*) to follow in order to attain the objective. BASIC, Pascal, COBOL, FORTRAN, and many other languages are in this category. With a nonprocedural language, we only specify *what* the task is. In this text, the primary prime example of a nonprocedural language is SQL. If we want to use SQL to retrieve all the orders placed at Premiere Products by all the customers of sales rep 12, for example, we don't have to write a complicated procedure. We merely indicate the task as follows:

```
SELECT ORDNUMB, ORDDATE
    FROM CUSTOMER, ORDERS
    WHERE CUSTOMER.ORDNUMB = ORDERS.ORDNUMB
    AND SLSRNUMB = 12
```

Then it is up to the computer to figure out *how* to get the job done. Clearly, if the computer takes over this responsibility, we can be much more productive.

Procedural Language

Why do we need a nonprocedural language to be available as part of the application generator? The answer is obvious: it makes us much more productive in the development of an application. Given this, why do we need a procedural language to be available as well? The reason is that nonprocedural languages can't do everything. Some of them are very powerful and are capable of handling a wide variety of tasks, but, no matter how powerful they are, we are still likely to encounter tasks that are beyond their capabilities. In such cases, we need to be able to resort to some good procedural language.

This isn't the whole story, however. We need the two languages, the nonprocedural and the procedural, to be closely integrated. In some environments that furnish both types of languages, once the task is beyond the capabilities of the nonprocedural language, that language becomes totally useless. Our only choice is to turn to the procedural language and give up all the benefits offered by its nonprocedural counterpart.

Ideally, commands from the nonprocedural language can be embedded in a program written in the procedural language. Then both languages can serve a role: the nonprocedural language handles everything it can, and the procedural language handles only those tasks which are left over. Many application generators furnish this good, close relationship between the two languages.

Menu Generator

Some application generators include a facility to assist us in putting all these pieces together in a convenient menu-driven environment. Since the application system we are developing ought to be menu-driven (because this is easiest for the user), such a facility is very important. Without it, we have to write programs to display menus, get choices from users, and take appropriate actions.

Help Facility

A **help facility** is simply a vehicle whereby users can obtain information about the system they are using without having to resort to a printed manual. Typically, a user presses a special key (Function key 1 is commonly used for this), and when he or she does, information about the part of the system he or she is using appears on the screen. After the user has viewed this information, he or she hits another key and resumes working with the system at the point where he or she left off.

When we talk about an application generator, we refer to two different help facilities. First, the application generator itself should have a help facility that can immediately provide us with information concerning the use of the application generator any time we seek it. Second, we would like to build a help facility into our application systems. Some application generators include facilities to assist us in doing this. In the absence of such facilities, we have to come up with the code to furnish this help to the users of our application system. This is not especially difficult to do, but it does take time; so it is far preferable for the application generator itself to possess this facility.

Program Generator

Basically, application generators work in one of two ways. Let's use the report writer facility to illustrate the difference. In the first approach, we describe a report; when the report is requested and printed, it will match the layout we have described. Sometime later, the report layout can be changed, and the next time the report is printed, it will match the new specifications. The drawback here is that we can't access the report description except through the report writer facility; in other words, this tool is essential for making changes in the report. This is reminiscent of the difficulty of working only with a nonprocedural language. If the report is beyond the ability of the report writer to produce, the tool is of no value to us; we have to write a complete program to do the job, just as if we had no report writer to assist us at all.

In the second approach, we also describe the layout of the report to the report writer, but then the report writer *creates a program which would be run to produce the report*. At this point, if we wanted to, we could make modifications in this program. We could still use such a report writer, even if the report to be produced is beyond its scope. We would specify a report that had as many of the characteristics we wanted as the report writer could handle. At this point, we would let the report writer create the program. Then we would make whatever changes were necessary in this program so that it would do the things that were beyond the scope of the report writer. Note that as with the first approach, a special program still has to be written. The difference here is that the report writer will create a large share of it and the portion left for us to write, we hope, is quite small.

This approach is often called a **program generator**. It actually creates, or *generates*, programs for us. As you can see, the benefits are considerable. Note that although we used the report writer feature to illustrate the concept, its application is not limited to that. The screen painter and the menu generator are examples of other facilities to which it is quite well suited.

9.4 APPLICATION GENERATORS: UTILIZATION

How do we utilize all these components to put together an application system? First, we need to determine the various reports, queries, and updates that make up the application system, and we need to design the underlying database. Then we can begin to use the application generator. (Actually, some of the work in the first step can proceed in parallel with some of the work with the application generator, but we will assume that the work in the first step has already been completed.)

We begin using the application generator by describing the contents of the database to the data dictionary. We indicate the columns that comprise each table, and the characteristics of these columns. For the *CUSTOMER* table in the *PREMIERE PRODUCTS* application system, for example, we might enter the data shown in Figure 9.12 on the next page. The other tables that make up the *PREMIERE PRODUCTS* database would be described in a similar fashion. If the system also allows us to describe relationships between tables, we will do so.

Figure 9.12

Layout of
CUSTOMER table

Structure for CUSTOMER table

Field	Field Name	Type	Width	Dec
1	CUSTNUMB	Numeric	3	
2	CUSTNAME	Character	15	
3	ADDRESS	Character	20	
4	BALANCE	Numeric	7	2
5	CREDLIM	Numeric	4	
6	SLSRNUMB	Numeric	2	

** Total **   52

Dec stands for decimal places, and the total includes a
decimal point in the *BALANCE* field.

After these descriptions have been entered, we use the screen painter to create the forms for all the on-line updates. After the forms have been created, we can begin to enter sample data. (Many systems automatically provide a simple form that can be used to actually enter some of the data even before these update forms have been created. An example of such a form is shown in Figure 9.13.)

Figure 9.13

System-created
update form for
CUSTOMER table

```
CUSTNUMB    999
CUSTNAME    XXXXXXXXXXXXXXX
ADDRESS     XXXXXXXXXXXXXXXXXXXX
BALANCE     9999.99
CREDLIM     9999
SLSRNUMB    99
```

We then move on to any reports and/or queries that are required by the system, and we use the report writer and query facility to create them. Since we have already entered some sample data, we can test these reports and queries as soon as they are created.

Whatever is beyond the scope of the screen painter, report writer, and/or query facility must now be added to the system. This will probably mean writing commands in either the nonprocedural or procedural language, either by adding commands to programs generated by the system or by writing entire programs from scratch. We always attempt to use the nonprocedural language first. Only when we encounter something that is beyond the capabilities of the nonprocedural language do we use the procedural language. Even then, we retain as much as we can of what has been accomplished with the nonprocedural language.

Programs also have to be written for functions that are not addressed by the application generator. This typically includes utility programs, automatic updates, and backup and recovery programs.

Finally, everything must be tied together in a menu-driven environment. We can do this either after we've prepared all the individual pieces or as they are being developed. Many application generators have facilities for developing menu-driven systems. If the one we are using does not, a program must be written to accomplish the same task.

Including a help facility is an option. As in other cases, if the application generator contains features that can assist us in developing this help facility, we will use them; if not, we must write the code to implement such a facility.

The steps involved in developing an application are summarized in Figure 9.14 on the next page.

Figure 9.14 Steps in developing an application system with an application generator	**Steps in Developing an Application**

1. Describe contents of the database to the data dictionary.
2. Use screen painter to create forms for on-line updates.
3. Enter sample data.
4. Create reports and/or queries. Test with sample data.
5. Add missing features, using nonprocedural language wherever possible. Use procedural language for anything else.
6. Create menus for all of the choices. Use menu-generation facility of the application generator if it has one; otherwise, write a menu program.
7. Consider including help facility. Again, use the tools provided by the application generator to do this if one is present. If not, modify programs you have written to include this facility.

9.5 APPLICATION GENERATORS: OTHER TERMINOLOGY

Unfortunately, there is no general agreement on terminology here. Some people use the term application generators, as we have done in this chapter. Others use the term **fourth-generation language (4GL)**. (Machine languages comprised the first generation of computer languages. Assembly language, a major advance in productivity for programmers, marked the second generation. The so-called *high-level languages* like FORTRAN, COBOL, BASIC, and Pascal, are part of the third generation; they provided another major increase in productivity.)

Since application generators made programmers still *more* productive, the term fourth-generation language is often applied to them. However, the term is used inconsistently. Some people use it to refer only to the nonprocedural part of an application generator, whereas others use it to refer to all its components; that is, to the whole "environment." This second usage has led to another term, the **fourth-generation environment**. This refers to an environment that is inclusive of all the various features we discussed with regard to application generators.

Thus, the terms application generator and fourth-generation environment are synonymous. Whether the terms application generator and fourth-generation language are synonymous depends on whom you ask.

One more note of interest on this subject. There is another definition for fourth-generation language which is even harder to get a handle on. Some people say it is any language or environment that makes a programmer ten or more times more productive than a third-generation language would allow him or her to be. While this definition accurately reflects the goal of a fourth-generation language, it is certainly not very easy to apply!

SUMMARY

1. An application system is a system that has been developed to handle some specific task. A microcomputer application system typically provides the following:
   a. A menu, from which users select the action to be taken.
   b. Interactive updates through forms on the screen that are filled in by the user (some of the updates may also be made automatically).
   c. A variety of reports.
   d. A simple way for users to query the database.
   e. Facilities for backup and recovery.
   f. Audit trails, that is, records of all the updates that have taken place in the database.
   g. A range of utility-type services.
   h. Various administrative-type services.

2. Application generators are tools that are used for the rapid development of an application system. The components of such tools are as follows:

   a. A programmer's workbench, which includes the various tools a programmer needs in order to use the other components of the system effectively.

   b. A data dictionary, which contains descriptions of the tables and columns that make up the database and which is integrated with the other components of the system.

   c. A DBMS, which manipulates the data.

   d. A screen painter, which facilitates the development of screen-oriented on-line update programs.

   e. A query facility, which provides a simple mechanism for querying the database.

   f. A report writer, which facilitates development of the reports required in the application system.

   g. A nonprocedural language, that is, one in which we specify *what* the task is rather than *how* it is to be done.

   h. A procedural language, which handles the tasks that are beyond the scope of the nonprocedural language; the two languages should be able to work together.

   i. A menu generator, which allows us to link the various operations in the application system with a menu of available choices.

   j. Two types of help facilities, one that gives us help on the application generator as we are using it, and one that provides us with a simple means of including a help facility of our own in the application system we are developing.

   k. A program generator, which creates programs we can then modify, if necessary, to accomplish a given task.

3. Some people use the term fourth-generation language as a synonym for application generator; others use the term to apply only to the nonprocedural language component. Another term for application generators is fourth-generation environment.

KEY TERMS

application	nonprocedural language
application generator	procedural language
application system	program generator
data dictionary	programmer's workbench
fourth-generation environment	query
fourth-generation language (4GL)	report writer
help facility	screen generator
menu generator	screen painter

EXERCISES

1. What is meant by an application system? Who creates application systems? Is a DBMS an application system?

2. What is an application generator? How does it differ from an application system?

3. List and briefly describe the components of a modern application system.

4. What do we mean when we say that an application system is menu-driven?

5. How are interactive updates usually achieved in a modern application system? Are any updates noninteractive?

6. Describe the type of backup and recovery procedures that are provided by a typical microcomputer application system. What problems, if any, are associated with these procedures?

7. What is an audit trail? What is its purpose?

8. What types of administrative services might be furnished as part of a microcomputer application system?
9. Describe a programmer's workbench.
10. What is a data dictionary? What role does it play in an application generator?
11. What is a screen painter? How is it used? What is another name for it?
12. What is a query facility? What is a report writer? Can you describe a way in which they differ?
13. What is a nonprocedural language? Give an example. What purpose does it serve?
14. What is a procedural language? Give an example. Explain its usefulness.
15. Does an application generator need both a nonprocedural language and a procedural language? If an application generator contains both types of languages, what relationship should exist between them?
16. What is a menu generator? What purpose does it serve? What action should be taken if an application generator does not contain one?
17. What is a help facility? What purpose does it serve? What should be done if an application generator does not contain one?
18. What is a program generator? Is it essential for an application generator to contain one? What advantage does a program generator offer?
19. List the steps involved in developing an application system by means of an application generator.
20. Describe the relationship between the following terms: application generator, fourth-generation language, fourth-generation environment.

1

Introduction to Microsoft Windows

OBJECTIVES

1. Understand a graphical user interface (GUI).
2. Describe Microsoft Windows.
3. Investigate how to use the mouse, menus, and dialog boxes.
4. Examine how to use Windows applications.
5. Use the online Help system.

M1.1 INTRODUCTION

In this module we discuss Microsoft Windows. Section M1.2 describes a graphical user interface (GUI) and discusses the advantages and features of a Windows-based environment. Section M1.3 covers the elements of the Windows desktop and some of the most useful Windows applications. In section M1.4, we look at how to work with Windows, including using the mouse, menus, and dialog boxes. Finally, in section M1.5 we conduct a brief tour of Microsoft Windows. We start and exit applications, enter and edit a document, save the document, print the document, and obtain online Help while using an application.

M1.2 MICROSOFT WINDOWS OVERVIEW

This module introduces the concept of using a GUI to assist you in communicating with the computer hardware. The user interface provides the computer's environment, or "look-and-feel." The user interface determines the "personality" of the computer as far as how you communicate what the computer is to do. Some computers have the user interface built into the operating system as an integral component (for example, Apple Macintosh and NeXTStep). Other user interfaces are available as additional software that runs on top of the operating system (for example, Microsoft Windows) to make the operating system easier to learn and use.

There are two primary types of user interfaces: a **character-based user interface (CUI)** and a **graphical user interface (GUI).** With CUIs, you type commands as words at a prompt. MS-DOS, the predominant example of the command-driven CUI for microcomputers, requires that you remember and then type in each command in order to use the operating system. For example, to copy a file from the floppy drive to a subdirectory on the hard disk, you would type in a command similar to: `copy a:myfile.ext c:\files\myfile.ext.` These interfaces provide greater access to the power of the operating system and yet are considered difficult to use and more difficult to learn. This module concentrates on the increasingly popular GUI.

Graphical User Interfaces

The graphical user interface (GUI) (pronounced "gooey") introduced by Xerox and popularized by the Apple Macintosh, combines window-oriented software, a pointing device (for example, a mouse), icons, and multitasking. A GUI adds graphical pictures (icons) and menus to the operating system to replace the entering of cryptic commands (Figure M1.1).

Figure M1.1

A graphical user interface

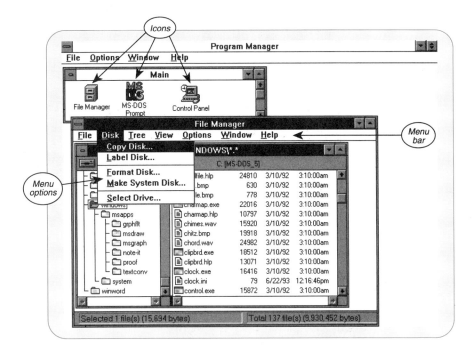

While these early GUIs were integrated into the operating system, Microsoft Windows was introduced to make a GUI available to IBM and compatible microcomputers. Here, Windows is referred to as an operating environment, or shell, which is loaded on top of MS-DOS, replacing the DOS command interface with typical GUI functions.

Whether the GUI is part of, or added to, the operating system, it typically includes the following characteristics:

- **Windows** that display multiple computer tasks simultaneously on the computer's monitor.
- **Icons** (graphical descriptions) to represent various computer tasks and applications. Icons may indicate the type of application, command, or file. These icons are then available to let you activate them to tell the computer what to do instead of entering a command.
- A **pointing device,** such as a mouse, used to select text or objects or to activate icons.
- **Pull-down menus** controlled either by the pointing device or the keyboard to select commands and options.
- **Multitasking,** which is the ability of the computer to run multiple computer tasks concurrently.

Advantages of a GUI

Although there are numerous reasons for the GUI's popularity, some of the most compelling advantages include, but are not limited to, the following:

- Ease of learning and use.
- Multitasking.
- An integrated environment.
- Memory management.

Ease of Learning and Use   The role of any user interface is to make the software product easier to use. Recent research studied which user interface, CUI or GUI, enabled workers to be the most effective. Studies revealed that workers who used GUIs improved their productivity. GUI users got more work done in less time, and with fewer errors. Workers using GUI also learned to use the system faster. Once users learned how to use the GUI for one application, they transferred the knowledge gained to other applications, thus learning subsequent applications in less time.

Multitasking   While it is a GUI's graphical nature that attracts the most attention, perhaps the most compelling reason to use such a system is the advantage of multitasking, which allows the integration of multiple application programs.

Multitasking lets you have several applications working at the same time in different windows. For example, you can be typing in a word processing program in one window and then, with the word processing window still running in that window, switch to another application, such as a spreadsheet, make some calculations, and then copy the results back into the word processing window. Some systems provide the ability to switch between applications but not to keep them running simultaneously. Depending on the system capabilities, however, other systems are able to not only have several applications open at the same time, but also to be performing operations simultaneously. With these systems, you could use the word processor, while at the same time a communications program could be downloading information from another computer.

Integrated Environment   In addition to allowing simultaneous multiple applications, under a GUI most applications can share information interactively. For example, you can copy, cut, and paste information between applications. Newer features include the ability to link objects between applications. To include data from a spreadsheet in a word processing document, you copy the data in the spreadsheet and then switch to the word processor and paste the data in that document. If the data in the spreadsheet is changed, however, you would have to re-copy the modified information into the word processing document.

With automatic linking, you do not have to copy and paste the data. Instead, an object is placed in the word processing document identifying the data in the spreadsheet that should be included. Whenever the data in the spreadsheet is changed, the change is reflected in the word processing document automatically.

Memory Management   Along with the ability to run multiple applications simultaneously comes the need for increasingly large amounts of memory. While single-tasking command-driven systems such as MS-DOS were able to run under limited memory, multitasking systems require access to large amounts of memory. Most GUI systems have integrated programs to manage the use of memory.

M1.3   HOW MICROSOFT WINDOWS WORKS

Now that you know the concepts behind Microsoft Windows, let's see how it actually works. Microsoft Windows, or Windows, makes it easy to learn and use application software. Because its graphical user interface (GUI) is consistent from one application to the next, you can focus on what the application *does* rather than *how* it does it. Word processing, spreadsheet, and database applications software designed to operate in the Windows environment can be integrated easily . Data can be shared among applications software, and learning a new application requires minimum training.

Starting Microsoft Windows

Many computers are set up to start Windows automatically when the computer is turned on. If this is not true for your system, then you need to start Windows from the DOS prompt. To start Windows, follow these steps:

- Type CD\WINDOWS
- Press the Enter key.
- Type WIN
- Press the Enter key.

Note 1: You can enter the commands in either upper- or lowercase letters.

Note 2: Your facility may have special instructions for accessing Windows. Check with your instructor before doing this tutorial.

The Windows logo and copyright information are displayed on the screen, followed by a blank screen with an hourglass-shaped icon in the center of the screen. Windows uses icons to depict various items. Finally, the Windows opening screen displays (Figure M1.2).

Figure M1.2

Microsoft Windows

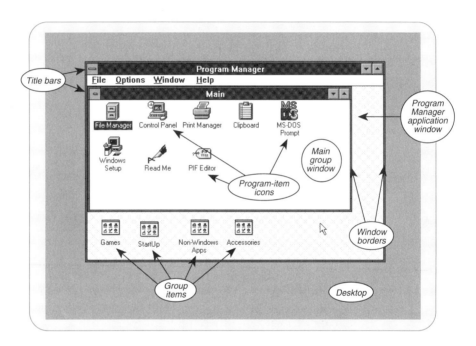

The Windows Desktop

The first time you start Windows, you will see a screen similar to the one shown in Figure M1.2. Don't be concerned if your screen looks different. Windows can be customized to include different icons, program items, and group items. The basic elements don't change, but they are simply rearranged.

The screen in Figure M1.2 contains the Program Manager window, with the Main group window open inside it. The Program Manager window represents the Program Manager application. A window that holds programs that are currently running is called an **application window.** Program Manager is in an application window.

The Main window contains a number of applications with each application represented by an icon. Since each icon represents a program, the icons are referred to as **program-item icons.** The Main window is called a **group window** because it holds a group of applications. Each window contains a *title bar* that identifies the window.

At the bottom of the screen are several group icons, which represent other groups of applications similar to the Main group.

The background on which the various windows display is called the **desktop.** You can change the desktop by moving items, adding new items and removing items you no longer need.

Window Elements

In this section, we describe the elements of a window. Most windows have certain elements in common, such as a title bar and menus. Not all windows, however, have every element. Figure M1.3 illustrates the common parts of a window.

Figure M1.3

Elements of a window

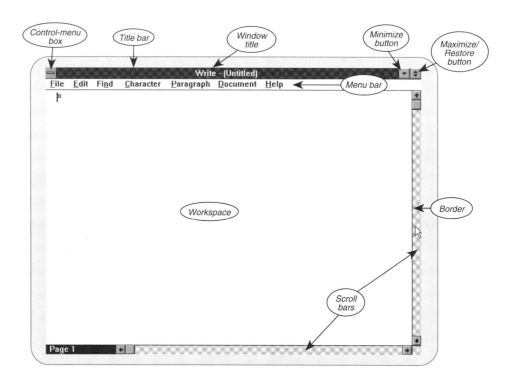

The **Control-menu box,** in the upper-left corner of a window, closes a window. It is also useful if you use your keyboard to work with Windows. There are Control-menu commands to resize, move, maximize, minimize, and close a window, as well as a command to switch to other applications.

The **title bar** shows the name of the application or document. If more than one window is open, the title bar for the active window (the one in which you are working) is shown in a different color or intensity.

The **window title** can be the name of an application and the name of a document, or the name of a group, directory, or file. The specific title depends on the type of window in which it appears.

The **menu bar** lists the menus available for a particular window. Most applications have a File menu, an Edit menu, and a Help menu.

The **scroll bars** allow you to move parts of a document into view when the entire document doesn't fit on the screen.

The **Maximize** button enlarges the active window so that it fills the entire desktop. The **Minimize** button reduces the active window to an icon. After you enlarge an active window, the Maximize button is replaced by a Restore button. The **Restore** button contains both an Up and a Down arrow and returns the window to its previous size.

The window **border** is the border around the outside edge of a window. You can lengthen or shorten the border on each side of the window.

You carry out actions in the workspace that occupies the center of the window.

Program Manager

The Program Manager is an application that is central to everything that you do in Windows. It starts whenever you start Windows and runs until you quit Windows. You use Program Manager to start other applications, to organize applications and files into groups, and to manage your desktop.

Windows has several predefined groups that display within the Program Manager window. Using Program Manager, you can add group windows, delete program items from existing group windows, and move an application from one group window to another.

The group windows that typically display when Program Manager is started are:

- **Main.** Contains the Windows system applications, including File Manager, Control Panel, Print Manager, Clipboard Viewer, MS-DOS Prompt, Windows Setup, PIF Editor, and Read Me.
- **Accessories.** Includes Write, Paintbrush, Terminal, Notepad, Recorder, Cardfile, Calendar, Calculator, Clock, Object Packager, Character Map, Media Player, and Sound Recorder.
- **Games.** Displays the games Solitaire and Minesweeper.
- **StartUp.** Holds programs that you want to start automatically when Windows starts.
- **Applications.** Contains programs found on your hard disk during installation.

While a complete discussion of these applications is beyond the scope of this text, there are three—File Manager, Clipboard Viewer, and Print Manager—that will be useful to you in future assignments.

File Manager provides a graphical representation of files and directories. You can use File Manager to view, copy, rename, move, and delete files. You can also create directories and format diskettes.

The Clipboard Viewer gives you temporary storage for information that you want to transfer from one application to another. Information that you place in the clipboard remains there until you place another piece of information into it or until you quit Windows. With Clipboard Viewer, you can view, save, retrieve, and delete the contents of the clipboard.

Print Manager controls printing for Windows applications. When you issue a Print command in a Windows application, it sends the command to the Print Manager rather than to the printer. Print Manager then controls the printing tasks, which allows you to continue working in your application while a file is being printed.

M1.4 WORKING WITH MICROSOFT WINDOWS

Using the Mouse

Because Windows is a GUI program, it works best with a mouse. Although almost all operations may be conducted with a keyboard, windowing software is written to maximize a pointing device. You use the mouse (or some other pointing device) instead of typing commands on the keyboard to select text or items, run applications, or navigate around windows.

You use the mouse by moving it around on a flat surface. As you move the mouse, a ball on the underside of the mouse transmits the movement to the computer, and the mouse pointer (usually an arrow) on the screen moves in the same direction.

There are normally two mouse buttons: left and right. The left button is the primary button used in Microsoft Windows. There are four basic mouse actions: pointing, clicking, double clicking, and dragging.

You **point** to objects by moving the mouse around on a flat surface, which simultaneously moves the mouse pointer on the desktop until it is pointing to the object you wish to choose.

Module One — Introduction to Microsoft Windows

After you point to an object, a **click** means you press and release the left button. A click will highlight an object or relocate your cursor from its current location to where it is clicked.

A **double click** means you move the mouse to an object and rapidly press and release the left mouse button twice. A double click runs a program or opens a document. It is usually the same as pointing to an object, clicking on it to select it, and then selecting Open from the File menu. Although double clicking is a shortcut for many operations, not all actions require the second click.

Note: If you move the mouse between a double click, Windows does not accept the action as a double click, but rather, as two separate clicks.

To **drag,** you first point to an object, hold down the left button, while dragging the mouse to a new location, and then release the left button. Dragging is used to move objects such as icons as well as to select text in applications software (such as a word processor or fields in a database) to copy, delete, move, or apply a new feature to the text.

To practice using the mouse, open an application, Windows Notepad, and manipulate the window to demonstrate the various uses of the mouse. To begin, you should have Windows running with the Program Manager active. Although your screen will probably look slightly different than the one presented here, it should look similar to Figure M1.4 with group icons for Accessories, Applications, Games, and so on, some of which may be open as group windows with program icons visible. In this example, all group windows have been minimized to icons at the bottom of the screen. If you cannot see the Accessories group icon or window, you may need to minimize other windows to clear the screen. (Alternatively, you may simply select the Windows menu from the Program Manager as Figure M1.4, which lists all group windows, indicates.) Once the Program Manager is running and the Accessories group icon is visible, you are ready to proceed.

Pointing   Let's practice moving the mouse around to point to an object.

- Move the mouse around on the desk or mouse pad and notice the corresponding movement of the pointer on the screen.
- Now hold the mouse a few inches above the desk and move it around.

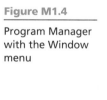

Figure M1.4

Program Manager with the Window menu

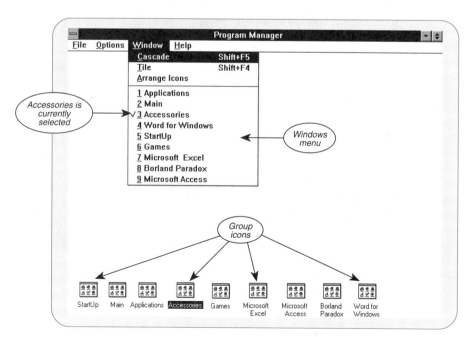

You will notice that the arrow doesn't move. To readjust the position of the mouse without moving the pointer, lift the mouse from the surface, and set it down at the new location.

Clicking   As you move the mouse around on the desk, the arrow does not change as it passes over icons. To select an item with the mouse, press the left button on the mouse as it points to that item.

■ Move the mouse until the arrow points to the Accessories group icon. As the arrow touches the icon, press and release the left mouse button.

The Accessories icon should now be highlighted. Highlighting an item indicates that any subsequent actions will be directed toward it.

Double Clicking   Double clicking (rapidly pressing and releasing the left mouse button twice) an object is a shortcut to open or run that object. For example, double clicking on a group icon will open it into a window; double clicking a program-item icon will open and run that program; and double clicking on the program's control-menu box will close the window.

■ Move the pointer to the Accessories icon and double click.

When you double click on the Accessories icon, a group window opens with program icons inside of the window. Although the icons may be arranged differently, your screen should resemble Figure M1.5.

Figure M1.5

Accessories group window

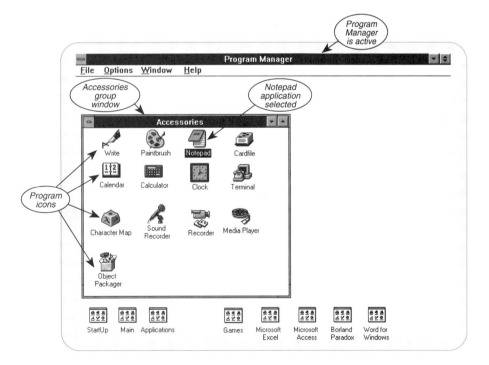

Dragging   To drag an item to rearrange the order of icons or to drag across text, hold down the left mouse button while moving the mouse and then release the left button upon reaching the destination. For example, to move the location of the Notepad icon to the bottom of the window, follow these steps:

■ Drag the Notepad icon to the bottom of the window to the location shown in Figure M1.6 by holding down the left button while moving the icon down and then releasing the button.

■ Move the Notepad icon back to its original location.

Figure M1.6

Accessories group windows with Notepad icon moved

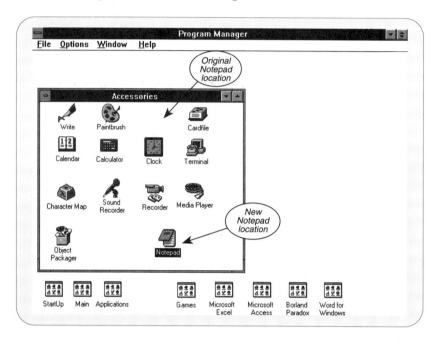

Double clicking on a program icon will open or run that program. To open and run Notepad,

■ Move the pointer to the Notepad icon and double click.

This will run Notepad and open a new document named Untitled, as Figure M1.7 shows.

Figure M1.7

Notepad application window

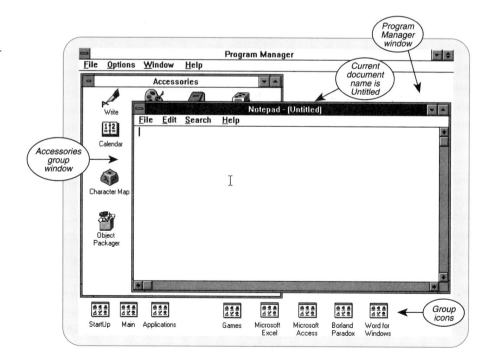

Working with applications such as Notepad will be covered later in the text. For now, let's look at how you can use the mouse to work with text in a document. You will notice in Figure M1.7 that the mouse pointer changes from an arrow to an insertion beam (in the shape of an uppercase I) when the pointer is moved to an application window. The cursor for text insertion, however, is at the blinking vertical bar. To change the cursor to another location, move the mouse pointer to that location and click.

To begin working with text, follow the steps outlined next.

■ Enter the following data in the Notepad window:
```
The mouse is used to...
1. Point
2. Click
3. Double-click
4. Drag
...items and text.
```

To see how the mouse is used to change the cursor location,

■ Move the I-beam pointer to the space before the word *mouse* in the first line, and press the left mouse button.

The cursor is now located where you clicked. To use the double-click feature with text,

■ Move the pointer to the word *mouse* and double click.

Your screen should now look like Figure M1.8.

The mouse may also be used to drag across text to select it.

■ Move the I-beam to the space before the word *mouse* and click.
■ Hold down the left mouse button.
■ Move the pointer to include the three words *mouse is used.*
■ Release the left mouse button.

You should have selected the three words, as in Figure M1.9.

We will exit Notepad now without saving our work. There are three ways to exit an application using the mouse: (1) double clicking on the application control box, (2)

Figure M1.8

Notepad application window with text selected

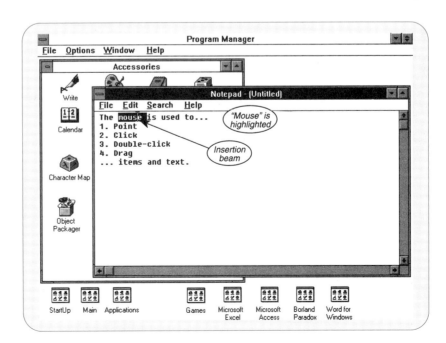

Figure M1.9

Notepad
application
window with
extended text
selected

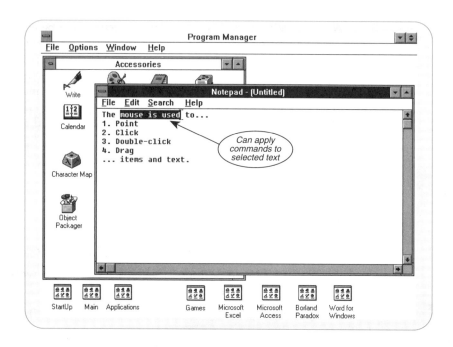

choosing Close from the application control box (Figure M1.10), or (3) choosing Exit from the File menu (Figure M1.11). Let's try using the double-click method. Be careful not to double click on the Program Manager's control box.

■ Double click on the Notepad application's control box.

Windows pops up a dialog box warning that the Notepad document has not been saved. To exit without saving,

■ Click on the No button.

Figure M1.10

Closing an
application from
the Control-menu
box

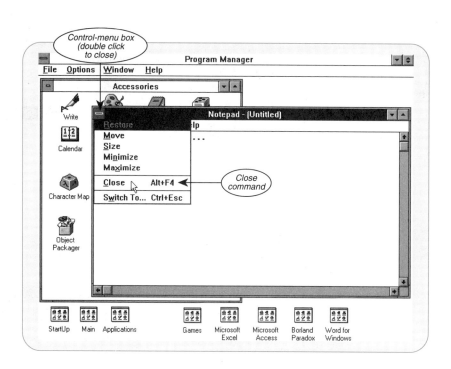

Figure M1.11

Exiting an
application from
the File Menu

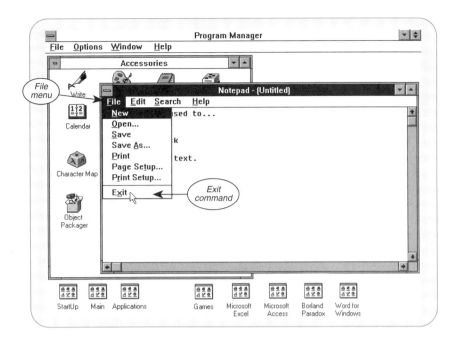

Note 1: Throughout this tutorial, an instruction to **select** an item means to click and highlight it, while a **choose** operation means to double click on an item or object.

Note 2: Some computers, most notably notebook computers, substitute the mouse with an alternate pointing device, such as a trackball. While the mechanics of the movement are different, all of the above actions are identical.

Using the Keyboard

Everything that you accomplish with a mouse you can also do with the keyboard by typing in the associated menu commands or shortcuts. Whether you choose to use the keyboard or mouse is a matter of individual preference. However, studies have shown that the mouse has proven most effective once mastered and is therefore recommended.

Using Windows Menus

You may select commands from Microsoft Windows by whichever method you prefer, since there are usually at least two or three different ways of accomplishing the same action. For example, there are three alternative methods to exit an application: (1) using the menu with the mouse, (2) using the menu with the keyboard, and (3) using keyboard shortcut keys.

Using the menu with the mouse.   The first method would be to select the option from the menu by clicking on each option with the mouse. To exit an application, click on the File menu and then on the Exit command.

Using the menu with the keyboard.   Each menu option indicates what letters can be selected with the keyboard by underlining a letter in each option. To activate these letters, first press the Alt key, followed by each underlined letter in succession. To exit an application, this would be shown as Alt + F + x. You can also use the arrow keys on the keyboard to select items on the menu and then press the Enter key.

Using keyboard shortcut keys.   When you first use Microsoft Windows, you will want to use the menus to view and select commands. However, once you are familiar with the commands, you will probably want to use the **keyboard shortcut keys.** Shortcut keys normally combine the Alt, Ctrl, or Shift key with a function key or one letter of a menu option. If a shortcut key is available, it will be listed on the right-hand side

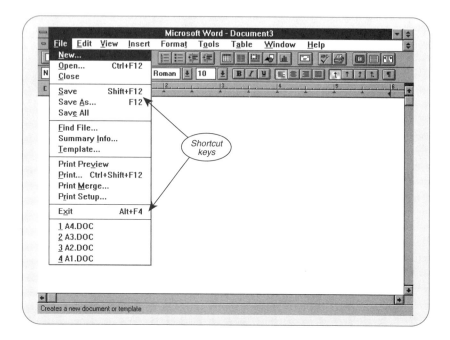

of the associated menu. In most application programs, the shortcut key to exit an application is Alt + F4 and will be indicated in the File menu (Figure M1.12).

Sometimes, menu commands are dimmed. This means that these commands are currently unavailable because they are not appropriate to the ongoing actions. For example, the Close option is not available under the File menu unless a window is open, and you cannot select Paste from the Edit menu until after something has been copied.

Using Dialog Boxes

You will also notice that some menu items are followed by an ellipsis (. . .). The ellipsis indicates that additional selections are available for that item. Windows uses a **dialog box** to access additional options or to provide information to the user. You may use a dialog box to give error messages and ask what action you want to take or to simply gather more information. Figure M1.13 shows a sample dialog box.

Dialog boxes vary in the amount of information provided or requested. Some dialog boxes may only ask you to select OK to confirm or Cancel to abort an operation, while others prompt you through a complex set of options. Dialog boxes typically contain one or more of the following elements:

Text boxes are used to enter information needed to complete a command. This usually involves providing a file, directory, or item name.

List boxes present lists of possible choices. An example of a list box would be a list of available fonts. Scroll bars allow you to scroll up and down through the list. Usually, a text box is associated with a list box and is linked with the list so that when you select an item from the list box, it displays in the text box.

Drop-down lists are single text boxes that have a Down-arrow button on the right side of the text box that is used to open a list of choices.

Option buttons present a group of choices from which you may choose only one. An example of an option button is one requesting whether you want to print in the Portrait or Landscape orientation. (Only one may be selected.) Option boxes are indicated as circles, with the circle appearing filled-in when it is selected. Clicking on an option box selects it, and clicking on a selected option box de-selects it.

Check boxes are similar to option boxes, except that these are options independent of other check boxes, and one or all of them may be selected. Check boxes are indicated

Figure M1.13

Dialog Box

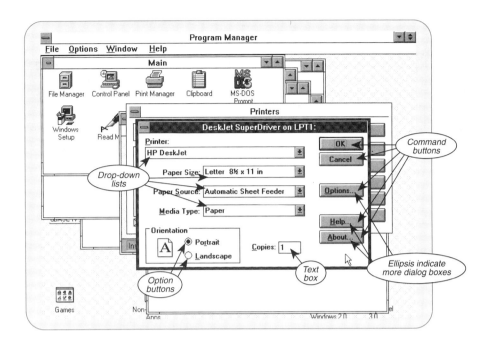

as a box, with an **X** displaying in the box when it is selected. Clicking on a check box selects it, and clicking on a selected check box de-selects it.

Command buttons carry out the Windows commands displayed in the button. For example, most dialog boxes will have a Cancel, Open, OK, or Done button. If the button includes an ellipsis, choosing it will open another dialog box much as a menu item with an ellipsis opens another menu.

M1.5   USING WINDOWS APPLICATIONS

Now that you know how to move around in Windows and how to start and exit an application, let's see how to use some of the various Windows applications. In this section, you will create a directory using File Manager, enter a document in the Notepad application, copy the document to the Write application, save the document in our new directory, and print the document. You will also use the Paintbrush application to illustrate the Windows online Help feature.

File Manager   You use File Manager to help manage directories and files. To start the File Manager program,

- Double click on the Main group window.
- Double click on the File Manager program-item icon.
- Click on the Maximize button.

A window similar to the one shown in Figure M1.14 displays on the screen. Your window will vary in that it may show a different drive and the directories and file-names will be different, but the design will be the same. The right side displays a directory tree listing the directories on the drive you are using. The contents list on the left side displays files and subdirectories in the selected directory.

Figure M1.14 shows the contents of the Windows directory. The Windows directory includes two subdirectories, MSAPPS and SYSTEM, and a number of files. The icons in front of the text graphically explain whether the contents are directories, data files, or program files.

Figure M1.14

Using File Manager

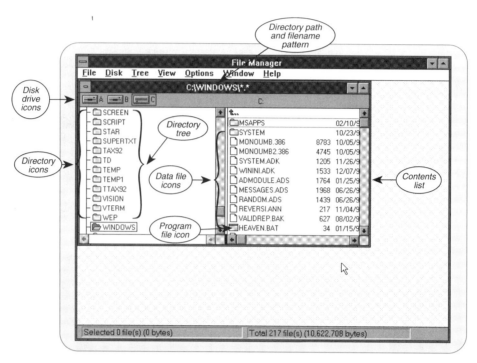

Let's create a directory on a disk in drive A.

- Insert your disk in drive A.
- Select the File menu.
- Select the Create Directory command.

The Create Directory dialog box shown in Figure M1.15 displays.

- Enter `a:\temp` in the Name text box.
- Click on the OK button.

Figure M1.15

Create Directory
dialog box

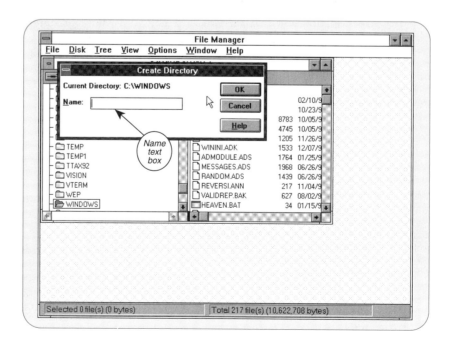

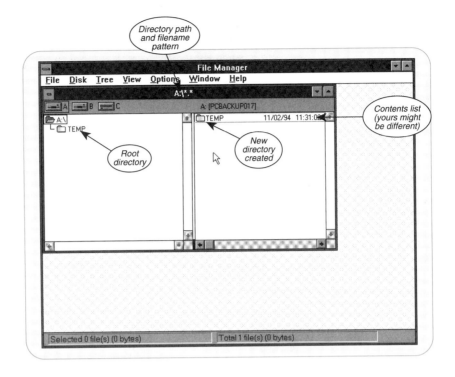

The directory will be created on the disk in your A drive. To confirm this,

- Click on the A disk drive icon.

The directory listing for drive A displays (Figure M1.16). Your listing may show additional files and directories.

To exit File Manager,

- Select the File menu.
- Select Exit.

You are returned to the Program Manager, and the Main group window displays on your screen. To close the Main group window,

- Click on the Control-menu box for the Main group window.
- Select Close.

The Main group window is closed and reduced to an icon.

Cut, Copy, and Paste between Windows

One of the more useful features of Windows is the ability to cut, copy, and paste data between application windows. To demonstrate, let's create a document in the Notepad application and then copy it to the Write application.

To open the Notepad application,

- Double click on the Accessories group window.
- Double click on the Notepad program icon.

The Notepad application window displays.

- Maximize the Notepad application window.
- Enter the text shown in Figure M1.17.

Now let's copy the text into the Windows Write application. To copy the text, you need to select it and then choose the Copy command from the Edit menu. This will place a copy of the text in the clipboard. You then need to open the Write application

Figure M1.17

Notepad
Application

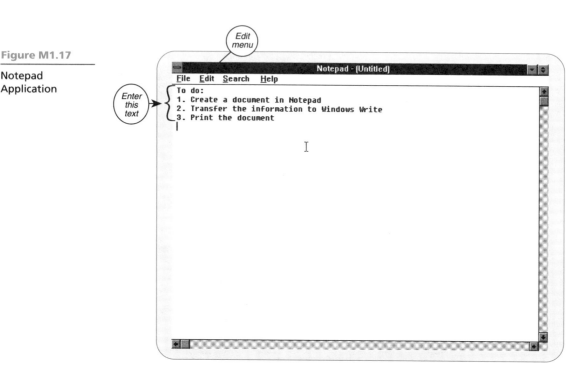

and place the text from the clipboard into the Write application. The clipboard acts as a temporary storage location for data that you want to copy or move. You can view the current contents of the clipboard by using the Clipboard Viewer application.

- Use the techniques covered in section M1.4 to select (highlight) the text in the Notepad application.
- Select the Edit menu.
- Select Copy.

The text is copied to the clipboard and can be placed in another Windows application. Now let's open the Write application without closing the Notepad application. To do this, use the Task List. Task List is a window that displays a list of all the applications you are currently running. You can use Task List to switch to another application. You can access Task List by selecting the Switch To command from the Control-menu box, or by using a keyboard shortcut. Use the keyboard shortcut here.

- Press Ctrl-Esc.

The Task List window displays, as Figure M1.18 shows. Notice that the list of applications in your Task List is different. You can now select the application you want by highlighting the application and clicking the Switch To button or by double clicking the application.

- Choose (double click on) Program Manager.

Program Manager displays with the open Accessories window.

- Double click on the Write program icon.
- Maximize the Write window.
- Select the Edit menu.
- Select the Paste command.

Your screen should now look like the one shown in Figure M1.19. The text is currently entered in the Notepad application, the Write application, and the clipboard. To print the document,

Figure M1.18

Windows Task List

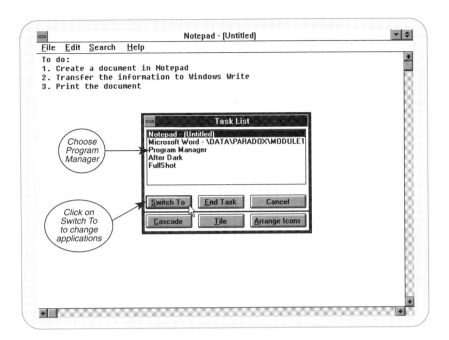

Figure M1.19

Windows Write
application

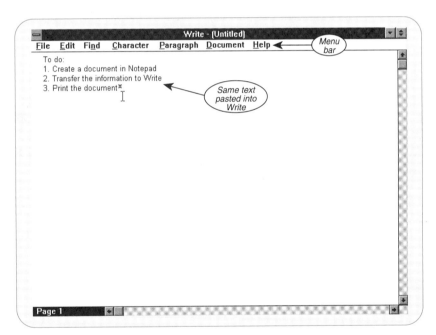

- Select the File menu.
- Select the Print command.

Before you close the Write application, save the document in the directory you created on drive A.

- Select the File menu.
- Select the Save As command.
- Enter `a:\temp\todo` in the File Name text box.
- Click on the OK button.

The Write program appends the extension .WRI to the file name to indicate that the document is in Windows Write format.

Now you need to close both the Write application and the Notepad application. To close the Write application,

- Select the File menu.
- Select the Exit command.

You are returned to the Program Manager. Now you need to close the Notepad application. You can switch quickly to another application by pressing and holding down the Alt key, and then pressing the Tab key repeatedly. As you continue to press Tab, you see the title of each open application. To close the Notepad application,

- Hold down the Alt key and press the Tab key until the words *Notepad-(Untitled)* display on the screen.
- Release the Alt key.
- Double click on the Control-menu box.

The dialog box shown in Figure M1.20 displays.

- Click on No.

Figure M1.20

Exiting the Notepad application

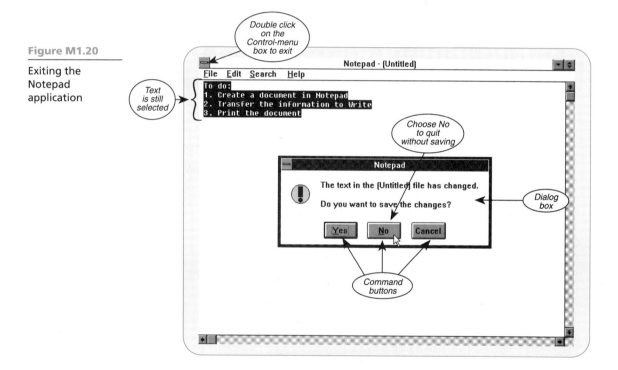

You are again returned to the Program Manager. Close the Accessories group window now.

- Double click on the Control-menu box in the Accessories group window.

Windows Online Help

If you need help on a task you are performing, you can use Windows online Help. Help is available for all applications except Clock. To illustrate Windows Help, let's start the Paintbrush application. (The Paintbrush application is a drawing program that allows you to create simple or elaborate color drawings.)

- Open the Accessories group window.
- Open the Paintbrush application.
- Maximize the window.

The window shown in Figure M1.21 displays. Paintbrush includes a number of drawing tools. Select the tool you want and use it to "draw" your picture in the workspace. Let's create a "happy face" drawing.

- Select the Circle tool.
- Draw a circle, as Figure M1.22 shows. (Hold down the left mouse button while moving the mouse, and a dotted line will outline the circle.)
- Select the Paintbrush tool.
- Fill in the "face," as Figure M1.23 shows.

Figure M1.21

Using the Paintbrush application

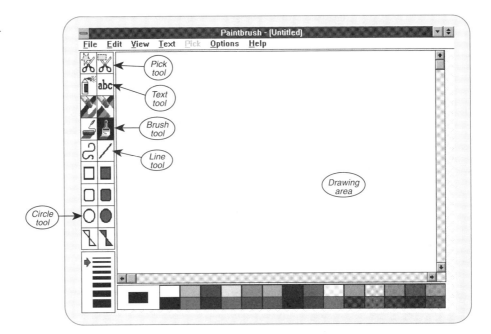

Figure M1.22

Using Paintbrush to draw an object

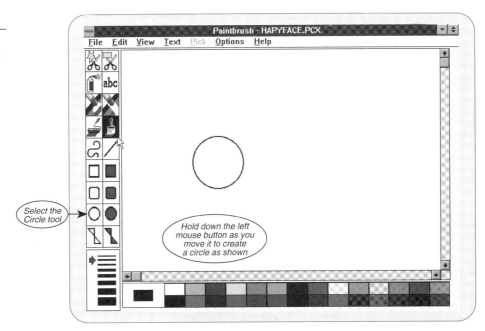

Figure M1.23

Using the Brush
tool

You have completed our drawing. At this point, you could print the drawing, or you could "cut" the drawing and paste it into another Windows application. To find out how to cut and paste the drawing, let's use the Help menu.

- Select the Help menu.
- Select Contents.

An alphabetical list of search topics displays in the Help window (Figure M1.24). To select a topic, scroll the list, point to the Help topic you want, and click the left

Figure M1.24

Windows Help
system

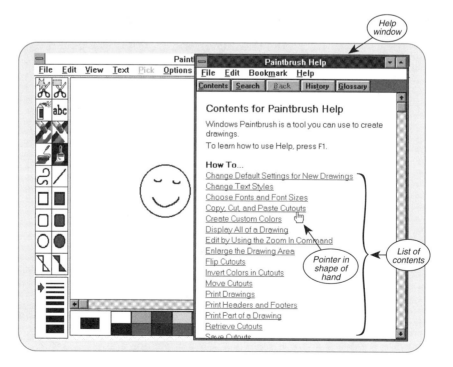

mouse button. When you point to a Help topic, the mouse pointer changes to a hand. To obtain Help on the cut and paste operation,

- Click on the Copy, Cut, and Paste Cutouts topic.

The window shown in Figure M1.25 displays. You can scroll the information in the Help window or print the information by selecting the Print Topic command from the File menu.

Figure M1.25

Windows Help
System

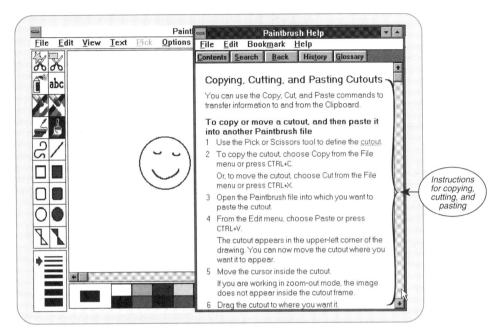

To exit the Help window,

- Select the File menu in the Paintbrush Help window.
- Select the Exit command.

Now, exit the Paintbrush application without saving your drawing.

- Select the File menu.
- Select the Exit command.
- Click on No.

The Paintbrush application is closed. Let's close the Accessories group window.

- Double click on the Control-menu box for the Accessories window.

Quitting Windows

Before you can quit Windows, you must be sure that all applications except Program Manager are closed. You then quit Windows from Program Manager. Remember that Program Manager starts when you start Windows and runs until you quit Windows.

By default, if you make changes to the Windows desktop (for example, moving icons) during a Windows session, the changes will be saved when you exit Windows. To cancel any changes you make to the desktop during a session, you can de-select (toggle off) the Save Settings on Exit command from the Options menu (Figure M1.26). A check mark preceding the command indicates that settings will be saved. To exit Windows without saving any changes to the desktop,

- Select the Options menu.

When the Options menu displays, look at the Save Settings on Exit command. If there is a check mark preceding the command, select the command to remove the check

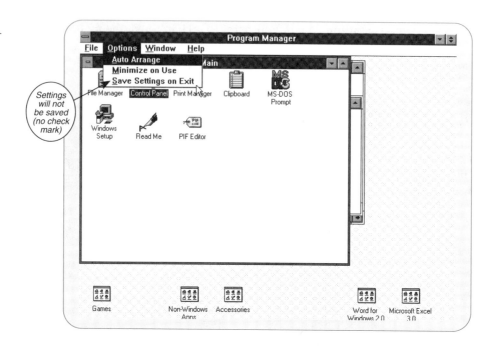

mark and toggle the option off. If there is no check mark, click on the word *Options* again to remove the menu from the screen.

- Toggle off the Save Settings on Exit command, if necessary.

Now, to quit Windows,

- Select the File menu.
- Select Exit.

The Exit Windows dialog box displays, asking you to confirm that you really want to quit Windows.

- Click on the OK button.

The Windows session is ended, and you are returned to the DOS prompt.

SUMMARY

1. The user-interface provides a means of communication between you, the user, and the computer.
2. A graphical user interface (GUI) uses graphic pictures or icons to replace the act of entering commands for common functions.
3. Multitasking is the ability to run multiple computer tasks concurrently.
4. Most GUIs, including Windows, provide an integrated environment that allows sharing of information among independent applications.
5. To start Windows from the DOS prompt, change to the Windows directory and type WIN.
6. Most windows have the following elements: control-menu box, title bar, window title, menu bar, scroll bars, maximize button, and minimize button.
7. Program Manager is an application that is central to Windows activities. You use Program Manager to start other applications, to organize applications and files into groups, and to manage your desktop.
8. The group windows that typically display when Program Manager is started include: Main, Accessories, Games, StartUp, and Applications.

9. File Manager provides a graphical representation of files and directories. You can use File Manager to view, copy, rename, move, and delete files. You can also create directories and format disks.

10. The Clipboard Viewer provides temporary storage for information that you want to transfer from one application to another. You can use the Clipboard Viewer to view, save, retrieve, and delete the contents of the clipboard.

11. Print Manager controls printing for Windows applications.

12. You can use the mouse to click on an object to select it, double click an object to open or run it, and drag the mouse to highlight or move an object.

13. Menus may be used to select commands. You activate Windows menus by pressing the Alt key, followed by the highlighted letter in the menu.

14. Keyboard shortcuts are often provided to speed the menu-selection process. An example of a keyboard shortcut is pressing `Ctrl + C` to copy the highlighted object or text.

15. To open an application, open the group window containing the application and double click the program-item icon. To close an application, select the Exit command from the File menu, select Close from the application's Control-menu box, or double click the application's Control-menu box.

16. To copy data from one application to another, select the data to be copied, select Copy from the Edit menu, transfer to the second application, and select Paste from the Edit menu.

17. You may use the Task List to move between applications running in different windows. Press `Ctrl + Esc` to show the Task List window.

18. The online Help system is available for all Windows applications except the Clock. To use the Help system, select the Help menu, select Contents, scroll the topics list, and select the topic of interest. You can print the topic contents.

19. To quit Windows, select Exit from the File menu in Program Manager. Before quitting, toggle off the Save Settings On Exit command from the Options menu.

EXERCISES

1. What is a graphical user interface (GUI)?
2. What is a group window? a program-item icon? an application window?
3. What is Program Manager?
4. How do you open a group window? How do you close a group window?
5. How do you start an application? How do you exit an application?
6. What does the Maximize button do? the Minimize button? the Restore button?
7. What do the terms *clicking, double clicking,* and *dragging* mean?
8. What does the File Manager program do? When would you use it?
9. What is a dialog box? When do dialog boxes display on the screen?
10. What is the Clipboard Viewer? When would you use it?
11. How can you copy data from one application to another?
12. What is the Task List? Why do you use it? How do you access it?
13. Why is it necessary to check the status of the Save Settings On Exit command before exiting Windows?

COMPUTER ASSIGNMENTS

1. Open the File Manager program and perform the following tasks:
   a. Create a directory on drive A called TEMP1.
   b. Copy the TODO.WRI file you created in section M1.5 to the TEMP1 directory.
   c. Rename the TODO.WRI file in the TEMP directory as DONE.WRI.
   d. Delete the TODO.WRI file in the TEMP1 directory.

e. Print the Help topic on naming files and directories. (**Hint:** Select Work with Files and Directories from the Contents list and then select Naming Files and Directories.)

2. Open the Calendar application in the Accessories group window. Add the following appointments to the current date:

8:00	Database Management Class
10:00	English Composition Class
12:00	Lunch with Database Project Team
4:00	Computer Club Meeting
8:00	Intramural Volleyball

Print the calendar for the current date.

3. Create a party invitation typed in Microsoft Write that includes a map drawn in Paintbrush.

a. Enter the following text in the Write application:

   PARTY!!!

   The Valley chapter of Beta Gamma Sigma invites all Junior and Senior students in the school of business to a ''Let's Get Acquainted Party.''

   When: Friday, Oct. 7, 1994 7:00 PM
   Where: Green Valley Arboretum

   If you are able to attend, please follow the map below and plan on an enjoyable evening.

b. Access the Paintbrush application and draw the roads, creek, and arboretum map shown in Figure M1.27. (**Hint:** To draw straight lines, use the Straight-line tool. To add place names, use the Text tool.)

c. Cut the map from the Paintbrush application and paste in the Write application. (**Hint:** Use the Pick tool to outline or cut out the map, and then select Cut from the Edit menu.)

Save the invitation as INVITE to the TEMP directory in drive A. Print the invitation.

Figure M1.27

Map created with Paintbrush

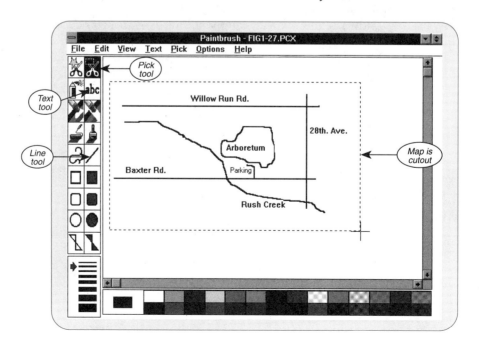

Introduction to Microsoft Access

OBJECTIVES

1. Describe the basic use of Access system.
2. Create and add data to a table.
3. Produce a report involving data in a table.
4. Back up a database.

M2.1 INTRODUCTION

In this module, we explore Microsoft Access, one of the leading microcomputer database management systems (DBMSs). We will investigate the basic mechanics of using Access screens and menus. Then we will use what you have learned by creating the *Branch* table in the *Books* database. Once you have created this table, you will add all the branches. At this point, we will briefly look at the manner in which you produce a quick report using ReportWizard. In later modules, we will see how to create reports with grouping and subtotals. Finally, in this section, we look at how you make a backup copy of your data.

Note: We will use the *Premiere Products* database in these examples. Make sure that this database is in the active directory or disk drive (normally Drive A) before you start using Access.

Note: In Access, you can use a mouse to accomplish many of the tasks you can accomplish with the keyboard. The procedures for using a mouse will be displayed in a special box like this one. When you see an instruction to "click on" some portion of the screen, it means move the mouse pointer to that portion of the screen, press the left mouse button, and then release it. An instruction to "double click" means that you should move the mouse pointer to that portion of the screen and then "click" twice in rapid succession. When you see an instruction to "drag" an item on the screen, it means to move the mouse pointer to the item, press and hold the left mouse button, move the item to the new location, and then release the mouse button.

To begin using Access, you must first ensure that the Windows Program Manager appears on the screen and that the Access icon is available. Refer to the procedures presented in Module 1 to accomplish this task. We also assume that your database is on a disk that will be in drive A. If this is not the case, substitute the letter for the appropriate drive. To begin using Microsoft Access, follow these remaining steps:

- Place your data disk in Drive A.
- Point to the Microsoft Access program-item icon.
- Double-click the left mouse button.

As soon as you have completed these steps, the Microsoft Access copyright message is displayed, then the initial Access window appears, as shown in Figure M2.1.

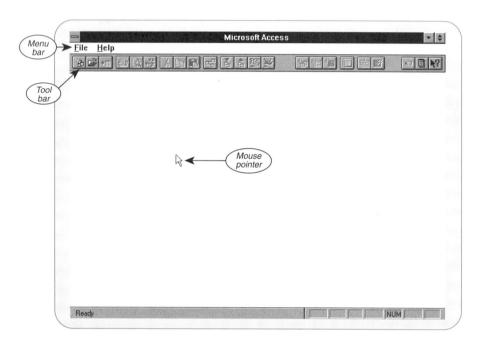

Note: Most relational DBMSs use the terms *table, row,* and *column.* Although Access uses the term *table,* it replaces the terms *row* and *column* with *record* and *field.* As you read through the Access material, be aware of this difference. The correspondence between the two sets of terms is as follows:

GENERAL TERM	ACCESS TERM
table	table
row	record
column	field

M2.2   WORKING WITH ACCESS

Access Objects

A Microsoft Access database is actually a collection of **objects.** Anything that you can give a name to is called an object. There are six main objects in a Microsoft Access database: tables, queries, forms, reports, macros, and modules. While many other microcomputer database systems use the term *database* to refer to the file of data, an Access database includes all the major objects related to that data in one file. Brief descriptions of the six major objects follow.

Note: Because Access database names are also the filenames, they must abide by an eight-character limit and a three-character extension of .MDB, which identifies the file as an Access database. Most database programs store tables, reports, forms, and so forth as separate files; therefore, they too must abide by the eight-character limit. Access, however, stores all objects as part of one database file. As such, they are not limited to the eight-character name and instead may be a descriptive 64 characters. Although long object names are possible, all objects will be abbreviated to eight characters throughout this text.

Tables.   A table contains information about a group of entities, such as customers or inventory items. Tables contain records comprised of fields that contain information about a part of that entity, such as the customer name or inventory item number. Each table can also have a primary key, which is one or more fields that give a unique reference for each record, and one or more indexes for each table.

Queries. A query is a custom view of data in one or more tables that shows selected fields in records that meet certain criteria forming a dynamic subset of the data called a *dynaset*. Queries may be defined either in a graphical query by example (QBE) or with standard SQL (Structured Query Language).

Forms. You may present data in a format other than a datasheet by creating a customized layout much like a familiar paper form but also providing electronic methods to input, edit, display, or print data.

Report. Like a form, data may be displayed on screen or printed in a customized report. A report includes the instructions for formatting, summarizing, and printing the data.

Macro. You may use macros to automate tasks that are used repeatedly. A *macro* is a set of commands you use to automate Access tasks, without the need to write computer-programming code. You accomplish this by selecting commands from menus, which are saved as a set of commands that you can execute with one command.

Modules. You may develop complete applications with Access modules that store statements and procedures. Procedures are written in Access BASIC, a programming language much like Microsoft Visual Basic, but including commands specific to the Access database.

Access Screens

We begin our study of Access by looking at some of the important screens you will use, as well as some of the keys you will use to accomplish various tasks. If you have not started Access, do so now in the manner described in section M2.1.

Your screen should now look similar to the one shown in Figure M2.1. This screen is where you will begin most of your work. Let's take a look at the various portions of this screen.

The top line on this screen is the Access Initial menu. A **menu** is simply a list of options from which you can choose. The words on this line (File and Help) are the options on the menu. As you work with Access, you will encounter situations in which the row of words at the top of the screen is different, but the menu line always plays the same role—a collection of options that you can use for a desired action. (You will soon see how to use these menus.)

The bottom line on the screen gives a description of the currently highlighted option. It also indicates the effect of certain special keys. Currently, for example, it tells you that Access is loaded in memory and ready to either create or load a database or to ask for help. We now turn our attention to the way you work with Access.

Pressing Escape

Sometimes, you might find that you inadvertently choose the wrong option, or you might not want to proceed with some action you have started, but you're not sure how to get out of it. When this happens, simply press the Escape key. In some cases, this immediately returns you to the Main menu. In others, Access provides specific instructions about the action you need to take to escape from your task.

To illustrate the use of the Escape key, select the File option from the Access menu (Figure M2.2). Let's assume you do not want to proceed with this option.

- Press Alt + F. (Press the Alt key and then the F key.)
- Press the Escape key twice.

Figure M2.2

Access Window
with the File menu
pulled down

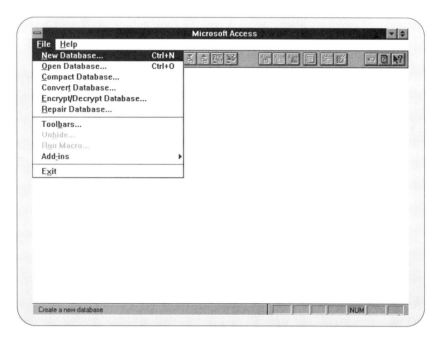

You return to the initial menu. Then the second Escape key cancels the Alt key selection that highlighted the menu.

Using Menus   There are two ways you can select an option from a menu. The first method is to use the mouse and click on the menu option you want. Alternatively, you can use the keyboard. You must first press the Alt key to activate the menu. Then you could either press the underlined letter for each menu option or you could use the Left or Right arrow keys to move to the desired option and then press the Enter key. Let's use this technique to select the Open option.

- Press the Alt key.
- Press the Down arrow until "Open Database..." is highlighted.
- Press the Enter key.

Once you have selected this option, you will see the Open Database dialog box (Figure M2.3). To select an option from it, first you would use the Tab key to move to the item you wish to change, and then press the Up or Down arrow to move the highlight to the option you want. Then press the Enter key. (You may need to first select Drive A.)

To use a mouse to select an option, click on the option; that is, move the mouse pointer to the option and press the left button on the mouse. For example, to accomplish the above process with a mouse you would click first on the File menu option and then on the Open Database option. Throughout this text, this process will be given as Select the File menu and then Choose the Open Database option. In this process, Select indicates action the user takes without immediate results, whereas Choose indicates immediate action.

Your screen should now look like the one shown in Figure M2.3. Access is asking you for the name of the database you wish to use. You can either type the name or you can press Tab, Down arrow, then the Enter key to select the name from a list. Let's use the Tab and Down arrow to select the *Books* database from a list.

- Press the Tab key and then the Down arrow to highlight the *Books* database.
- Press the Enter key.

Figure M2.3

Open Database
dialog box

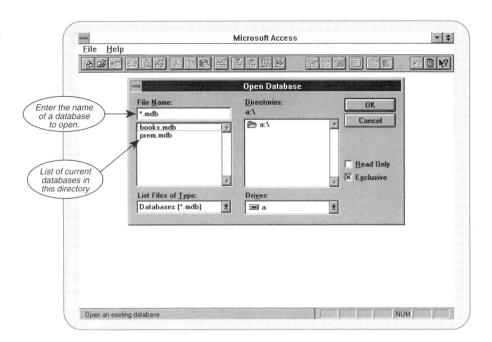

On any screen containing a box labeled OK, clicking on OK is equivalent to pressing the Enter key. On any screen containing a box labeled Cancel, clicking on Cancel is equivalent to pressing the Escape key.

Your screen should now show a list of the tables in the *Books* database, as Figure M2.4 shows. In this database, there are currently five tables. To the left of the list of tables are icons for the six objects with the table icon selected. Just as these five tables are a part of the *Books* database, any queries, forms, reports, and so forth are also part of this one database file.

Viewing a Table

We begin by looking at one of the tables, a list of publishers.

- Select *Publshr* from the list (Figure M2.4) by moving the highlight to it with the arrow keys or by clicking on it with the mouse.
- Press the Enter key or click on OK.

There is a shortcut when you are using a mouse to select an item from a list. Rather than clicking on the item and then clicking on OK, you can click on the item twice (called double clicking). You have to do this very rapidly, however. If you double clicked but the selection has not been made, you probably waited too long between clicks. In this case, you could either try it again or simply click on OK.

You will now see the data in the *Publshr* table (Figure M2.5). This table is comprised of fourteen records, each with three fields. This is the same as saying that this table is fourteen rows by three columns. Although all fourteen records are visible on the screen, the system is always working with one field of one record. The current record is indicated by an arrow pointing to the first row. The current record number is listed on the control bar at the bottom of the window.

Navigating between Fields and Records

There are different methods to move between fields. To move to the next field in a record, press the Tab key; to move to the previous field, press Shift + Tab. To move to the first field of the current record, press Home; to move to the last field of a record, press End. You also could use the mouse to click in the field name selector in the toolbar and select the field to which you wish to move.

Figure M2.4

Books database
window showing
current tables

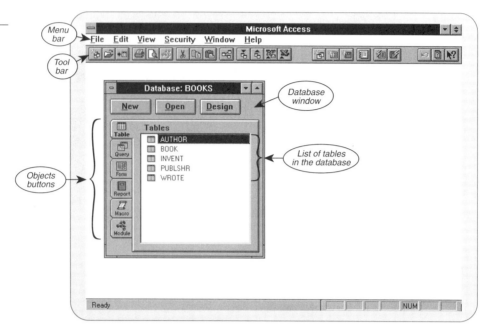

Figure M2.5

Publshr table

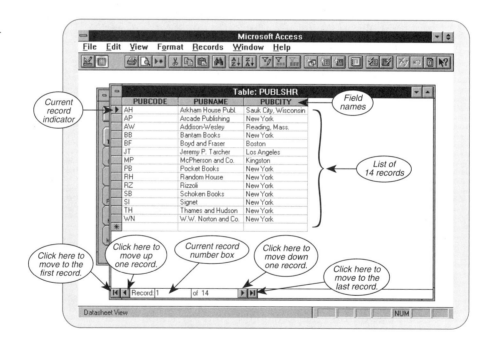

There are numerous options to move between records. With the keyboard, you could press the Up and Down arrows to move between records, Ctrl + Home to move to the first record, and Ctrl + End to move to the last record. You could also use the Go To command from the Records menu. With the mouse, you could click anywhere on the row to which you wish to move, or you could use the navigation bar in the lower-left corner of the datasheet window (Figure M2.5). In the navigation bar are buttons for the first, previous, next, and last records. Also on the navigation bar is the record number box that displays the current record number. With the mouse, you could double click on the record number and enter the number of the record to which you wish to move and then press the Enter key.

Manipulating Windows

In this section, we examine the ways we can manipulate windows on the screen. To begin, we will activate both the *Publshr* and *Book* tables.

- Select Window.
- Use the Down arrow key to highlight "1 Database: *BOOKS*" (shown in Figure M2.6).
- Press the Enter key.

The screen now shows the *Books* database window with the list of tables (Figure M2.7).

- Use the Up arrow to select the *Book* table.
- Press the Enter key.

Figure M2.6

Using the Window menu to open another table

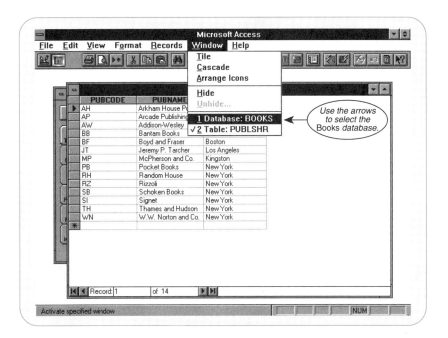

Figure M2.7

Opening a second table

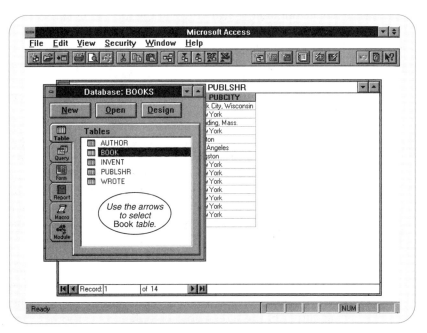

Your screen should look like the one shown in Figure M2.8. Notice that the *Publshr* window is almost completely hidden by the *Book* window. We will see ways of changing this shortly. First, however, let's examine the special features on this screen. The window on top is called the active window (Table: BOOK, in this case). This is the window you can manipulate.

Figure M2.8

Working with a second table

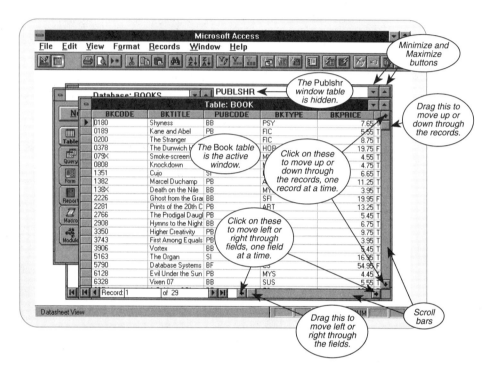

The small boxes near the upper-right-hand corner of the active window are called the Minimize and Maximize buttons. Clicking on the maximize button will make the window as large as possible. Refer to the discussion of windows in Module 1.

The strip along the right-hand side of the active window is called the **vertical scroll bar.** Clicking on the arrowhead at the top of the vertical scroll bar will move the cursor up one row. Clicking on the arrowhead at the bottom will move you down one row. Dragging the small rectangle in the scroll bar up or down in the strip moves the cursor up or down several rows at a time. Moving it all the way to the bottom moves the cursor to the last row. Moving it back to the top moves the cursor to the first row. Moving it to approximately the middle of the strip will move the cursor to approximately the middle row.

The strip along the bottom of the active window is called the horizontal scroll bar. It is similar to the vertical scroll bar except that it moves the cursor to the left or right rather than up or down. Clicking on the arrowhead at the right of the horizontal scroll bar moves the window to the right one column. Clicking on the arrowhead at the left will move the window one column to the left. Dragging the small rectangle in the scroll bar left or right in the strip moves the window to the left or right several columns at a time.

You don't need a mouse to work with Windows, however. You can use the Tab and arrow keys to move around within the active window. Also, you can use the window control menu, which is represented by the hyphen symbol in the upper-left-hand corner of the screen, to accomplish several other things.

To select the control menu from the keyboard, press Alt + hyphen. Press the Down arrow to highlight Maximize and then press Enter. To select it with a mouse, simply click on the hyphen symbol, and then click on the maximize option. You could also press Alt + hyphen and then press the letter x to maximize.

■ Select the window control menu (Figure M2.9).
■ Select Maximize.

Figure M2.9

Using the window
control menu

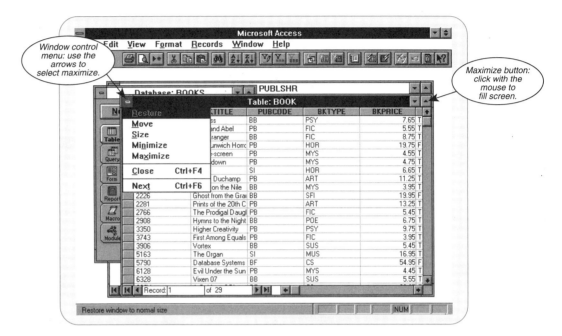

Your screen should now look like the one shown in Figure M2.10. The active window has been expanded to the maximum possible size. Let's now return it to its original size.

■ Select Restore from the control menu or click on the Restore button.

Figure M2.10

Maximized
window

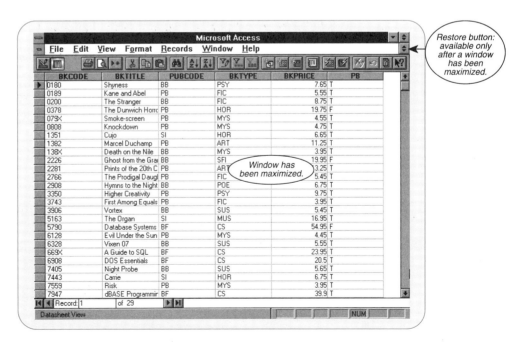

The window resumes its original size. Next, let's change the size of a window and also move a window.

To change the size of a window with a mouse, you use the lower-right-hand corner, called the Resize corner. You change the size by dragging the Resize corner to a new position. The size of the window changes as you move the corner. To move a window, you drag the top line of the window. The entire window moves as you move the line.

Now let's change the size of the window to match the one shown in Figure M2.11. We will use the mouse in the method just described. Alternatively, you could use the window control menu.

■ Use the mouse to change the size of the window to match Figure M2.11.

Figure M2.11

Resized window

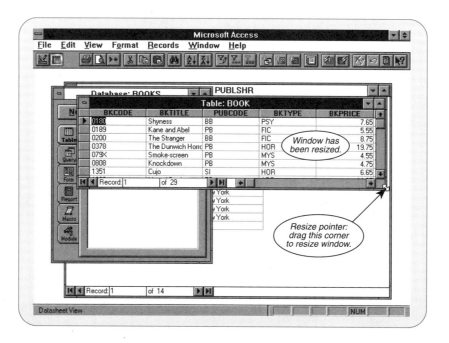

Next we will move the window to the position shown in Figure M2.12. Again, you can use the mouse in the manner described by dragging the top line of the window.

■ Use the mouse to move the window to match Figure M2.12.

When you have more than one window on your screen, you frequently need to move back and forth between the windows. You can use the Next option of the window control menu to do this, but pressing Ctrl + F6 accomplishes the same thing.

To use a mouse to move to a different window, click anywhere within the window.

■ Either press Ctrl + F6 or click on some portion of the *Publshr* window to make it the active window.

The *Publshr* window is now the active window (Figure M2.13). You know which window is the active window by the highlighted title bar. Notice that the *Publshr* window has now overlapped the *Book table* window.

Now let's resize the *Publshr* window so that both windows are completely visible. You may use either the mouse to resize the window or the control menu to select Size.

Figure M2.12

Book window
moved

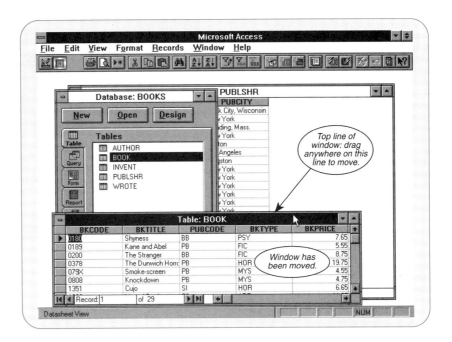

Figure M2.13

Publshr table made
active window

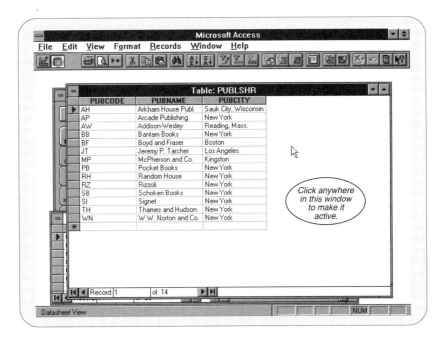

■ Change the size of the *Publshr* window to the one shown in Figure M2.14.

Finally, let's practice moving around within the active window. Use the *Book* window to do this.

■ Move to the *Book* window.
■ Repeatedly press the Down arrow to move down to record 15. (You can also use the vertical scroll bar with the mouse.)

Now your screen should look like the one shown in Figure M2.15. Notice as you move the record indicator box shows which record you are in. The scroll bar on the

Figure M2.14

Working with
multiple windows

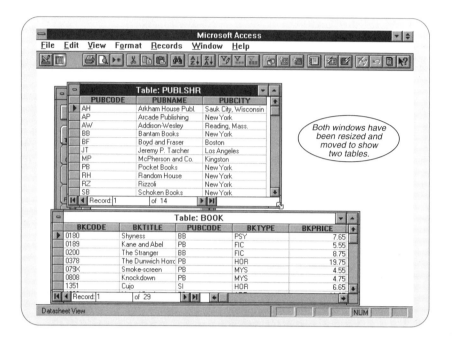

Figure M2.15

Moving around
within the active
window

right side of the active window also shows the relative position in the file. You will move through the fields now. First you will move back to the first record.

- Repeatedly press PageUp until you have moved back to the first record in the *Book* table. Notice that you move back several records at a time using this method.
- Repeatedly press the Right arrow to move to the *PB* field. (With a mouse, you can also use the horizontal scroll bar.)

Note: If, when you press the Right arrow key, the cursor only moves one character at a time, you are in Edit mode. Press the Tab key to move one field to the right.

Notice that some of the fields moved off the left-hand side to make room for the extra fields on the right. The scroll bar on the bottom of the screen indicates the relative position among the fields. You can press the End key to return to the last field and the Home key to move to the first field. Now return to the first field of the first record.

■ Press Ctrl + Home to return to the top-left corner of the *Book* window.

Changing Column Width

You may need to adjust the width of columns to better match the width of the data in each column. For example, the data in BKTITLE is wider than the width of the column, while the BKPRICE column width does not need to be as wide as it is. To change the width of each column, you may either use the mouse or select the Column width option from the menu. Begin by changing the first column. Move your cursor to the BKCODE field. The *Field* box in the toolbar should indicate in which field you are working. To use the menu to change the width, select Column width from the Format menu (Figure M2.16).

■ Select Format and then Column width.
■ Change the value in the Column Width box to 10.
■ Press the Enter key.

Figure M2.16

Changing Column Width

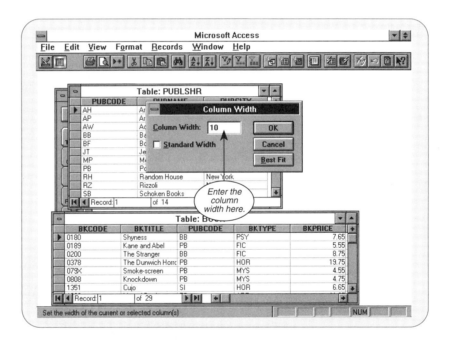

To change the width of a column with the mouse, move the mouse to the line separating the names of each column. When you move the mouse pointer to one of the separating lines, the pointing arrow changes to a double arrow (Figure M2.17). Hold down the left button on the mouse while dragging to the left to shrink the column width and dragging to the right to widen a column.

Practice changing column widths using either the mouse or the menu (Format and Column width) to match the approximate widths shown in Figure M2.17.

This concludes our practice with windows so now you will remove both windows from the screen.

■ Select Windows.
■ Use the arrows to highlight Database: *BOOKS.*
■ Press the Enter key. (This makes the database window active.)

Figure M2.17

Changing column
width with the
mouse

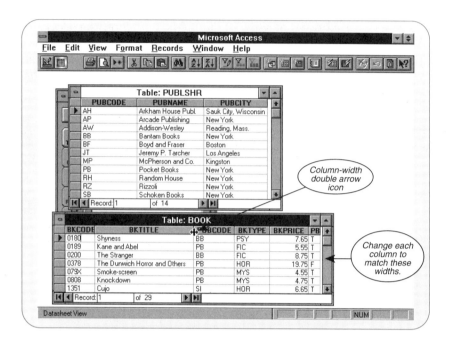

- Select File and then Close Database.
  (Answer Yes to, "Save layout changes to Table: *BOOK*?")

(You could also close each window with the mouse by using the Close buttons.)

Getting Help   Access has an extensive Help facility. You can obtain help on a variety of topics by pressing the F1 key. (Help is also available by clicking on the Help button on the toolbar or by choosing Help from the menu.)

- Press F1, or use the mouse to click on the Help button on the toolbar.

You will see a screen similar to the one shown in Figure M2.18. This is the top-level Help screen. (If you had been in the middle of a specific task, you would instead

Figure M2.18

Using Access Help

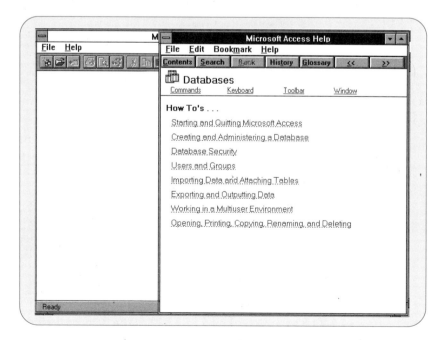

have received help on the task.) The window contains appropriate Help options that allow you to move directly to specific topics. For example, to move to Help on using the keyboard, you could either click on the keyword "Keyboard" or press the Tab key to highlight the keyword and press the Enter key.

- Press the Tab key three times to highlight the word *Keyboard*.
- Press the Enter key.

Help changes to the first screen on using the keyboard (Figure M2.19). You will notice a line of navigation buttons display at the top of the screen. These buttons may be used to go to a table of contents, move back to the previous Help screen, move forward, or move backward through Help screens. Now we want to move to the next keyboard Help screen. The last button on the control bar is the >> button and is used to move forward one screen. You can activate this key by either clicking on it with the mouse, using the Alt key combination, or pressing the tab key to highlight it and then pressing the Enter key.

- Press the Alt and > keys.

Figure M2.19

Database window keyboard guide

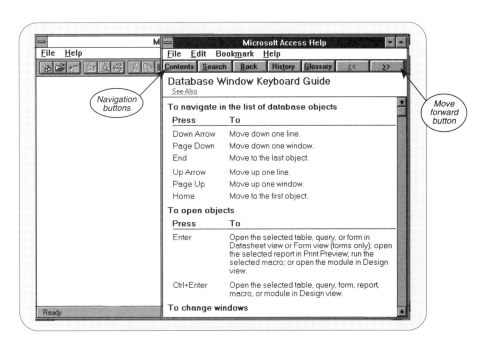

The Table Window Keyboard Guide (Figure M2.20) provides a list of options to jump to. For further help on using the function keys, you could either again use the Tab key to select the option you want, or you could use the mouse. You will notice that as you move the mouse arrow pointer over a green highlighted word, the arrow changes to a pointing hand. When the arrow changes to a hand, it indicates that you have moved to a hot spot. **Hot spots** are mouse-sensitive areas used to jump to another screen. To see how this works, click the mouse on the words *Function Keys*.

- Click on the Function Keys option.

You should now see the screen shown in Figure M2.21, which gives specific information on function keys. At this point, let's say that you want to return to a previous Help screen. The Back button returns to the screen that was seen just prior to the current one.

Figure M2.20

Table Window
Keyboard Guide

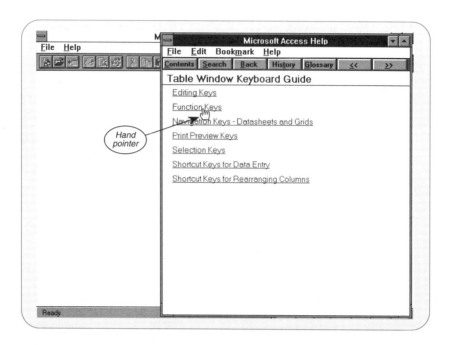

Figure M2.21

Function Keys Help
screen

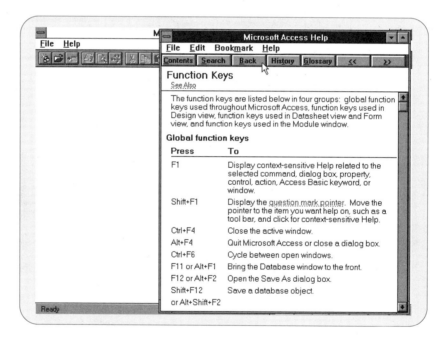

■ Click on the Back button.

Now you have returned to the Keyboard Guide screen. You could continue pressing the Back option to retrace your steps through the screens previously viewed. Instead, at this point, let's examine the Search feature. Click on the Search button on the toolbar.

■ Click on the Search button.

The Search dialog box (Figure M2.22) has two main sections. The first section provides a text box to enter a search criteria. Once a criteria has been entered, a list of matching topics is displayed. Select one of these topics and then activate the search

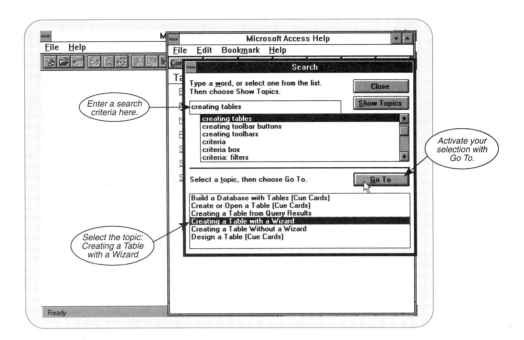

with Go To. To demonstrate the Search feature, let's search for information on creating a table, the next section that will be covered.

In the criteria box, you will enter "Creating tables," which will display a set of matching topics. Then you will select "Creating a Table with a Wizard" and activate it with the Go To button.

- Enter `Creating tables` in the Search criteria box.
- Press the Enter key.

A list of topics displays in the lower list box.

- Click and select "Creating a Table with a Wizard" from the topic list.
- Click on the Go To button.

A Help screen on creating a table will display (Figure M2.23). On this screen, you will notice some of the words highlighted in color. There are words such as *table*, *records*, and *fields*, for instance. These are special words known as hot spots.

As mentioned earlier, the Help system of Access provides special hot spots that jump, or link, to other Help topics or more information about the current topic. Hot spots are displayed with an underline to identify them. A solid underline indicates the hot spot will jump to another screen, while a hot spot with a dashed underline will display additional information that appears in a pop-up window. To close a pop-up window, click anywhere on the pop-up screen, or press any key.

- Click on the *table* hot spot with the pointing hand.

A pop-up box displays a description of a table. To close this box, click anywhere in the box. At this time, you may find it useful to browse through these Help screens. Use the forward/backward buttons or the jumps to investigate the Help system. When you have finished, exit Help as instructed below.

Exiting Help   To leave the Help system, you must be at the File menu.

- Select File and then Exit (or press Alt + F and then x).

This closes the Help screen and returns you to the Access main menu.

Figure M2.23

Help topic
"Jumps"

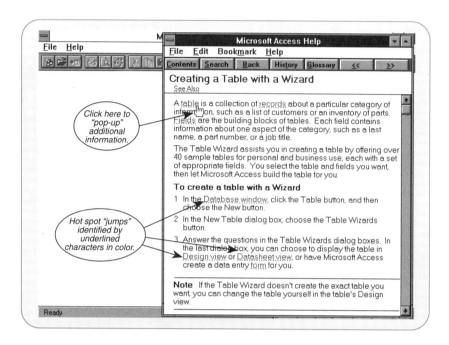

M2.3   CREATING A TABLE

Now that you have seen the basic mechanics of working with Access, let's get to work. First, you need to create a new table. The table you will create is the *Branch* table in the *Books* database. When you create a new database, you create a new file. This file contains all the tables and other objects related to the database. When you add another table, you simply add it to the existing database file.

Beginning the Table Creation Process

To create a new table, open the *Books* database and add *Branch* to the list of tables.

- Select File and then Open.
- Type Books for the name of the database to open. (Alternatively, if the database is visible in the dialog box, you can click on it with the mouse.)

A list of the current tables is displayed. (If the database window does not list the tables, click on the table object icon.) We want to add a new table.

- Click on the New button.

Access provides two methods for creating database objects, Wizards and custom design. For example, when you select New while in the Table view, Access offers the choice of the Table Wizard, or New Table. The same procedure is followed for each new table, query, form, or report. An Access Wizard is like having an expert who prompts you with questions and offers suggested samples and then builds the object based on your responses. We will use Access Wizards in later modules. For now, we will use the custom method to create a new table.

- Click on the New Table button.

Your screen should now look like the one shown in Figure M2.24. You will define the structure by entering the names and types of all the fields in your table. The rules for field names are:

1. They must be no more than 64 characters long.
2. The first character must not be a blank.
3. The same name cannot be used for two different fields in the same table.

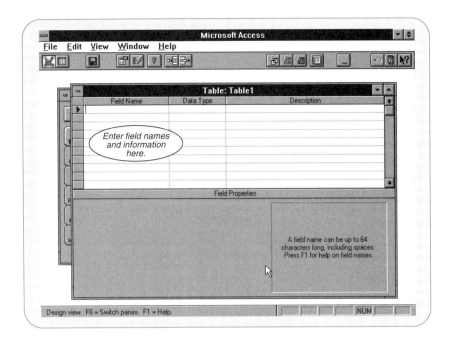

Each field has a field type. This indicates the type of data that you can store in the field. The possibilities are:

1. *Text.* The field can contain up to 255 alphanumeric characters.
2. *Memo.* The field can contain large (32,000 bytes) amounts of alphanumeric characters.
3. *Number.* The field can contain only numbers. The numbers can be either positive or negative. You can use fields of this type in arithmetic operations.
4. *Date/Time.* The field can contain a legitimate date stored in one of the date forms *short* (mm/dd/yy), *medium* (dd-mon-yy), or *long* (day, mon. dd, yyyy) such as in 11/05/93 or 05-Nov-93 or Friday, November 5, 1993. Time formats also are available in *short* (16:30), *medium* (4:30 pm), or *long* (4:30:00 pm).
5. *Currency.* The field contains monetary values. The values will be displayed with dollar signs. As with numeric fields, you can use currency fields in arithmetic operations.
6. *Counter.* This field is a numeric value that Access automatically increments for each record you add.
7. *Yes/No.* Boolean values, True or False.
8. *OLE Object.* OLE objects are graphics or other binary data from other Windows applications.

Defining Fields

The first step in defining a field is to type its name in the column labeled Field Name. Since the cursor is already in the column for the first field, enter `Brnumb` to create a field for the branch number.

- Type `Brumb`
- Press Enter.

The cursor is now in the column labeled Data Type. Since branch numbers are numeric but are not dollar amounts, we will use Number as the field type. You will notice that when you press Enter, the cursor moved to the Data Type field and the default type Text displayed. To see a list of the valid data types, use the mouse and click on the Down arrow in the Data Type field. (The keyboard alternative is to press Alt + Down Arrow.)

■ Click on the Down arrow in the Data Type field.

This pulls down a menu to list the types available for selection. Your screen should look like the one shown in Figure M2.25. To select the Number type, just click on the word *Number*, or press the Down arrow twice and press the Enter key.

■ Choose Number.
■ In the Description column, enter `Branch number` to describe the field.

Figure M2.25

Adding fields

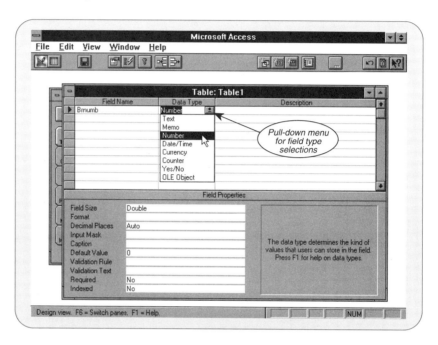

In addition, we will designate this field as a **key field.** This means that Access will maintain the data in this table in *Brnumb* order and reserve the data as a unique entry. To designate a field as a **Primary Key** field, click on the Primary Key icon in the toolbar (see Figure M2.26).

Figure M2.26

Branch number field information and primary key

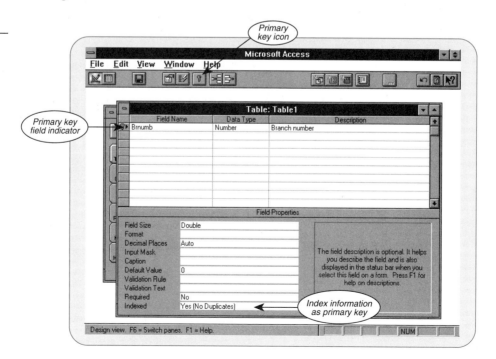

■ Click on the Primary Key icon.

You will now make the remaining entries. If you make any mistakes, you can use the keys shown in Table M2.1 to correct them.

■ Enter the information for the remaining fields as shown in Figure M2.27.

Table M2.1

Special keys used when working with tables.

KEY	PURPOSE
↑	Moves the cursor up one row.
↓	Moves the cursor down one row.
→	Moves the cursor one column to the right.
	Moves the cursor one character to the right when editing data.
←	Moves the cursor one column to the left.
	Moves the cursor one character to the left when editing data.
Home	Moves the cursor to the first column.
End	Moves the cursor to the last column.
Ctrl-Home	Moves the cursor to the first record.
Ctrl-End	Moves the cursor to the last record.
Backspace	Moves the cursor one position to the left and erases the character that was in that position.
Enter	Completes the current entry and moves the cursor to the next column.

Figure M2.27

Branch field list

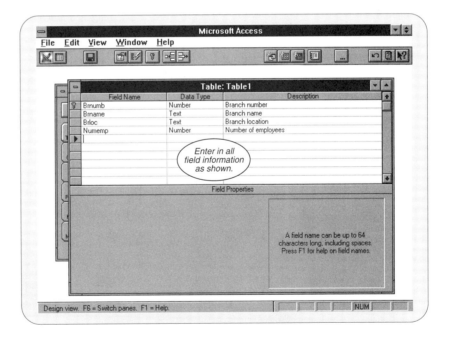

Revising a Field's Attributes

You entered the field type for *Brname* as Text. The default field size for text fields is 50 characters. The actual size we want for this entry is 20 characters. We will use this change to illustrate the process of revising field attributes.

■ Click anywhere on the *Brname* field row.
■ Click in the Field Size box at the bottom of the screen.
■ Type 20 and press Enter (Figure M2.28).

Repeat this process to change the width of the *Brloc* field to 20 characters.

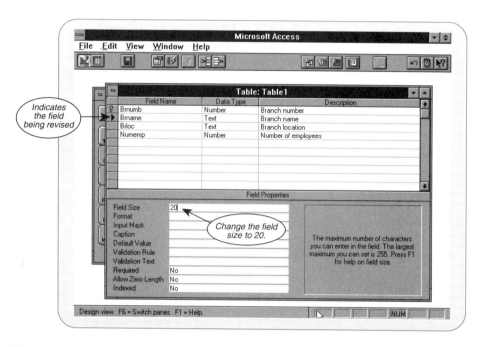

Finishing the Process

Now that you have described all the fields, you can save the table. Select Save As from the File menu.

■ Press Alt + File and then choose Save As.

Access next prompts for the table name, as Figure M2.29 shows.

■ Type **Branch** for the table name and press the Enter key.

Viewing the Table as a Datasheet

You have been working with the *Branch* table in the design view. To work with the table in the Datasheet View, either click on the Datasheet View button on the toolbar, or select Datasheet from the View menu.

■ Press Alt + View and then select Datasheet.

Your screen should now look like Figure M2.30 and is ready to enter data.

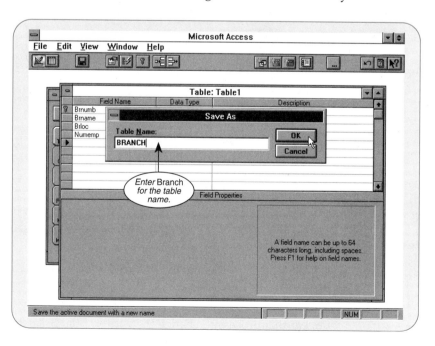

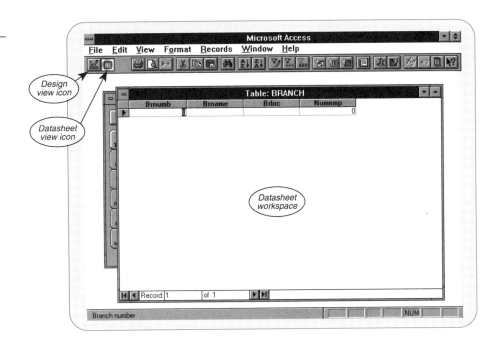

M2.4 UPDATING A DATABASE

**Entering
Initial
Records**

In this section, you will add data for the branches to your database. As you will see, Access gives you several different ways of accomplishing the task. You can enter data using either a form or directly into the table. Once you have learned the various methods, you can select the one that seems most natural.

We will first use the method that adds records directly into the table. Begin by entering the branches shown in Table M2.2. We will make some mistakes to illustrate the process of making corrections.

Table M2.2

Data for the
Branch Table

BRNUMB	BRNAME	BRLOC	NUMEMP
1	Henry's Downtown	16 Riverview	10
2	Henry's On the Hill	1289 Bedford	6
3	Henry's Brentwood	Brentwood Mall	15
4	Henry's Eastshore	Eastshore Mall	9

- Type 1
- Press Enter.
- Type `Hnryy's Dawntown`
- Press Enter.
- Type `16 Riverview`
- Press Enter.

At this point, your screen should look like the one shown in Figure M2.31. Before proceeding, let's fix the branch name.

- Use the Left arrow key to move back to the *Brname* field.

There are two ways to make the correction. As you move back each field, you will notice that the entire field is highlighted. The first method of editing is to simply enter the new data and it will replace the current contents. With this method, however, you must retype all of the data. If the field is long and there are only a few changes, this method is not the one you want to use. Instead, you can enter the Edit mode by pressing

Figure M2.31

Entering data

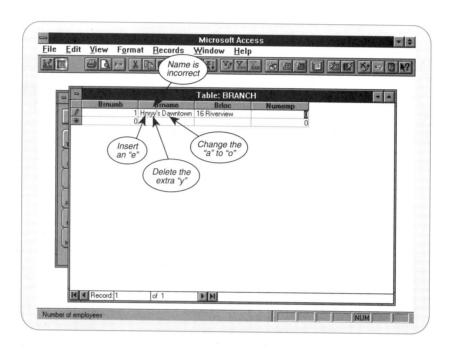

the F2 function key. You can now use the arrow keys to move the cursor within the field. You can delete the character to the left of the cursor by pressing the Backspace key. You switch between the two modes by pressing the F2 key.

- Press F2 to enter the Edit mode.
- Change the data in the Brname field to `Henry's Downtown.`
- Press Enter and to leave the Edit mode and move to the next field.

Now finish entering the rest of the first record.

- Use the Right arrow key to move to the *Numemp* field.
- Type `10`
- Press the Enter key.

You have finished the first record and are ready to enter the second.

Note: Access automatically saves your changes as soon as you move to another record or close the form or datasheet. You do not have to save the changes. However, if you want to save the changes without leaving the record, choose Save Record from the File menu.

- Enter the remaining records exactly as Figure M2.32 shows.

Notice that you have made intentional mistakes on the second record. The third record should actually be the fourth record. The fourth record you entered should not be in the file. We will correct these mistakes in the next section.

Changing Records

You can change records using either the Datasheet (Table) View or Form View. We will illustrate the use of the Form View later in the text. For now, we will edit the mistakes in the Datasheet View. Make sure your screen looks like the one shown in Figure M2.32. You need to first correct record 2. You can move to the record you want to change by using either the arrow keys to move to the record you want to edit or click on the record with the mouse. Since the number of records currently on the screen is very small, this is not a problem. (It would be cumbersome if your file were large.) Fortunately, there are other ways to accomplish this task. For now, however, just follow one of these procedures: Either click on the second record or move there with the cursor keys.

Figure M2.32

Entering data

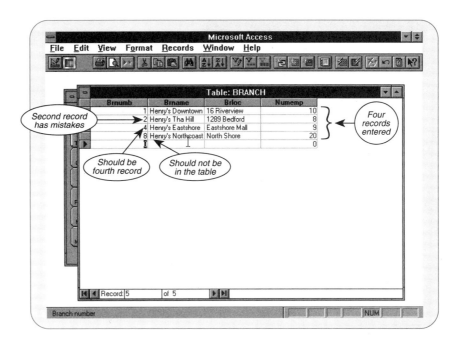

When you have selected the second record, press the Right (or Left) arrow key until the cursor is on the field to be changed. You can correct the data in the field by pressing the F2 function key and entering the Edit mode. You can use the arrow keys in Edit mode to move left or right in the field until you reach the incorrect character; then press Backspace to erase the character to the left of the cursor and correct the error.

- Press the arrow keys to move to the *Brname* field of record 2.

You can also click on the *Brname* field with the mouse.

- Correct the name. The correct entry should be `Henry's on the Hill.`
- Now correct the value in the *Numemp* field. The correct value is 6.

Once you have made the corrections, Access will save your work when you move to another record.

- Press the Down arrow to the next record to save the changes to record 2.

Deleting Records

Deleting records is similar to changing them. Let's delete the record for branch 8 since it does not belong in the database. Move to the record to be deleted by clicking with the mouse on the record handle on the left edge of the row or choosing Select Record from the Edit menu (Figure M2.33). To delete the record,

- Move the cursor to the record for branch 8.
- Select Edit, and then Select Record.
- Press the Delete key.

The record has now been removed from your image of the *Branch* table and an Access dialog box requests verification of the deletion (Figure M2.34).

- Press the Escape key or click on the Cancel button.

The deleted record is reinserted in the datasheet and remains highlighted. This verification dialog box is your safety net for accidental deletions. Once the record has been deleted, the Undo option in the Edit menu is not available. Since we really want to delete the record, let's repeat the process.

Figure M2.33

Editing data

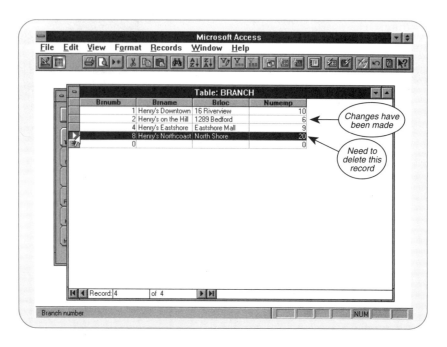

Figure M2.34

Deletion dialog box

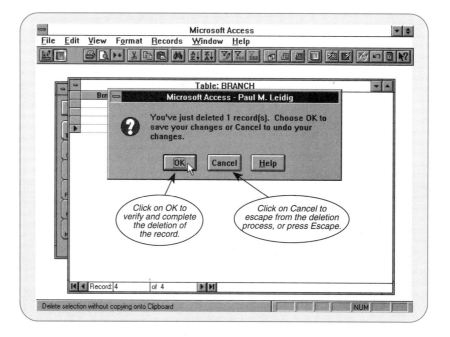

- Press the Delete key.
- Press Enter or click on OK.

Adding Records

You can add records in the Datasheet View. If you want to add a record at the end, move past the last record and type the data for the new record. The data is saved once you have pressed Enter after the last field or moved to another record. Typically, though, you will use the DataEntry mode; that is the way we will add Branch 3 to our table. The first record in Figure M2.35 is the record we want to add to the table. The second one is actually an error since the branch number duplicates the number of a branch already in the table. Let's enter this data and see what happens.

Figure M2.35

Using DataEntry to add records

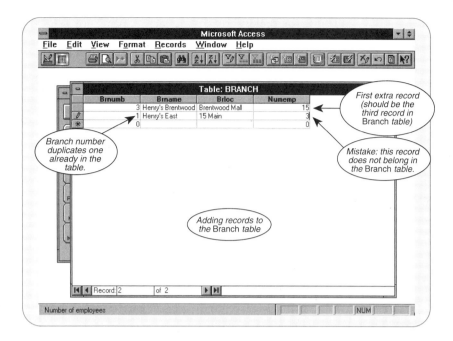

- Select Records and then DataEntry.
- Enter the records shown in Figure M2.35.

When you press the Enter key after the *Numemp* field for the first record, the record is accepted and moves to the next record. However, when you attempt to accept the second record with the duplicate *Brnumb,* an Access dialog box informs you of a "duplicate key error" and does not update the index (Figure M2.36). The reason for this is that when you created the *Branch* table, you indicated that the *Brnumb* field was the primary key and could not accept duplicate values. To acknowledge the error message, either click on OK or press the Escape key.

- Press the Escape key.

Figure M2.36

Duplicate key error

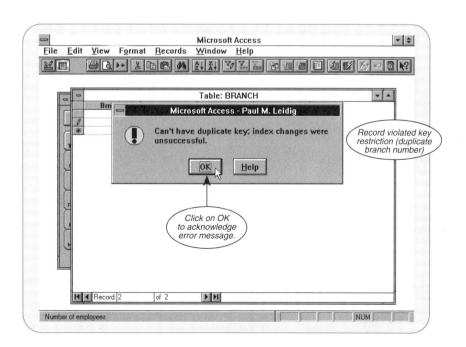

The Error dialog box disappears, waiting for you to correct the error. Since you do not want to enter this second record, you can press Escape again to exit entering the second record.

■ Press the Escape key.

You now only have the branch 3 record in the DataEntry screen. To return to the full datasheet, select Show All Records from the Records menu.

■ Press Alt + Records and then Show All Records.

You will now see an image of the updated *Branch* table (Figure M2.37). Notice that not only has branch 3 been added to the table, but it is in the correct location as well.

We will now clear the workspace.

■ Close the *Branch* table window by pressing Alt + File and then Close.

Figure M2.37

Viewing the image of the *Book* table

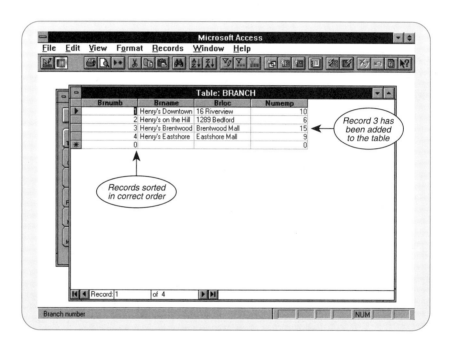

M2.5   INTRODUCTION TO REPORTS

We continue this module by looking briefly at the process of creating reports. Later in the text, we will examine the process in detail. For now, you simply want to be able to quickly and easily get a report from the data in your database. We will illustrate the process by producing a report of all books. Although you may view reports on the screen, they are intended primarily for producing printouts.

Beginning the Report Creation Process

Creating a report is easy with the aid of ReportWizard. ReportWizard prompts you through a series of questions in a step-by-step method, asking for the fields to be printed, the order in which the records are to be sorted, and how the records should be grouped. ReportWizard provides seven predefined styles of reports to choose from. In this section, you will create a simple Groups/Totals report.

Begin creating a report by selecting ReportWizard: Either (1) click on the Report Tool on the left side of the Database window and then on the New button, click on the Report Button on the toolbar, or (2) select File, New, and then Report from the menu.

- Select File, New (**Note:** not New Database), and then Report.

Your screen should now look similar to Figure M2.38. ReportWizard is asking which table the report should be created from. Select the *Book* table.

- Press the Down arrow twice to highlight the *Book* table. (Do *not* press Enter yet.)
- Select ReportWizard.

Figure M2.38

Creating reports with Report Wizard

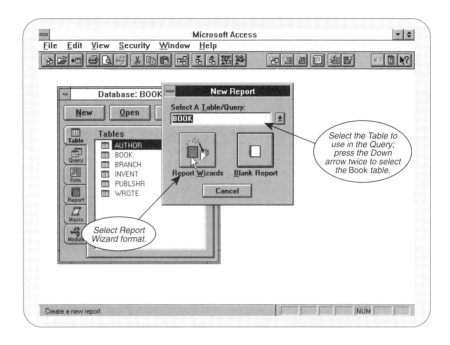

ReportWizard next asks which style of report you want to create. Choose Groups/Totals and click on the OK button (Figure M2.39).

- Click on Groups/Totals.
- Click on OK.

Figure M2.39

ReportWizard style box

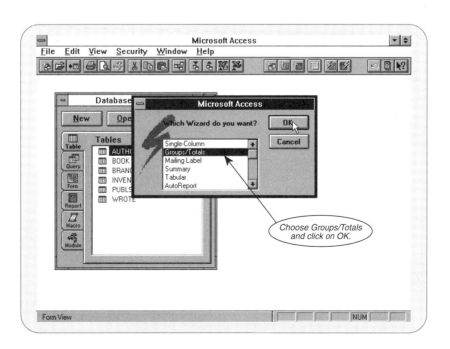

Figure M2.40

Adding fields to a report

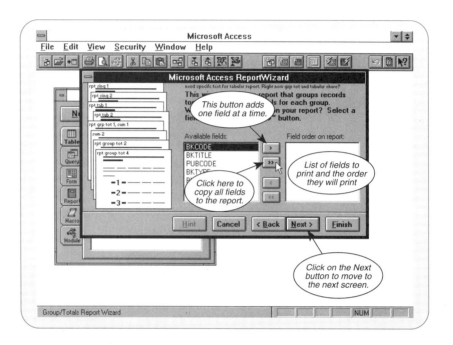

The next screen (Figure M2.40) prompts for the fields to print. You click on each field and then the copy (>) button in the order that you want the columns printed from left to right. To select all of the fields in the current order, you simply click on the Copy All (>>) button.

■ Click on the Copy All button (>>)
■ Click on the Next> button or press Alt+N.

The next two screens prompt for the fields to group or sort by and a report style. To bypass these screens, simply click on the Next > button for each.

■ Click on the Next> button (or press Alt+N) to bypass the group by prompt.
■ Click on the Next> button (or press Alt+N) to bypass the style prompt.

The last of the prompts (Figure M2.41) lets you enter a title for the report. Enter the title "Book Listing". We also want to de-select the "Calculate percentages . . ." option.

■ Enter Book Listing for the report title.
■ Click on Calculate percentages . . . to de-select the option.
■ Click on the Finish button to view the report.

At this point, ReportWizard processes the information provided and creates a report for you. This report is then displayed on the screen (as Figure M2.42 shows) and allows you to view, modify, or print the report. As you move the mouse pointer over the page image, the arrow turns into a magnifying glass icon. You may use this to zoom in on a selected portion of the report or zoom out to the reduced view. When you click on the Zoom button, you display an enlarged view page (Figure M2.43).

■ Press Alt+Zoom or click on the Zoom button.

You may use either the arrow keys or the horizontal and vertical scroll bars to view the remainder of the report not displayed on the screen. Once you have finished looking at the report, close the file without saving it.

■ Select File and then Close to close the report.
■ Click on No when prompted for whether you want to save the report.

Figure M2.41

Creating reports
with ReportWizard

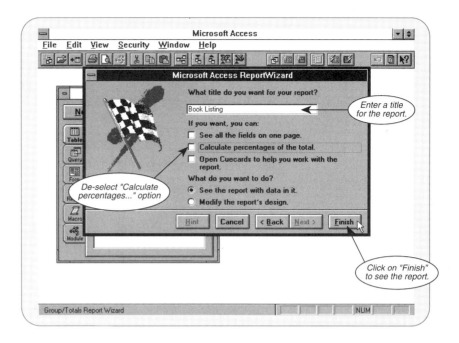

Figure M2.42

Viewing a report

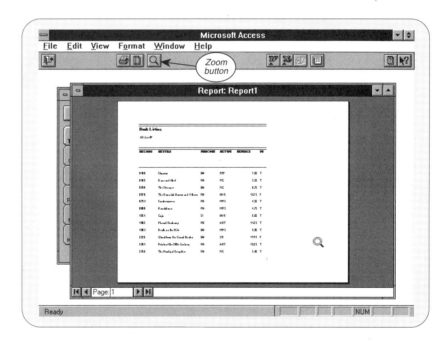

Note: If you want to print the report and the printer has been set up, you may click on the Print icon to send a copy of the report to the printer.

This short introduction to reports tells you how to print out customized reports with little effort. A detailed discussion of report design is provided later in the text. If, however, you merely want to print out a simple report of one table (with all of the fields printed in columns, as we have just done), you may accomplish this much more quickly by choosing Print Preview from the File menu while highlighting the desired table on the database window. Figure M2.44 gives an example of such a report. You will notice that this method does not give the means to give a report title, select the

Figure M2.43

Print Preview in
Report Wizard

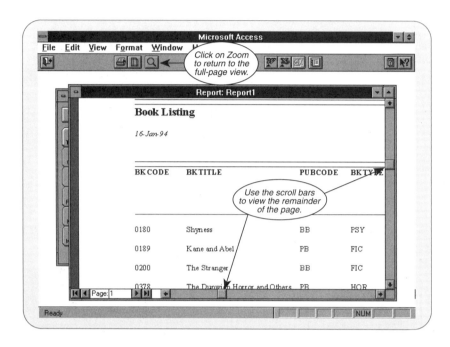

Figure M2.44

Using the default
Print Preview

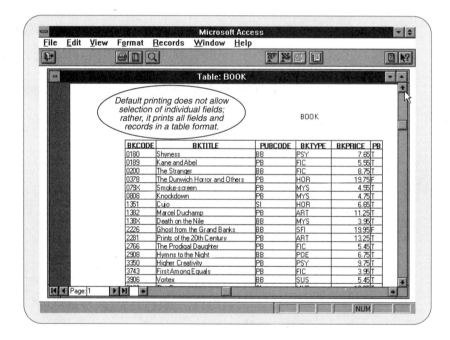

appearance of the report, fields to print, or the order in which reports are printed. It is, however, a quick method of printing out the contents of a table for a hardcopy listing of the data.

M2.6   BACKING UP YOUR DATABASE

The files you are currently using are called your **live files.** It is important to periodically make copies, called **backup** or **save copies,** of these files. (In the process, we say that you back up your files.) This protects you in case a problem occurs in your live files.

**Copying
Tables**

There is an Access menu choice to assist you in copying tables in a database. We will investigate the option in a later module. For now, we will simply look at how we can use it to make backup copies of a table in a database.

To make a backup copy of a table, use the Edit-Copy and Edit-Paste and options. Here, we will illustrate the process by making a copy of the *Book* table.

First make sure that you have closed all reports and tables and that the *Books* database is open with the tables listed in the database window. (If this is not the case, open the *Books* database, and click on the Table button.)

■ Highlight the *Book* table by clicking on it or moving to it with the arrow keys.
■ Select Edit and then Copy (Figure M2.45).

Figure M2.45

Copying an
individual table

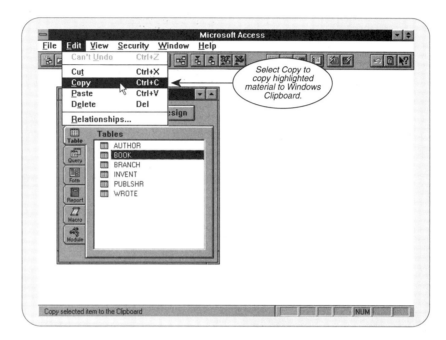

This places a copy of the entire table (structure and data) in the Windows Clipboard. You then can paste the clipboard back into the database.

■ Select Edit and then Paste.

Access will open the Paste Table As dialog box and request a name for the new table. Call the backup copy BOOK backup and append the current date.

Note: Remember that object names in an Access database may contain up to 64 characters, including spaces.

■ Enter BOOK backup and the current date (Figure M2.46).

Access places a backup copy of the *Book* table in the database and lists it as another table (Figure M2.47). You can then use this table as you would any other table.

**Recovering
Tables**

If you discover that a table has been damaged, you can return the table to its condition as of your last backup by deleting the damaged table and renaming the backup copy with the original name. To do this, select File and then Rename.

**Backing Up
the Entire
Database**

To make a backup of the entire database, including all tables, reports, forms, and so forth, use the Copy command in the Windows File Manager. For example, to make a backup of the *Books* database, you would close all of the *Books* files, switch to the Windows File Manager, and choose Copy from the File menu.

Figure M2.46

Pasting the table from the Clipboard into the current database

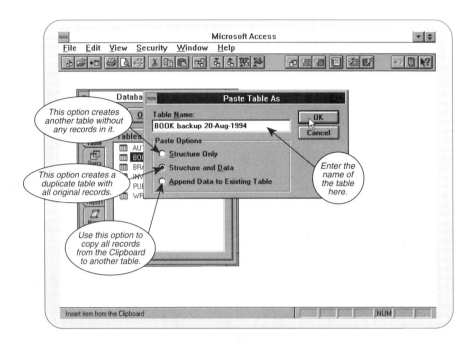

Figure M2.47

Backup table added to the database

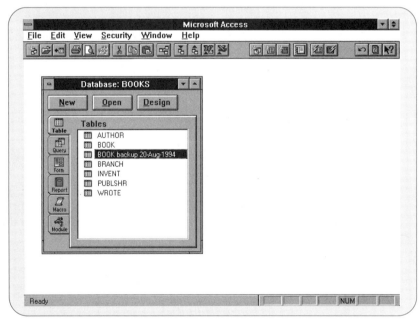

When Should You Back Up a File?

You should back up each file when you first create it. After this, you need to back it up only when you change it in some way. If, for example, no activity occurs today that affects the *Customer* table and you made a backup copy yesterday, it is not necessary to back it up again, although there is certainly nothing wrong with doing so. In backing up files, it pays to be a little *extra* careful.

SUMMARY

The following list summarizes the material covered in this module:

1. An Access database is a collection of objects. An object can be a table, query, form, report, macro, or module. All related objects are stored in one complete file with an extension of .MDB.

2. To select an option from the menu using the keyboard, press the underlined letter for each option on the menu. If none is highlighted, press Alt prior to selecting from the menu. Alternatively, you can use the arrow keys to move to the option and then press the Enter key. To select an option with a mouse, click on the option.
3. To escape from your current task, press Esc.
4. To view a table, select the View option of the Main menu and then select Table. A list of tables in the database is listed, select the table to be viewed.
5. To maximize or restore a window, click on the Maximize/Restore button.
6. To move a window with a mouse, drag the top border of the window. To resize a window with a mouse, drag the resize corner (the lower-right-hand corner).
7. To move to a different window, select the window from the Window option on the menu, or click on the other window.
8. To get help, press F1 or select Help from the menu.
9. To leave Access, select Exit from the File menu and then select Yes.
10. To create a new table, click on the Table icon and then the New button. Type the name of the table, and then enter the fields that make up the table.
11. When you are viewing or updating data, you can use Datasheet View, in which the data is displayed as a table, or Form View, in which the data is displayed using a form on the screen.
12. To delete a record, move to the record to be deleted, select the record from the Edit menu, and press Delete. Press Enter or click on OK to verify the deletion.
13. Select Undo from the Edit menu to reverse the most recent change.
14. To produce a report without having to specify details about the report layout, select New from the File menu, select Report, choose ReportWizard, and answer the questions to the prompts.
15. To make backup copies of a table, copy and then paste a table within a database to create a backup copy. To make a copy of the entire database, use the Windows File Manager to copy the database file.

EXERCISES

1. How do you make a selection from a menu in Access?
2. Describe objects in an Access database.
3. How do you view a table? What different ways can you view a table?
4. Describe the process of creating a table.
5. What field types does Access support?
6. Explain how to add records to a table.
7. Explain how to change the contents of records in a table.
8. Explain how to delete records from a table. When are the records permanently removed?
9. Explain how to create a report listing all the records in a table.
10. How can you make a backup copy of a table? A copy of the entire database? How can you use this copy if your database is damaged?

COMPUTER ASSIGNMENTS

1. You will be working with the *Books* database, so make sure that you are in Access with the *Books* database window open.
2. Create the *Bookcust* and *Bookord* tables shown in Table M2.2. The details you need for the fields are shown in Table M2.3. The *Bookcust* table contains the names and

Table M2.3

Bookcust and
Bookord tables

BOOKCUST

CSTNUMB	CSTNAME	CSTADDR	CSTCTY	CSTST	BRNUMB
1	Allen, Donna	21 Wilson	Carson	In	2
2	Peterson, Mark	215 Raymond	Ceder	In	2
3	Sanchez, Miguel	47 Chipwood	Mantin	Il	3
4	Tran, Thanh	108 College	Carson	In	1
5	Roberts, Terry	602 Bridge	Hudson	Mi	4
6	MacDonald, Greg	19 Oak	Carson	In	1
7	VanderJagt, Neal	12 Bishop	Mantin	Il	3
8	Shippers, John	208 Grayton	Ceder	In	2
9	Franklin, Trudy	103 Bedford	Brook	Mi	4
10	Stein, Shelly	82 Harcourt	Hudson	Mi	1

BOOKORD

BKCODE	CSTNUMB	REQDATE
0378	5	05/12/94
0378	10	05/20/94
0808	2	04/28/94
0808	8	05/15/94
1382	1	04/26/94
1382	10	05/20/94
3743	1	04/26/94
3743	5	05/12/94
3743	9	05/18/94
5163	4	04/29/94
7443	3	04/28/94

addresses of customers who have placed special orders at one of the branches of Henry's bookstore. The *Bookord* table indicates which books customers have ordered as well as the dates on which the customer placed the order.

3. Add the first eight customers shown in Table M2.3 to the *Bookcust* file. Add at least the five rows shown in Table M2.4 to the *Bookord* file. Print a report of all customers. Print a report of all rows in *Bookord*.

4. Change the name of `Peterson, Mark` to `Peterson, Mike`. Add the last two records to *Bookcust*. Add yourself as the eleventh customer. Print a report of all customers.

5. Change the name of `Peterson, Mike` back to `Peterson, Mark` and delete the eleventh record in *Bookcust*. Print a report of all customers.

BOOKCUST

FIELD NAME	FIELD TYPE	FIELD WIDTH	KEY FIELD	FIELD DESCRIPTION
CSTNUMB	Number		Y	Customer number
CSTNAME	Text	16	N	Customer name
CSTADDR	Text	12	N	Customer address (street)
CSTCTY	Text	10	N	Customer city
CSTST	Text	2	N	Customer state
BRNUMB	Number		N	Number of customer's branch

BOOKORD

FIELD NAME	FIELD TYPE	FIELD WIDTH	KEY FIELD	FIELD DESCRIPTION
BKCODE	Number		Y	Book code
CSTNUMB	Number		Y	Number of customer requesting the book
REQDATE	Date/Time		N	Date book was requested

Queries

OBJECTIVES

1. Retrieve data from an Access database using query forms.
2. Create a query using multiple tables with the JOIN function.
3. Update data in an Access database using an update query.

M3.1   INTRODUCTION

In this module, you will learn to use the powerful Query-by-Example (QBE) feature of Access. We use this feature to create **queries.** A query is really just a question, the answer to which is found in the database.

When you define a query, Access creates and displays a **dynaset,** which is a dynamic set of selected records. You create a query by clicking on the Query button in the database window, and then click on the New button. (Alternatively, you could select New, Query from the File menu.) The form that displays on your screen is called a **query form** (Figure M3.1). There are two main parts of a query form: The top half of the window is where you can identify the tables and fields to be listed, and the bottom half is the QBE section. This is where you enter the data that you want to select. To create your query, simply fill in the form. A completed query form might look like the one in Figure M3.1, which shows the relationships between fields of different tables, and an "X" in the Show box that indicates which fields to print, and the criteria of the search.

In this module, you will see the various types of options you can use when you create queries. You will also see that you can use these query forms to update data in a database.

M3.2   TIPS

Before you begin using query forms, you should be familiar with the concepts discussed next.

Working with Character Data

You need to know about values you enter in text fields. Text fields in a query must be enclosed with quotation marks. If the text consists of one continuous word, Access will place the quotation marks around the text for you. If the text includes any punctuation (such as a comma) or spaces in the value, you must provide the quotation marks. Thus

SMITH

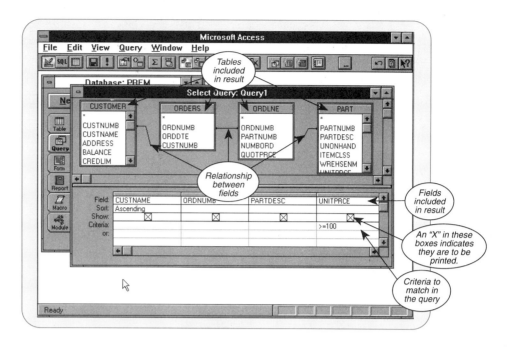

would not need to be enclosed in quotation marks, but

```
JOHN SMITH
```

would, since it contains a space. The correct format for the latter example would be

```
"JOHN SMITH"
```

In addition, conditions involving character fields are *case-insensitive*. This means that you do not need to worry about entering the right combination of uppercase and lowercase letters. If a name is stored in the database as Jones, you may search for jones and still find the record.

Restoring the PREMIERE PRODUCTS Database

In some of the examples in this module, you will make changes in the data of the *PREMIERE PRODUCTS* database. You can make any changes to this database that are indicated in the module and the exercises. Feel free to make any of the changes because you can return the *PREMIERE PRODUCTS* data to its original state at any time by exiting to either DOS or Windows and copying the file *premback.mdb* to replace *prem.mdb*. To restore the database using DOS, exit Access and Windows and copy the file from the C:> DOS prompt.

■ COPY A:PREMBACK.MDB A:PREM.MDB

Although you can take this action at any time, you should be sure to do it when you have completed the module so that the data is in the appropriate state for later modules.

M3.3   DATA DEFINITION

Example 1: Creation of a database.

Statement: Describe the layout of the *Sales Rep* table to the database management system (DBMS).

To create a table, you use the Database Design screen you saw in Module 2. Remember, Access prompts you to enter the name and physical characteristics of all the fields that make up the table.

Let's assume that you have described the following tables and fields for the *Premiere Products* database:

Slsrep:

Field Name	Field Type	Field Width	Key Field
Slsrnumb	Number		Yes
Slsrname	Text	15	No
Slsraddr	Text	22	No
Totcomm	Currency		No
Commrate	Number		No

Customer:

Field Name	Field Type	Field Width	Key Field
Custnumb	Number		Yes
Custname	Text	15	No
Address	Text	20	No
Balance	Currency		No
Credlim	Currency		No
Slsrnumb	Text	2	No

Part:

Field Name	Field Type	Field Width	Key Field
Partnumb	Text	4	Yes
Partdesc	Text	10	No
Unonhand	Number		No
Itemclss	Text	2	No
Wrehsenm	Number		No
Unitprce	Currency		No

Orders:

Field Name	Field Type	Field Width	Key Field
Ordnumb	Number		Yes
Orddte	Date		No
Custnumb	Text	4	No

Ordlne:

Field Name	Field Type	Field Width	Key Field
Ordnumb	Number		Yes
Partnumb	Text	4	Yes
Numbord	Number		No
Quotprce	Currency		No

Let's also assume that you have entered all the *Premiere Products* data already (this has been done on your data disk). Now let's move on to the remaining examples.

M3.4 SIMPLE RETRIEVAL

Example 2: Retrieval of certain columns and all rows.

Statement: List the number, name, and balance of all customers.

To indicate that a column is to appear in the result, we place a mark ("X") in the Show box. We do this by moving the cursor to the column and clicking in the Show box. First, you need to get a query form for the *Customer* table on the screen.

- Open the *PREM* database if is not already open.
- Click first on the Query button and then on the New button.

As with creating a new table, Access provides the choice of using Query Wizards or a custom New Query. For now, we will continue to use the custom format. (Later modules will address the use of Wizards).

- Click on the New Query button.

You should have a screen similar to Figure M3.2 now. Access first prompts for the first table to include in the query.

Figure M3.2

Creating a new query

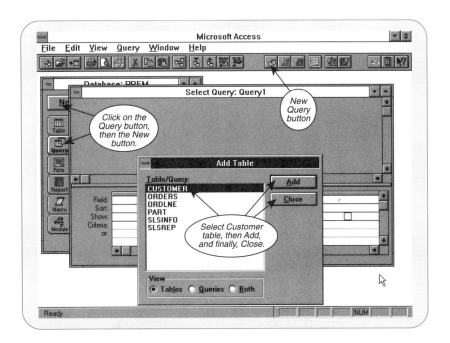

- Select the *Customer* table.
- Click on the Add button.
- Click on the Close button.

Access needs to know which fields you want to include in the query. You now identify the fields *Custnumb, Custname,* and *Balance.* To do so, with the cursor blinking in the first column, press Alt and the Down arrow (or click on the drop-down list icon), move to the field that you want to add, and press Enter. The field name is entered in the column with an ''X'' in the Show box. The cursor automatically moves to the next column.

- Press Alt and the Down arrow.
- Use the Down arrow to move the cursor to the *Custnumb* field.
- Press the Enter key.
- Repeat the process to select the *Custname* field.
- Repeat the process to select the *Balance* field.

Your screen should now look like the one shown in Figure M3.3. To see the results of the query, click on the Datasheet button in the top-left corner, or Select Datasheet from the View menu.

- Select Datasheet from the View menu.

Figure M3.3

Completed query

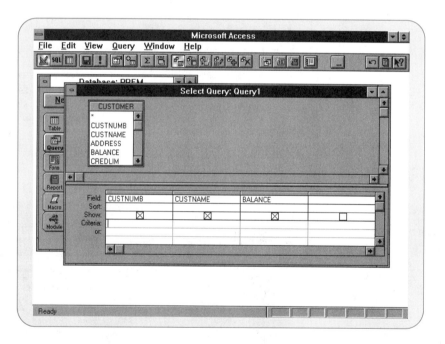

You should now see the resulting query dynaset on your screen (Figure M3.4). Close this query window either by clicking on the Control-menu box in the upper-left corner and then clicking on Close, or select Close from the File menu.

- Select Close from the File menu.
- Answer *No* to the Save Query? prompt.

Example 3: Retrieving all columns and all rows.

Statement: List the complete *Orders* table.

The only difference between this example and the previous one is that this one has all columns. You could select each column individually as you did in the previous

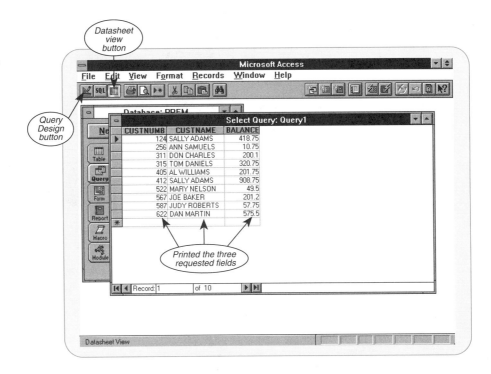

column. Since the *Orders* table has only three columns, this would not be a problem. In tables with many columns, however, this could become tedious. Fortunately, there is a shortcut. When you select the * under the name of the table, Access will use all columns in the table and will list it as ORDERS.*. The asterisk is a wildcard, which represents all field names. Let's use this shortcut.

- Click on the Table icon.
- Double click on the *Orders* table, or click on the *Orders* table, then click on the Open button.
- Select New, then Query from the File menu.
- Click on the New Query button to create a custom query.
- Double click on the * under the *Orders* table name.

You should now see the screen shown in Figure M3.5. The window containing the query fields will indicate ORDERS.*, meaning that all fields from the *Orders* table will be shown.

- Click on the Datasheet icon, or select Datasheet from the View menu.

Figure M3.6 shows the result of the query dynaset. Notice that all rows and all columns are included in the query results. When you have finished, close this query window by clicking on the Control-menu box in the upper-left corner and then clicking on Close, or select Close from the File menu.

- Select Close from the File menu.
- Answer *No* to the Save Query? prompt.

M3.5 CONDITIONS

Example 4: Use of a condition.

Statement: What is the name of customer 124?

Figure M3.5

Displaying all fields
for every record

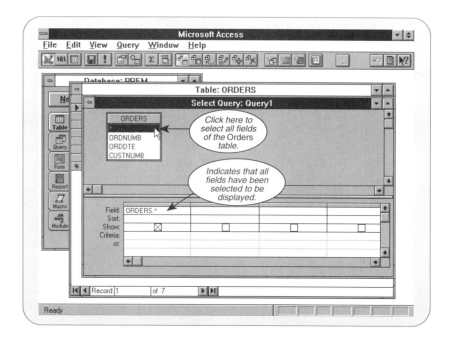

Figure M3.6

Dynaset that
includes all of the
records

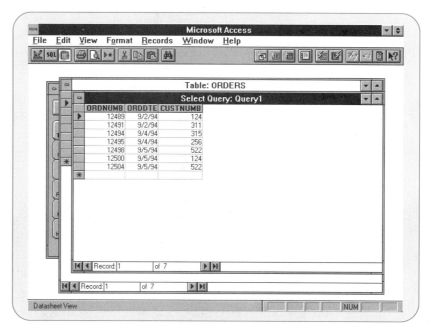

In this query, we want the value in the *Custname* field. We only want those rows in which the value in *Custnumb* is 124. To implement this restriction, we simply place the 124 in the *Custnumb* column.

- Select the *Customer* table. (Click on the Table icon, then on Customer.)
- Select New, then Query from the File menu.
- Click on the New Query button to create a custom query.
- Press Alt and then the Down arrow to select the first field.
- Use the Down arrow to move to the *Custnumb* field.
- Press the Enter key.

This selects the *Custnumb* field to be shown and moves to the second column. We do not want this field shown; rather, we want to use it to apply a criteria.

- Press the Left arrow to move back to the *Custnumb* column.
- Use the Down arrow to move down to the Show option.
- Press the space bar, or click on the box to remove the "X".
- Press the Down arrow to move to the Criteria option.
- Enter 124
- Press the Enter key to move to the second column.
- Move up to the field row.
- Press Alt and the Down arrow to select the second field.
- Use the Down arrow to move to the *Custname* field.
- Press the Enter key.

The resulting query form is shown in Figure M3.7. Notice that there is an "X" in the *Custname* field but not in the *Custnumb* field. In addition, the criteria option under *Custnumb* specifies 124.

- Select Datasheet from the View menu.

Figure M3.7

Specifying conditions in a query

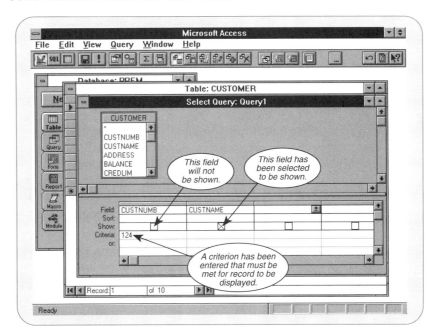

Figure M3.8 shows the resulting query dynaset. Notice that only the *Custname* column is shown and that only customer 124, "Sally Adams", is shown. When you have finished, close this query window by clicking on the Control-menu box in the upper-left corner and then clicking on Close, or select Close from the File menu. Then close the *Customer* table, also.

- Select Close from the File menu. (This will close the query window.)
- Answer *No* to the Save Query? prompt.
- Select Close from the File menu. (This will close the table windows.)

Conditions in query forms can involve any of the operators shown in Table M3.1. The Equal to operator (=) does not need to appear; that is, if no operator is included, Access assumes you want Equal to. That is why you could type the condition as 124 rather than =124.

Example 5: Use of a compound condition.

Statement: List the descriptions of all parts in warehouse 3 and for which there are more than 100 units on hand.

Figure M3.8

Dynaset of one
specific record

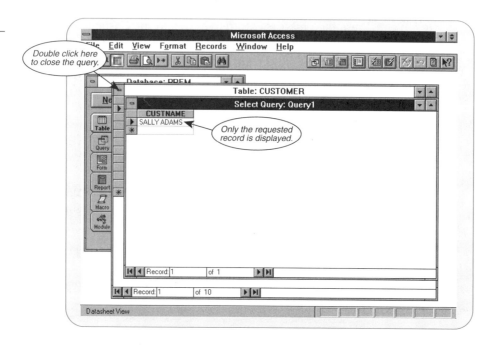

Not only can you use the operators shown in Table M3.1, but you can also combine conditions to create *compound* conditions. In many situations, you would do so by placing the word AND or the word OR between the conditions. In Query-by-Example (QBE), it is done a little differently. If you want records that meet one criterion *and* another criterion, as is the case in this example, place the conditions on the same line. If you are looking for records that meet one criterion *or* another criterion, place the criteria on different lines. The next few examples illustrate this rule.

- Select the *Part* table. (Click on the Table icon, then on Part).
- Select New and then Query from the File menu.
- Click on the New Query button to create a custom query.
- Press Alt and the Down arrow to select the first field.
- Use the Down arrow to move to the *Partdesc* field.
- Press the Enter key. (This will place an "X" in the Show box and move to the next field.)
- Press Alt and the Down arrow to select the next field.

Table M3.1

Special operators
used with queries

Operator	Meaning
=	Equal to.
>	Greater than.
<	Less than.
>=	Greater than or equal to.
<=	Less than or equal to.
?	Wildcard—stands for any single character.
*	Wildcard—stands for any series of characters.
And	Used to meet both of two criteria.
Or	Used to meet either of two criteria.
Not	Must be used before another criterion. Will then find items that do not match the criterion.
Between	Used to specify a range of values.
In	Used to specify a list of values.
Like	Used to search for patterns in text fields.

- Use the Down arrow to move to the *Unonhand* field.
- Press Enter.

This will place an "X" in the Show box and move to the next field. Since we need to enter a criteria and do not wish to print this column, we need to go back to the column and make the changes.

- Use the Left arrow to move to the *Unonhand* field.
- Move down to the Show box and press the space bar to remove the "X".
- Press the Down arrow to go to the criteria line.
- Enter >100
- Use the Right arrow to move to the next column. Move up to the Field line.
- Press Alt and the Down arrow to select the next field.
- Use the Down arrow to move to the *Wrehsenm* field.
- Press the Enter key.

This will place an "X" in the Show box and move to the next field. Since we need to enter a criteria and do not wish to print this column, we need to go back to the column and make the changes.

- Use the Left arrow to move to the *Wrehsenm* field.
- Move down to the Show box and press the space bar to remove the "X".
- Press the Down arrow to go to the criteria line.
- Enter 3
- Press the Enter key.

Figure M3.9 shows the query form.

- Select Datasheet from the View menu.

Figure M3.9

Using AND with a compound condition

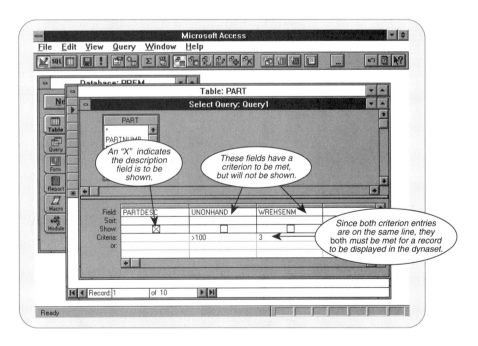

The resulting query dynaset is shown in Figure M3.10. Notice that only the Part-desc column is shown and that there are two records that meet both the Wrehsenm and Unonhand criteria.

If you want to combine conditions with OR, the conditions must go on separate lines. As an example, let's find the part description of all parts that are in warehouse 3

Figure M3.10

Dynaset of a query
using AND

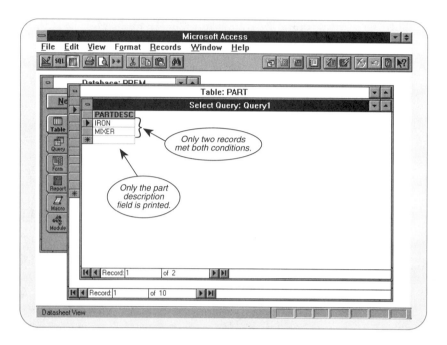

or the parts which exceed 100 units on hand. We'll put the condition for units on hand on the first line in the query form and the condition for warehouse number on the second. In addition, let's check the Show box for both the *Unonhand* and *Wrehsenm* fields.

- Select Query Design from the View menu.
- Use the arrow to move to the *Unonhand* column.
- Move to the Show line and press the space bar.
- Use the arrow to move to the *Wrehsenm* column.
- Move to the Show line and press the space bar.
- Press the Down arrow to move to the first criteria row.
- Press the Delete key.
- Press the Down arrow to move to the second criteria (or:) row.
- Enter 3

The query form as it should now appear is shown in Figure M3.11. As you can see, the criteria for *Unonhand* and *Wrehsenm* are on separate lines. This means that either of the criteria may be met in order for the record to be displayed.

- Select Datasheet from the View menu.

The resulting dynaset is displayed in Figure M3.12. With this query, there are four additional records that now meet the criteria.

When finished, close this query window by clicking on the Control-menu box in the upper-left corner and then clicking on Close, or select Close from the File menu.

- Select Close from the File menu.
- Answer *No* to the "Save Query?" prompt.

This closes the query window. You need to also close the *Parts* table because the next example will use another table.

- Select Close from the File menu.

Example 6: Use of computed fields.

Statement: List the number, name, and available credit for all customers who have at least an $800 credit limit.

Figure M3.11

Using OR with a
compound
condition

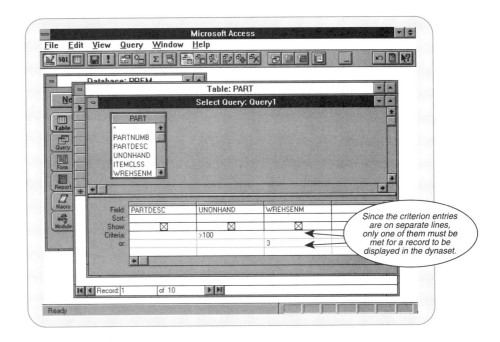

*Since the criterion entries
are on separate lines,
only one of them must be
met for a record to be
displayed in the dynaset.*

Figure M3.12

Dynaset of a query
using OR

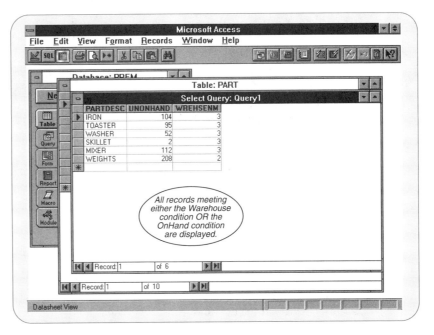

*All records meeting
either the Warehouse
condition OR the
OnHand condition
are displayed.*

There is no field for available credit in the *Customer* table. We can compute it, however (available credit = credit limit minus balance). We call such a field a *computed* field. To include computed fields in queries, you need to use expressions. Before we do, however, let's try to find just the number and name of all customers who have at least an $800 credit limit.

- Select the *Customer* table. (Click on the Table icon and then on Customer.)
- Select New and then Query from the File menu.
- Click on the New Query button to create a custom query.
- Press Alt and Down arrow to select the first field.
- Use the Down arrow to move to the *Custnumb* field.
- Press the Enter key.
- Press Alt and Down arrow to select the next field.

- Use the Down arrow to move to the *Custname* field.
- Press the Enter key.
- Press Alt and Down arrow to select the next field.
- Use the Down arrow to move to the *Credlim* field.
- Press the Enter key.

This will place an "X" in the Show box for each of the three fields. Since you do not wish to print the Credlim column and you need to enter a criteria, you need to go back to the column and make the changes.

- Use the Left arrow to move to the *Credlim* field.
- Move down to the Show box and press the space bar to remove the "X".
- Press the Down arrow to go to the criteria line.
- Enter >=800

Figure M3.13 shows the query.

- Select Datasheet from the View menu.

Figure M3.13

Computed fields in a query

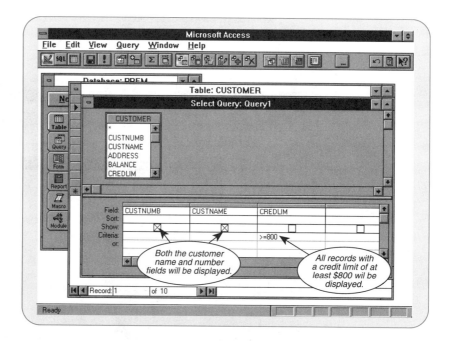

The results of the query are shown in Figure M3.14. Now you need to be able to identify values for the available credit, which is the balance minus credit limit. You do so by creating calculated fields. To enter a calculated field, enter an expression that combines names of the fields. This example displays the available credit, which is entered as [CREDLIM]-[BALANCE].

To enter the new column, you need to insert it after the customer name and before the credit limit column.

- Select Query Design from the View menu.
- Move the cursor to the Credlim column.
- Select Insert Column from the Edit menu.

This inserts a column to the left of the Credlim column.

- Use the arrows to move to the new column's field name line.
- Enter the expression using brackets [ ] around field names.
  Enter as: =[CREDLIM]-[BALANCE]
- Press the Enter key.

Figure M3.14

Dynaset for a
computed field

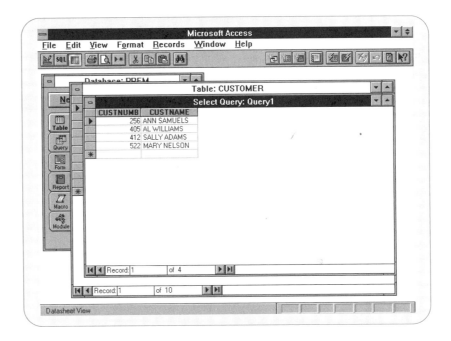

As Figure M3.15 shows, Access places the word *Expr1:* in front of the expression. If you leave the field name as it is, it would use *Expr1* as the column heading for the results. You can replace the *Expr#* by either moving the cursor to the field name and editing the name or by placing the name you want prior to entering the field names. In this case, you could have entered AVAIL CREDIT:[CREDLIM]-[BALANCE] to use AVAIL CREDIT as the column heading. Since the default Expr1 has been used, you can edit the heading.

- Move to the field name and replace Expr1 with AVAIL CREDIT
- Press the Enter key.
- Select Datasheet from the View menu.

The results are on your screen (Figure M3.16). Notice that the computation does indeed give the credit limit minus the balance. You are not restricted to subtraction in

Figure M3.15

Computed fields in
a query

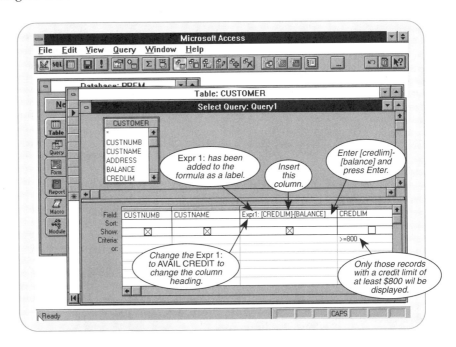

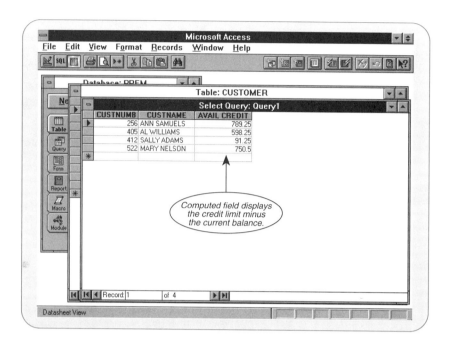

computations. You can use addition (+), multiplication (*), or division (/). Also, you can include parentheses in your computations.

Example 7: Use of a wildcard.

Statement: List the number, name, and address of all customers who live in Grant.

If the cities were stored in a separate field, this query would be similar to queries we have seen before. We would just enter *GRANT* in the city column. In this database, however, street, city, and state are all combined into the single field called *Address.* Thus, you need to find those records that contain *GRANT* somewhere within the address. To do so, we use one of two special symbols called wildcards.

The first of the two wildcards is the asterisk (*). It represents any collection of characters. Thus *GRANT* represents any collection of characters followed by GRANT followed by any other collection of characters. The other wildcard symbol is the question mark (?). This represents any individual character. Thus T?M represents the letter T followed by any single character followed by the letter M. In this problem, the asterisk is appropriate.

We will demonstrate the use of wildcards by modifying the current query. For this query, we want to delete the Avail Credit and Credlim columns and then add the Address column.

- Select Query Design from the View menu.
- Move to the Avail Credit column.
- Select Delete Column from the Edit menu.

This will delete the available credit column and place the cursor in the credit limit column.

- Select Delete Column from the Edit menu.

This deletes the Credit Limit column. You now want to add the Address column.

- Press Alt and the Down arrow to select the field name.
- Select the *Address* field.
- Press the Enter key.

Move the cursor back into the Address column, and then move down to the Criteria line.

- Enter `*grant*`
- Press the Enter key.

Because you did not provide specific details, Access automatically changed the field to *Like''*grant*''* (See Figure M3.17). Although you could have entered ='*grant*' in the field, letting Access automatically enter the proper syntax is simpler and faster.

- Select Datasheet from the View menu.

The resulting dynaset is displayed in Figure M3.18. As requested, all three records have "grant" in the address field.

Figure M3.17

Using wildcards in text searches

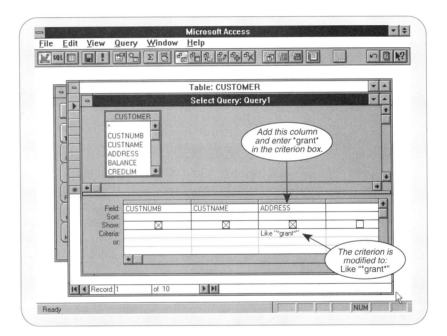

Figure M3.18

Dynaset for records meeting a criterion with a wildcard

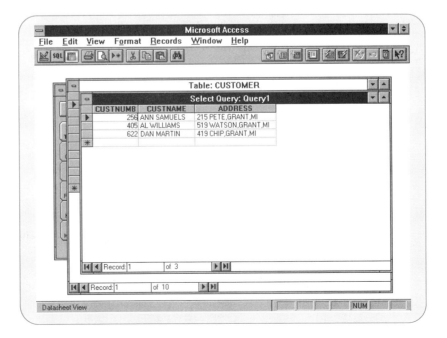

M3.6   SORTING

Example 8: Sorting.

Statement: List the name and address of all customers. The report should be ordered by customer number.

Since you already have the current query available, let's begin by modifying the current query. First delete the criterion for Grant in the Address field, set the *Address* field to Show, and then set the customer number as the column to sort.

- Select Query Design from the View menu.
- Use the arrows to move to the Criterion row for the Address field.
- Press Delete, and then Enter.
- Use the Left arrow to move back to the Custnumb column.
- Move the cursor to the Sort row.
- Press Alt and the Down arrow to select from the options.
- Select Ascending.
- Press the Enter key.

The current Query Design screen is displayed in Figure M3.19. Now switch to the datasheet view.

- Select Datasheet from the View menu.

Figure M3.19

Sorting a listing

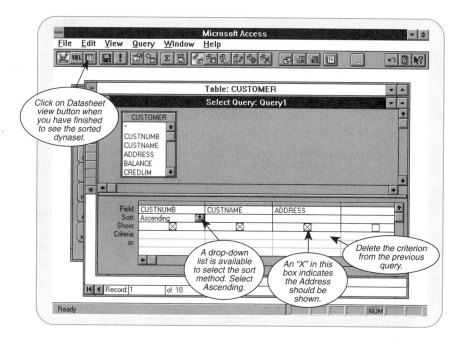

The resulting dynaset, one that shows all records, now lists all customers in ascending order by customer numbers (Figure M3.20). To sort on another field, perhaps by customer's name, you need only to remove the sort setting from the customer number column and select ascending order for the customer name column.

- Select Query Design from the View menu.
- Move to the Sort row of the Custnumb column.
- Press Delete, and then press the Enter key.
- Move to the Sort row of the Custname column.

Figure M3.20

Sorted query
dynaset

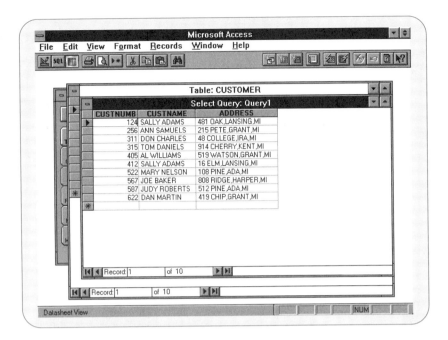

- Press Alt and the Down arrow to select from the sort options.
- Select Ascending.
- Press Enter.

Your query should now look like Figure M3.21.

- Select Datasheet from the View menu.

Figure M3.21

Sorting on the
name field

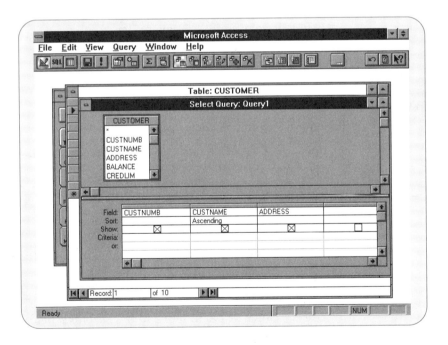

Figure M3.22 shows the results. Notice that the names are stored as first name followed by last name; therefore, they are actually sorted on first name. In order to sort on the last name followed by first name, you would have to enter the data as "Lastname, Firstname", or better yet, use separate fields for first and last name.

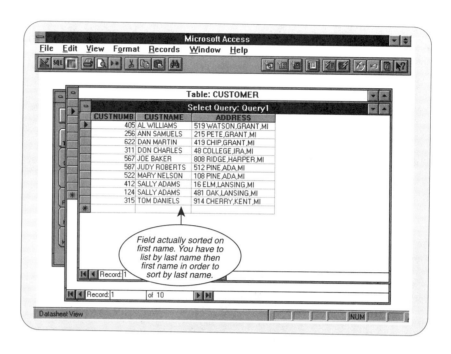

When finished, close this query window by clicking on the Control-menu box in the upper-left corner and then clicking on Close, or select Close from the File menu.

■ Select Close from the File menu.
■ Answer *No* to the Save Query? prompt.

This closes the query window. You need to also close the *Customer* table because the next example will use another table.

■ Select Close from the File menu.

M3.7   BUILT-IN FUNCTIONS

Example 9: Use of the Count command.

Statement: How many parts are in item class HW?

Most database systems, including Access, support the same collection of built-in functions: Count, Sum, Avg, Max, and Min.

In this case, Count is the appropriate operation. We can put Count in any column. (It doesn't matter whether we count part numbers, part descriptions, or anything else. We should get the same answer.) Let's use the *Part* table and put it in the Partnumb column.

■ Select the *Part* table. (Click on the Table icon and then on Part.)
■ Select New and then Query from the File menu.
■ Click on the New Query button to create a custom query.
■ Press Alt and the Down arrow to select the first field.
■ Use the Down arrow to move to the *Partnumb* field.
■ Click on the Totals button. (The Σ symbol will insert a Total row.) You could also select Totals from the View menu.
■ Press Alt and the Down arrow to select the *Count* function.
■ Press the Enter key.

The query form should now look like Figure M3.23.

■ Select Datasheet from the File menu.

Figure M3.23

Using the Totals command in queries

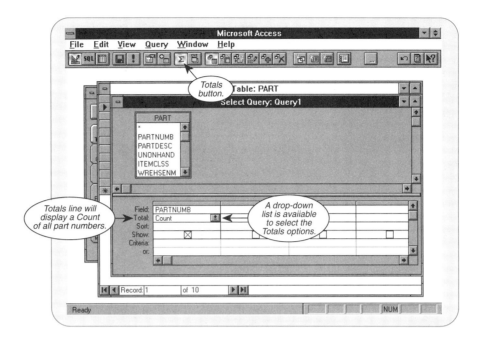

The result of the Count function is that it displays the count of all the records that do not have a null value in *Partnumb* (Figure M3.24). To extend the Count function, let's continue with this example. The requested information was the count of records with an item class of HW. To do this, we need to add another column with a "Where" condition as Figure M3.25 displays. To interpret this query form, it could be read as: Show the count of all partnumb records where itemclss values are equal to "HW".

■ Select Query Design from the View menu.
■ Move the cursor to the second column, on the field name row.
■ Press Alt and the Down arrow to select the *Itemclss* field.
■ Press the Enter key.
■ Move to the Total row, press Alt and the Down arrow to select Where. (You will have to scroll down to select Where.)

Figure M3.24

Using the Count command in queries

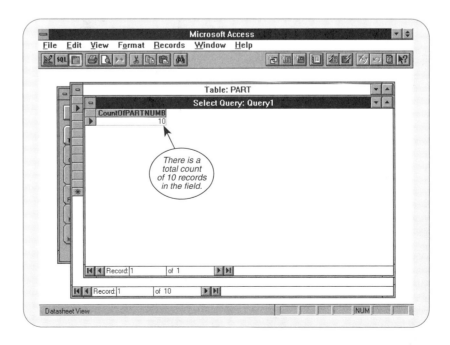

Figure M3.25

Using a Where
condition with a
Totals command

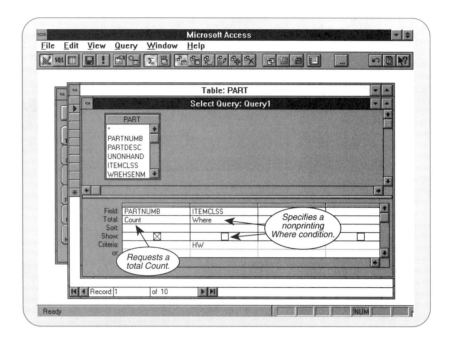

- On the Criterion row, enter the text HW.
- Press the Enter key.
- Select Datasheet from the File menu.

The results are shown in Figure M3.26 to indicate there are four such parts. Close this query window by clicking on the Control-menu box in the upper-left corner and then clicking on Close, or select Close from the File menu.

- Select Close from the File menu.
- Answer No to the Save Query? prompt.

This closes the query window. You need to also close the *Parts* table because the next example will use the *Customer* table.

- Select Close from the File menu.

Figure M3.26

Results of a Where
condition with a
Totals command

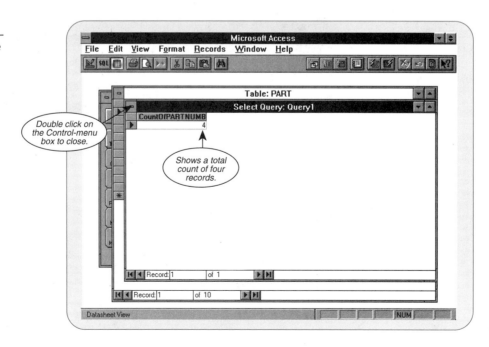

Example 10: Use of the built-in function Sum.

Statement: Count the number of customers and find the total of their balances.

Here, we want to use both Count and Sum. Let's put Count in the *Custnumb* column, and Sum in the *Balance* column since that is the column we want totalled.

- Select the *Customer* table. (Click on the Table icon and then on Customer.)
- Select New, then Query from the File menu.
- Click on the New Query button to create a custom query.
- Press Alt and the Down arrow to select the *Custnumb* field.
- Click on the Totals button. (The Σ symbol will insert a Total row.)
- Press Alt and the Down arrow to select the *Count* function.
- Press the Enter key.
- Move the cursor to the field name of the second column.
- Press Alt and the Down arrow to select *Balance*.
- Move to the Total row and press Alt and the Down arrow to Sum.
- Press the Enter key.

The result should look like Figure M3.27.

- Select Datasheet from the File menu.

Figure M3.27

Using the Sum totals

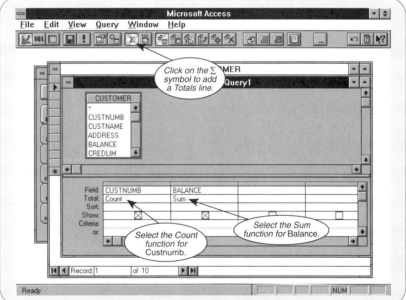

The results are shown in Figure M3.28. There are ten customers and the total of their balances is $2,944.80. In addition to Count and Sum, you could use other Access functions in a similar fashion.

Example 11: Use of the built-in function Sum.

Statement: Find the number of customers represented by each sales rep as well as the total of their balances.

You need the same functions as in the previous example. In addition, you need to specify that the records are to be *grouped* for the calculations. You want the calculations performed for the group of customers represented by sales rep 3, the group represented by sales rep 6, and the group represented by sales rep 12. For-

Figure M3.28

Using the Sum
totals

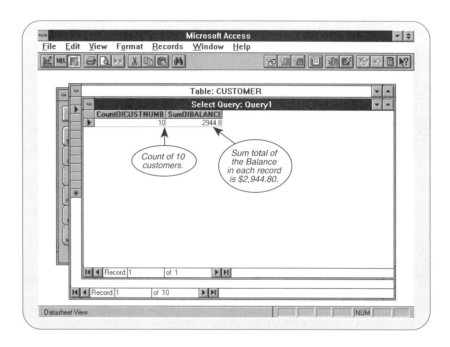

tunately, requesting such grouping is a simple process. You just need to insert a column for the sales rep number used for grouping—in this case, *Slsrnumb*.

- Use the arrow keys to move to the *Custnumb* column.
- Select Insert Column from the Edit menu.
- Use the Alt and the Down arrow to select the *Slsrnumb* column.
- Press the space bar on the Show row to remove the ''X''.
- Press the Enter key.

The query form should now look like Figure M3.29, and the results are shown in Figure M3.30. Notice that they are exactly what we wanted: The three sales reps 3, 6, and 12 are each listed with the number of customers and their balances.

Figure M3.29

Grouped sum
totals

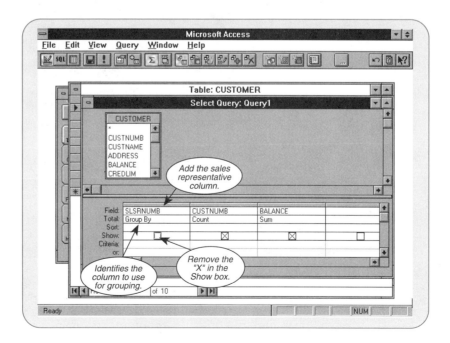

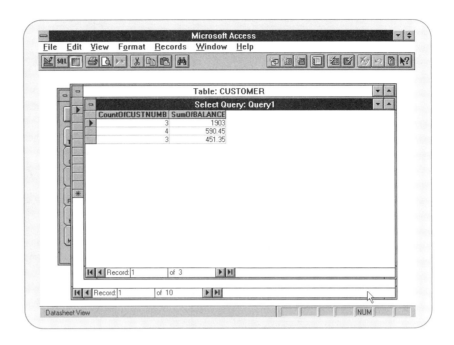

Since this set of queries is now finished, you need to close the query and the table. Close this query window by clicking on the Control-menu box in the upper-left corner and then clicking on Close, or select Close from the File menu.

■ Select Close from the File menu.
■ Answer *No* to the "Save Query?" prompt.

This closes the query window. You need to close the *Customer* table also.

■ Select Close from the File menu.

M3.8   JOIN

Example 12: Joining two tables.

Statement: List each customer's number and name, along with the number and name of the corresponding sales rep.

This query cannot be satisfied using a single table. The customer name is in the *Customer* table, whereas the sales rep name is in the *Slsrep* table. You need to **join** the tables. To do so, we need to add both tables to the query. With all tables and queries closed, and the *Prem* database open, you will create a new query.

■ Click on the Query object button and then New to create a new query.
■ Click on New Query to create a custom query.

Access first needs to know what tables to add to the query. In this example, you will need both the *Customer* and the *Slsrep* tables. Figure M3.31 shows the dialog box with the list of tables. We will add the *Customer* table and then the *Slsrep* table. This adds the tables to the top half of the query design form.

■ Click on *Customer* and then on the Add button.

This added the *Customer* table to the top half of the form.

■ Click on *Slsrep* and then on the Add button.

Figure M3.31

Adding tables to a
query

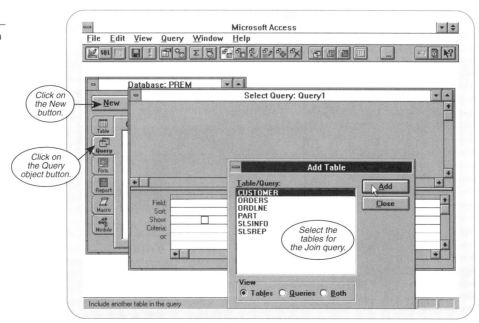

This added the *salesrep* table to the query.

■ Click on the Close button to end adding tables to the query.

Note: If you need to add additional tables after the query is designed, you can add more tables by selecting Add Table from the Query menu, and selecting from the list of tables.

Both tables should now be on the query form (Figure M3.32). (You could have selected the tables in reverse order.) Next you need to relate the two tables. Connect the two tables by using the mouse and dragging a field from one table to the matching

Figure M3.32

A query with
multiple tables

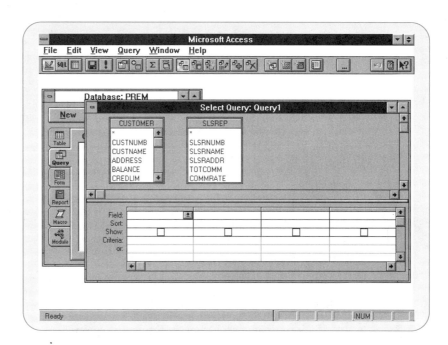

field in the other table. As you drag the mouse, the arrow pointer changes from the arrow to either a small icon of a field or an icon representing the *not* symbol (meaning that you are not in a proper location to make the connection). When you release the mouse button, a *join* line is drawn between the two matching fields/tables.

For example, we want to tie the *Slsrnumb* field in *Customer* table with the *Slsrnumb* field in the *Slsrep* table.

- Point (with the mouse) to the *Slsrnumb* field in the *Customer* table.

Note: You may need to scroll down the list of fields to reach this field.

- Hold down the left button and drag the mouse to the *Slsrep* table.

Notice that as you drag the mouse to the right, the arrow changes to a small circle with a line (the standard icon for *not*). This means that you cannot release the mouse button at this point. As you move the mouse to a valid field in the second table, the pointer changes to a small rectangle, symbolizing a field.

- Move the mouse to the *Slsrnumb* field in the *Slsrep* table and release the mouse button.

Note: A join line should now indicate that the two fields are linked (Figure M3.33). If you did not make the connection as shown, try again. If the join line is in the wrong location, you can click on the line to select it and then press the Delete key to remove the line.

The results are shown in Figure M3.33. Notice the *Slsrnumb* fields in the two table boxes are joined. Access interprets this to mean that *the value in the Slsrnumb column of the* Customer *table must match the value in the Slsrnumb column of the* Slsrep *table*.

To add fields to the query in previous examples, we selected fields from the drop-down lists. Another method of adding fields to the query is to drag the field name from the table box to the field name in the lower portion of the screen. We want to see the customer number, customer name, sales rep number, and sales rep name.

- Point to the *Custnumb* field in the *Customer* table and drag the icon to the field name row of the first column.

Figure M3.33

Joining tables in a query

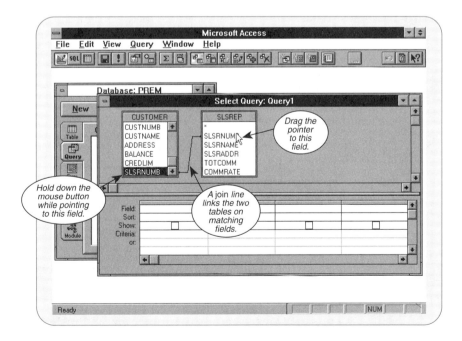

- Point to the *Custname* field in the *Customer* table and drag the icon to the field name row of the second column.
- Point to the *Slsrnumb* field in the *Customer* table and drag the icon to the field name row of the third column.
- Point to the *Slsrname* field in the *Slsrep* table and drag the icon to the field name row of the fourth column.

Your screen should now match Figure M3.34. To view the datasheet, click on the datasheet icon.

- Click on the View Datasheet icon.

Figure M3.34

Adding fields to a query

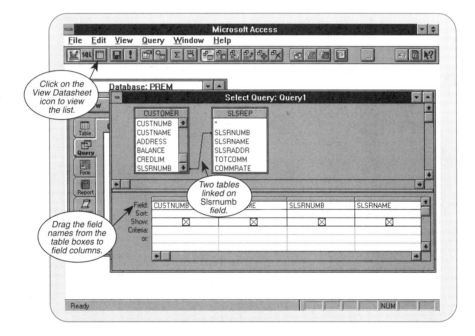

The datasheet view (Figure M3.35) includes data from both the customer and sales rep tables, linked by the matching sales rep numbers of each record.

The next query is similar to the current one, which saves you some work since you modify the current query.

- Click on the Query Design icon.

Example 13: Restricting the rows in a Join.

Statement: For each customer whose credit limit is $800, list his or her number and name, along with the number and name of the corresponding sales rep.

The only difference between this query and the previous one is that there is an extra restriction: the credit limit must be $800. To accommodate this, we put 800 in the *Credlim* column. (**Note:** The following steps assume the images from the previous query are still on your screen. If not, you will need to make the necessary adjustments to the process.)

First, we need to add the *Credlim* column, and the $800 criterion.

- Drag the *Credlim* field from the *Customer* table to the field name of the fifth column.
- Press the Down arrow to the Show row, and press the space bar to remove the ''X'' because you do not need to print this column.

Figure M3.35

A query of two tables

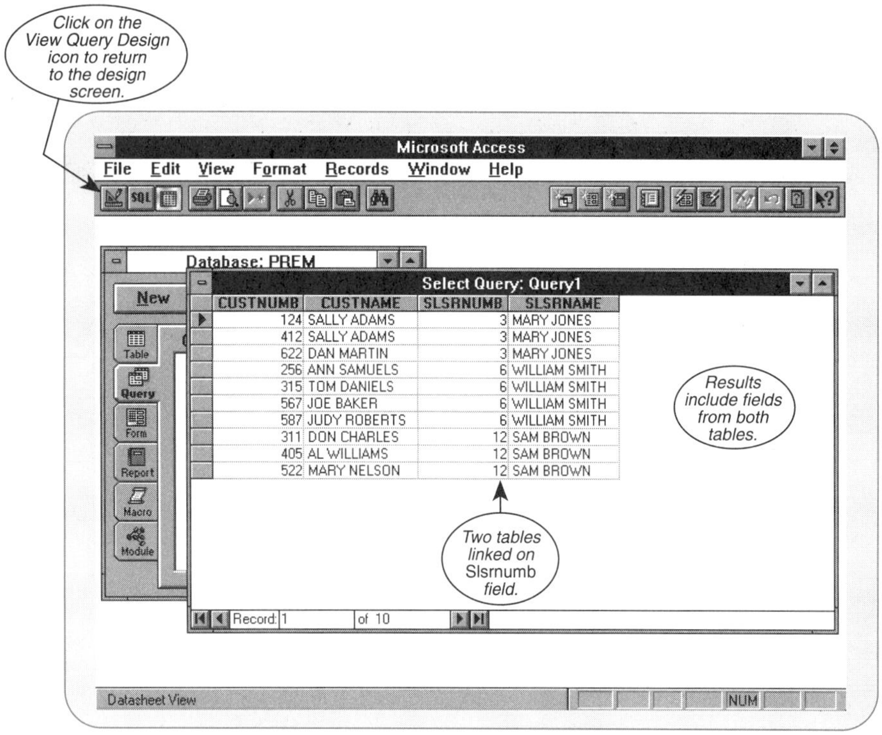

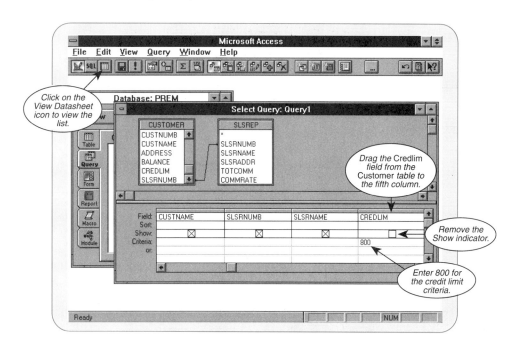

- Press the Down arrow to the Criteria row.
- Enter 800

The query design form should now match Figure M3.36. Let's look at the result.

- Click on the View Datasheet icon.

Your results should resemble ones shown in Figure M3.37. Notice that only the customers with an $800 credit limit are included.

- Click on the Query Design icon.

Figure M3.36

A query form with two tables and a condition

Figure M3.37

A query with two
tables and a
condition

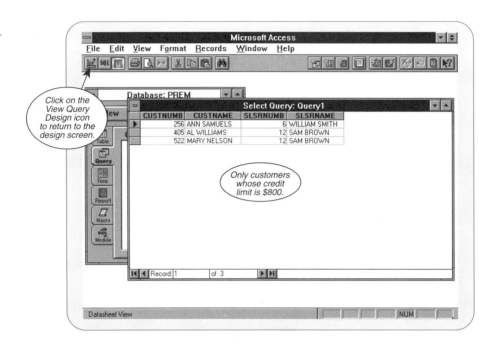

Example 14: Joining more than two tables.

Statement: For each order, list the sales rep number, the sales rep name, the customer number, the customer name, the order number, and the order date.

This query involves three tables, *Slsrep*, *Customer*, and *Orders*, so we add the `Orders` table to the query design screen. First, you need to remove the Credlim column.

- Click anywhere in the Credlim column.
- Select Delete Column from the Edit menu.

Next, you need to add the *Orders* table to the query.

- Select Add Table from the Query menu.

A dialog box lists the tables in the *Prem* database.

- Click on the *Orders* table.
- Click on the Add button.
- Click on the Close button.

The *Customer* and the *Slsrep* tables are linked by the *Slsrnumb* field. Similarly, you need to relate the *Customer* and *Orders* tables by indicating that the values in the *Custnumb* columns in both tables must match. As you would expect, you do so by dragging the *Custnumb* field in the *Customer* table to the *Orders* table.

- Drag the *Custnumb* field name from the *Customer* table to the *Orders* table.

The join line between the two tables passes under the *Slsrep* table and indicates the link between *Customer* and *Orders* table. Next, you need to add the Order number and the Order date columns to the lower part of the query screen.

- Drag the *Ordnumb* and the *Orddte* fields from the *Orders* table to the fifth and sixth columns.

Figure M3.38 shows that the *Orders* table is added, along with the two new columns. Although the first and second columns have scrolled off the left side of the screen, you may use the scroll bar arrows to view the entire form.

Figure M3.38

A query form with
three tables

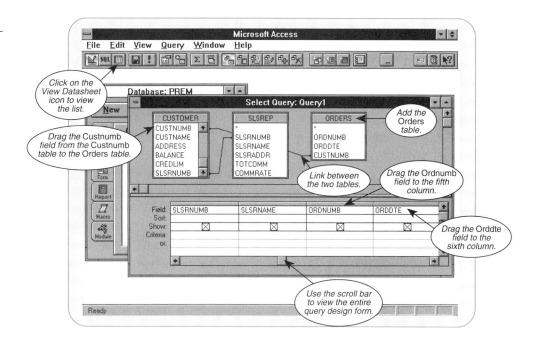

■ Click on the View Datasheet icon.

The results are shown in Figure M3.39. We now have a query with fields from three separate tables.

You can print an instant report. To do this, select Print from the File menu. To preview the report, however, you may select Print Preview instead.

■ Select Print Preview from the File menu.

Figure M3.40 shows the preview with a small view of what the report would look like. To see the report on the screen (Figure M3.41), you can zoom in.

■ Click on the Zoom button.

Figure M3.39

A query form with
data from three
tables

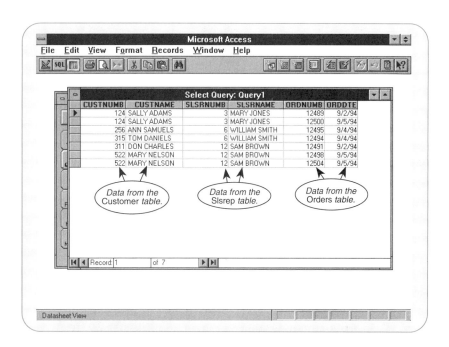

Figure M3.40

Print Preview of a
query report

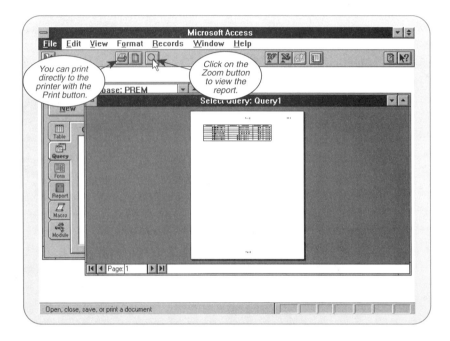

Figure M3.41

Print Preview of a
query report in the
Zoom view

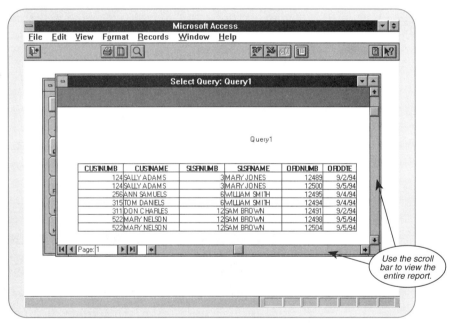

If you have a printer connected and set up properly, you can print the report to the printer, by clicking on the Print button. Since this section is finished, close the print preview and the query.

■ Select Close from the File menu.
■ Click on No in response to the Save query? prompt.

M3.9   UPDATE

Example 15: Changing existing data in the database.

Statement: Change the name of customer 256 to ANN JONES.

In the previous sections, you learned how to use queries to precisely select the data you want—even from multiple tables. These queries in Access are called *select queries*. You can also use *action queries* or *update queries* to update data in the database. In this example, you want to change the value in one of the fields. You only want to make this change on the record on which the value of *Custnumb* is 256. To restrict the condition to customer 256, you enter a criteria as you have done before. To indicate the change to be made, you change the query mode to Update and enter the new value.

- Click on the Table object button.
- Select the *Customer* table.
- Click on the Open button.
- Select New and then Query from the File menu.
- Click on the New Query icon to create a custom query.
- Drag the *Custnumb* field to the first column.
- Drag the *Custname* field to the second column.
- Enter 256 in the Criteria row under the Custnumb column.

So far, you have been using the select query. Now we want to change the query to an *update query* next.

- Select Update from the Query menu.

When you change to an update query, Access changes the title bar to an "Update to" line in the QBE lower portion of the form. Use this line to specify any changes you wish to make. Since we have already entered our customer number = 256 criteria, any changes will only take place to that record. Therefore, you can enter the new name as shown in Figure M3.42. The quotation marks are necessary because the entry is a text string.

- Enter ANN JONES in the Update To: row of the *Custname* column.

To run your query, you may choose the Run from the Query menu, or click the Run button on the tool bar. Access first checks to see how many records will change

Figure M3.42

Using a query form to update a database

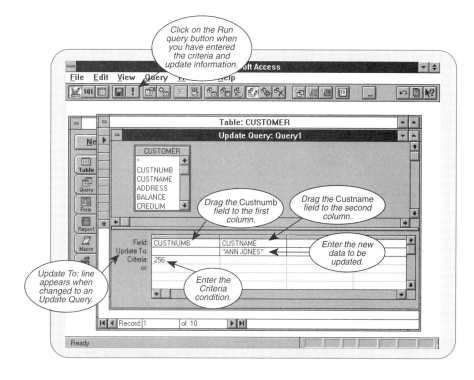

Figure M3.43

Update query
verification dialog
box

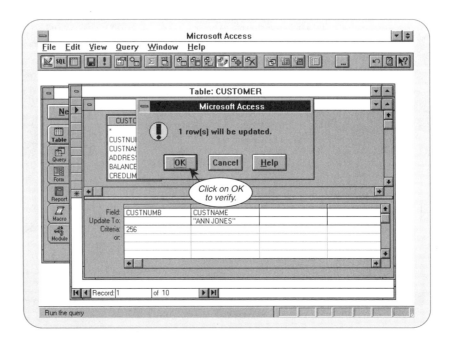

based on the selection criteria. You will see a dialog box like the one in Figure M3.43. Since you know that there is only the one record you want to change, verify the change.

- Click on Run Query button to make the change.
- Click on OK in the update dialog box.

You may check that the update actually took place by viewing the *Customer* table.

- Select window 2, Table: Customer from the Window menu.

As expected, Figure M3.44 shows that customer number 256 has been changed to Ann Jones. Let's return to the query design screen now.

- Select window 3, Update query: Query 1 from the Window menu.

We will continue using a query of the *Customer* database so we can save time by modifying this query.

Example 16: Changing data in the database on the basis of a compound condition.

Statement: For each customer with a $500 credit limit whose balance is less than $500, increase the credit limit to $800.

It is perfectly permissible to use a compound condition in an update query. Here, you need to enter <500 in the Balance column and 500 in the Credlim column. You also need to enter the Update To: expression to 800 in the Credlim column.
First, delete the current query.

- Select Clear Grid from the Edit menu. (This clears all entries in the query rows.)

You want to add the fields with criteria to be matched first. Then enter the criteria to be matched and, finally, enter the instructions for the update that is to occur.

- Drag the *Credlim* and *Balance* fields to the first and second columns.
- Enter 500 in the Criteria row under the Credlim column.
- Enter <500 in the Criteria row under the Balance column.
- Enter 800 in the Update To: row under the Credlim column.

Figure M3.44

Updated *Customer*
table

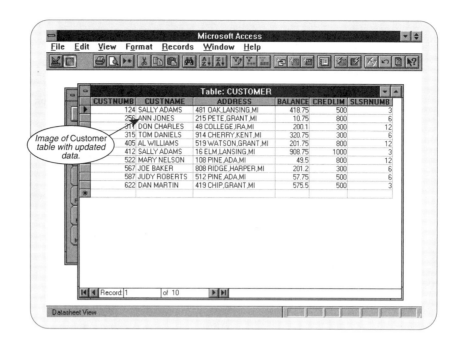

Your screen should now match Figure M3.45. To translate this screen, it reads; *For all records with a credit limit of 500, and a balance less than 500, update the Credlim to 800.*

■ Click on the Run button.

Access checks the records for a match of the criteria, and a dialog box requests verification that two records will be updated (Figure M3.46).

■ Click on the OK button.

You can check that the changes have been made by selecting window 2, Table: Customer from the Window menu.

Figure M3.45

Changing data on
the basis of a
compound
condition

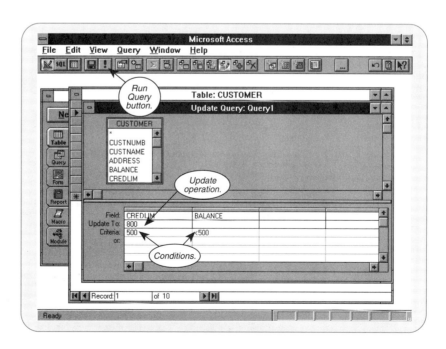

Figure M3.46

Changing data on
the basis of a
compound
condition

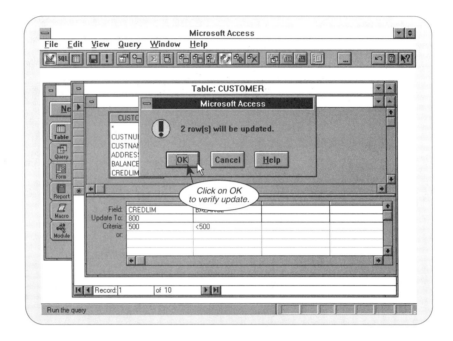

- Select Close from the File menu to close the query.
- Answer No to the Save Query? prompt.
- Select Close from the File menu to close the *Customer* table.

Q & A

Question: Suppose you wanted to change the credit limit of everyone who has a credit limit of at
most 500 and whose balance is less than the credit limit. You want to add 200 to the
credit limit of these customers. How would you do it?

Answer: Using an update query, you would fill in the query form as Figure M3.47 shows. Click
on the Run button to make the change.

Figure M3.47

Changing records
with a condition
and computation

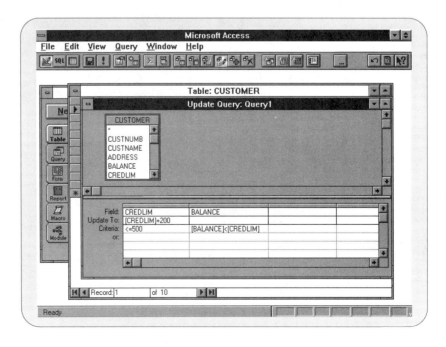

SUMMARY

The following list summarizes the material covered in this module:

1. You create queries to access data in an Access database by filling in **Query-by-Example** (QBE) forms. To produce a query form on the screen, click on the New Query button and then select the table to be queried.
2. Once a query form has been filled in, click on the Datasheet View button to view the results.
3. To include a column in the results of a query, drag the field from the table box to a column in the QBE form and place a check mark in the Show box. To include all columns, drag the (*) asterisk field to the QBE form.
4. To use a condition, type the condition into the Criterion of an appropriate column. To use a compound condition involving AND, type the individual conditions on the same line. To use a compound condition involving OR, type the conditions on separate lines.
5. To include a calculated field in a query, specify the computation, using the field-names in [brackets] for the computation.
6. The * wildcard indicates any collection of characters. The ? wildcard indicates any single character.
7. You can use the LIKE operator to find entries that are similar to the value, or have the entry as part of the value, rather than match the value exactly.
8. Output will be sorted on the field with a check mark in the Sort row. Use the drop-down list to select either ascending or descending order.
9. Access has many built-in functions available for statistical calculations, including: COUNT, SUM, AVERAGE, MAX, and MIN.
10. To join tables, first bring tables to the Query Design screen by selecting Add Table from the Query menu. Then drag matching fields from one table to the second table to indicate the join relationship.
11. To use a query to change data in a table, use an Update Query.
12. To use a query to delete records from a table, use a Delete Query.

EXERCISES

1. What is a Query-by-Example (QBE) form? How do you produce one?
2. How do you see the results of a query once you have filled in the query form?
3. How can you print the results of a query?
4. How do you include a field/column in a query? What is the simplest way to include all columns in the results of a query?
5. How do you enter a condition in a query? How do you enter a compound condition involving AND? How do you enter a compound condition involving OR?
6. How do you include a computed field in a query?
7. How do you use a wildcard? What wildcard symbols are available in Access?
8. What is the function of the LIKE operator? When might you use it?
9. What built-in functions are available in Access for statistical calculations? How do you use them?
10. How do you join tables in Access?
11. How can you use a query to change data?
12. How can you use a query to delete records? If you discover your deletions were wrong, how can you undo them?

COMPUTER ASSIGNMENTS

1. Use the Query-by-Example (QBE) feature of Access and the *Premiere Products* database to complete the following exercises:

a. Find the part number and description of all parts.
b. List the complete sales rep table.
c. Find the names of all the customers who have a credit limit of at least $800.
d. Give the order numbers of those orders placed by customer 124 on September 5, 1994.
e. Give the part number, description, and on-hand value (units on hand * price) for each part in item class AP. (On-hand value is really units on hand * cost, but we do not have a cost column in the *Part* table.)
f. Find the number and name of all customers whose last name is NELSON.
g. List all details about parts. The output should be sorted by unit price.
h. Find out how many customers have a balance that exceeds their credit limit.
i. Find the total of the balances for all the customers represented by sales rep 12.
j. For each order, list the order number, the order date, the customer number, and the customer name.
k. For each order placed on September 5, 1993, list the order number, the order date, the customer number, and the customer name.
l. Find the number and name of all sales reps who represent any customer with a credit limit of $1000.
m. For each order, list the order number, the order date, the customer number, the customer name, together with the number and name of the sales rep who represents the customer.

2. Use the QBE feature of Access and the *Books* database to complete the following exercises:
a. List the numbers, names, and addresses of all customers.
b. List the codes and titles of all books.
c. List the complete *Publshr* table.
d. List the names of all publishers located in New York.
e. List the names of all branches that have at least 10 employees.
f. List the codes and titles of all books whose type is HOR.
g. List the codes and titles of all books whose type is HOR and that are in paperback.
h. List the codes and titles of all books whose type is HOR or whose published code is PB.
i. List the codes and titles of all paperback books whose type is ART and whose price is under $12.00.
j. Customers who are part of a special program get a 10% discount. To see what the discounted prices would be, list the code title and discounted price of all books.
k. Find the numbers and names of all publishers whose name contains "and."
l. List the codes and titles of all books whose type is FIC, MYS, or ART.
m. List the names of all authors. Sort the output by name.
n. List the names of all authors. Sort the output by name in reverse order.
o. Find out how many books are of type MYS.
p. Find out the average price for each type of book. Repeat the process, but this time only consider paperback books.
q. For each book, list the book code, title, publisher code, and publisher name.
r. For each branch, list the branch number as well as list the book code, book title, and number of units on hand, of each book currently in stock at the branch.
s. For each customer who has a book on order, list the customer number, customer name, book code, book title, and the date the book was requested.

Advanced Relational Features

OBJECTIVES

1. Discuss how to examine the structure of a database.
2. Present the use of indexes in Access.
3. Investigate the process of searching for records that satisfy specific criteria.
4. Explain how to change the structure of a table.
5. Discuss the use of views (stored queries).
6. Introduce the use of macros.
7. Discuss the deletion of files as well as the copying of families of files.

M4.1   INTRODUCTION

In this module, we begin our investigation of some of the advanced features of the relational model as implemented in Access. We start in section M4.2 by discussing the way you can determine the structure of a particular database and a particular table. In section M4.3, we look at the uses of indexes in Access. We view how indexes can be used in place of sorting; the effect of updating a table when an index is in use; how to create single- and multiple-field indexes; and how to remove indexes. Searching, the process of locating a record that satisfies some specific condition, is the topic for section M4.4. In section M4.5, we examine how we can change the structure of a table. In section M4.6, we look at saved queries, the Access version of views. Finally, in section M4.7, we examine how to use the Access file operations to delete unwanted files and also to make a backup copy of an entire family of files.

Note: If you didn't return to the original *Premiere Products* database at the end of Module 3, you should do so now. This will ensure that the data in your *Premiere Products* database is correct. To do so, exit to either DOS or Windows and copy the file *premback.mdb* to *prem.mdb*, replacing the modified *prem.mdb* file. To restore the database using DOS, exit Access and Windows and copy the file from the C:> DOS prompt.

■ COPY A:PREMBACK.MDB A:PREM.MDB

To restore the database using the File Manager of Windows, simply select *premback.mdb* and choose Rename from the File menu. Rename the file *prem.mdb* to replace the current file.

M4.2   EXAMINING STRUCTURE

To use a database effectively, you need to know its structure. Fortunately, in Access, determining the structure of a database is simple. You need to be able to determine

259

which tables exist for each database. Also, you need to be able to determine the fields and associated field types that comprise each table. Let's begin by looking at the database files available to Access when selecting Open Database from the File menu, as Figure M4.1 shows.

- Select Open Database from the File menu.
- Select the *PREM* database.

Figure M4.1

Open database
dialog box

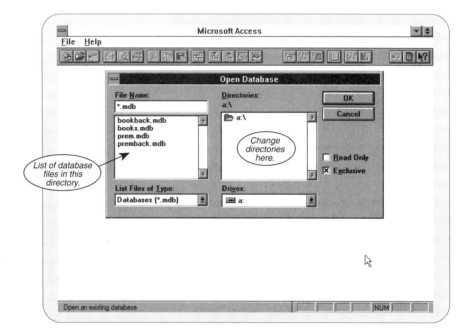

In Access, all related tables, queries, reports, forms, macros, and modules are stored in a single database file. When you open a database, the Database window has a column of object icons down the left side of the window, with the table objects listed by default. To view a listing of other objects, click on one of the object buttons.

**Listing
Tables**

You will now see a list of tables in the *PREM* database, as Figure M4.2 shows.

**Determining
Table
Structure**

To determine the structure of a table, switch to the Table Design mode. You will now be asked to identify the table whose structure you wish to see. Let's look at the structure of the *Customer* table.

- Click on the *Customer* Table.
- Click on the Open button.

When you open a table, by default the table opens in the datasheet view. To switch to see the structure of the table, click on the Design View button or select Table Design from the View menu.

- Select Table Design from the View menu.

Your screen should now look like the one shown in Figure M4.3. This shows the current structure of the table. All of the fields in the table are listed by name, along with the data type and a description. In addition, as you move from one field to another, the field properties for each field are listed at the bottom of the screen. If a primary key

Figure M4.2

Prem Database window

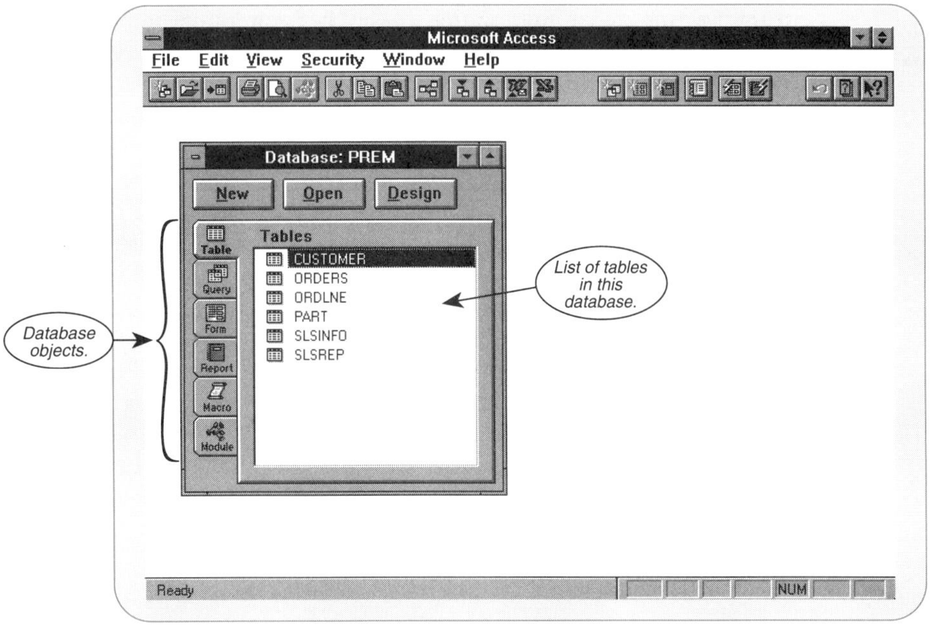

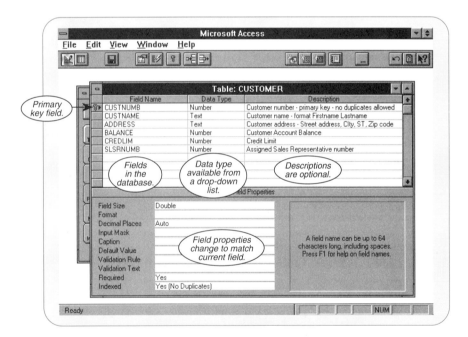

Figure M4.3

Table Design view

has been identified, as Custnumb has been in this table, the primary key icon will be displayed to the left of the field name.

We need to close the *Customer* table now so that we may use another one for the next section.

- Select Close from the File menu.

M4.3 INDEXES

We use an index to provide rapid access to records on the basis of the value of some field or combination of fields. Consider, for example, the *Customer* table. Without an index, finding a customer on the basis of his or her number would require us to examine each record in the file until we found the particular customer. On the other hand, if we use an index on the *Customer number* field, finding a customer on the basis of his or her number can be done very quickly since Access can go directly to this customer without having to examine others in the file.

If you want to list the records in a table in a certain order, you can sort them. The records are rearranged, so that they occur in the desired order. Although this does work, it is usually far less efficient than using an index, especially for large tables.

We can also use an index to effectively order the records in a file according to the values of some field or combination of fields. Consider again the *Customer* table. Without an index, if you were to produce a report that listed the records in the file, they would appear in whatever order in which they happened to be (typically, the order in which they were entered). On the other hand, if you were to use an index built on the *Customer number field*, customers would be listed in customer number order, regardless of the order in which they actually occurred in the table. Using such an index does not in fact rearrange the records in the table, but it does make it appear to a user as if they had been rearranged.

Primary Key / Indexes

The **primary index** is the main index for the table. As you have seen, it controls the order in which the records are displayed. It also prohibits the addition of a record whose values duplicate those on a record already in the database.

Creating a Primary Key

Creating the primary key in Access is simple. With the table open, switch to the Table Design View, then click on the field you wish to make the primary key, and click on the Key icon on the tool bar. Alternatively, select Set Primary Key from the Edit menu while on the field and in the table design view.

You can create the primary key on a combination of fields. To create a primary key with multiple fields, you must hold down the Ctrl key while clicking on the leftmost column, or field handle for each field to make up the primary key (each row will be highlighted), and then click on the primary key icon. The *Ordlne* table, for example, could have the *Ordnumb* and *Partnumb* combined to be the primary key. This would be necessary because both the order number and the part number could have duplicates. Therefore, you would have to combine the two in order to make each entry unique.

The first two fields are designated as key fields, since it is the combination of these two fields that uniquely identifies any record in the table. In this case, Access does not create two separate primary indexes but instead creates a *single* primary index on this combination. There is always only one primary key.

For Ordline, we will need to combine *Ordnumb* and *Partnumb* to form the primary key.

- Open the *Ordline* database.
- Click on the Table Design button.
- Hold down the Ctrl key and click on the leftmost column of the *Ordnumb* field.

Note: The arrow pointer changes to a right arrow when it is on the leftmost column.

- Continue pressing the Ctrl key and click on the leftmost column of the *Partnumb* field.

Both fields should be highlighted, as Figure M4.4 shows. We now will identify both fields as the primary key.

- Click on the Primary Key button.

Both fields will now have a small key icon to the left of the field name that identifies the fields as the primary key (Figure M4.5). Now, whenever you add or change a record, the record will be rearranged according to the primary key when you leave

Figure M4.4

Selecting multiple fields

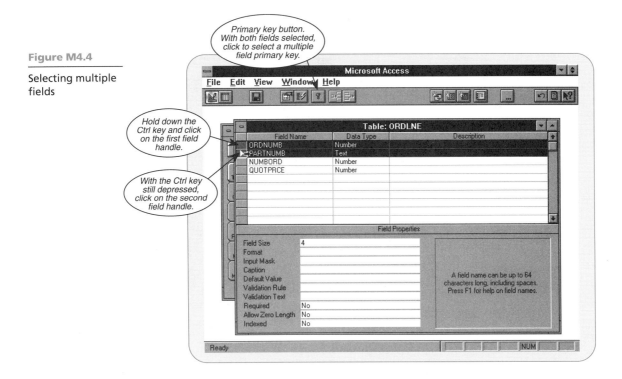

Figure M4.5

Selecting multiple field primary keys

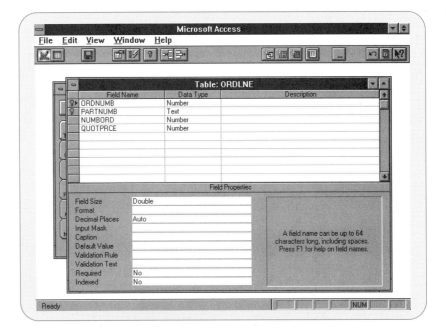

and then reopen the table. To view or modify primary keys, you can access them through the Indexes view.

■ Select Indexes from the View menu.

Figure M4.6 shows that the primary key is a combination of *ordnumb; partnumb*. In the table properties, you can change, or delete the primary key, or add secondary indexes. For now, let's change back to the Datasheet view and close the table.

Figure M4.6

Viewing or
modifying indexes

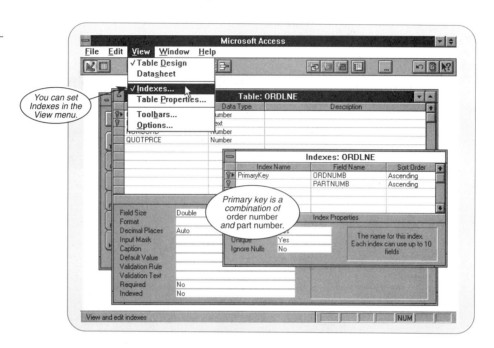

- Click on the Datasheet view icon.
- Click on OK to save the changes before switching to the Datasheet view. (See Figure M4.7.)
- Select Close from the File menu.

Secondary Indexes

At any given time, only one primary index can exist for a table. You should create the primary index for the unique identifier, or *primary key*, of the table. This means that using a primary index for sorting works only if the field or combination of fields on which you want to sort happens to be the primary key.

If you want an index for sorting or searching on another field or combinations of fields, you can use another type of index, called a **secondary index.** As an example of a secondary index, Figure M4.8 demonstrates the use of a warehouse number as a

Figure M4.7

Modifying table
properties

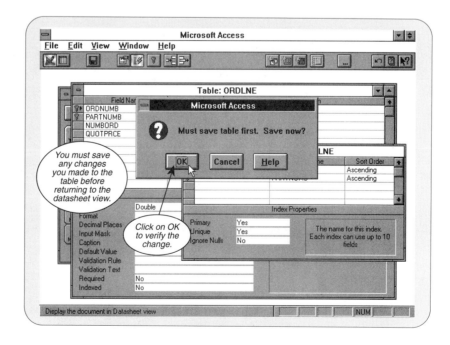

Figure M4.8

Example of a
secondary index

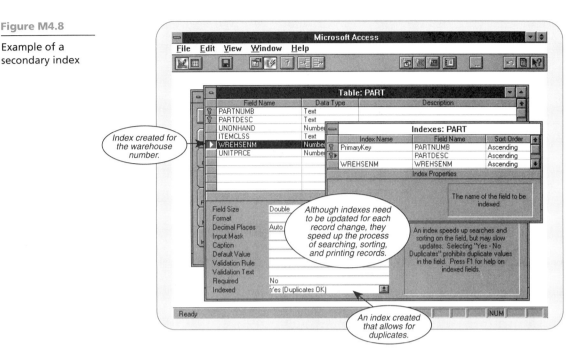

secondary index so that printing of reports and searching for records could be done faster.

The more ways that you might need to view your data, the more you need indexes to help Access search your data efficiently. An index is simply an internal table that contains two columns: the value in the field or fields being indexed and the location of each record containing that value in your table. In the above example, the index would contain the warehouse number and the location of each record recorded in the order of record numbers.

Multiple-Field Indexes

Most of the indexes you need to define will probably contain the values from only a single field. Sometimes you must create an index on a combination of fields. Suppose, for example, you would like the data sorted by warehouse as you did in the last section. This time, however, you would like all the parts in a given warehouse to be sorted by item class. In this case, we say that the output is to be sorted by item class *within* warehouse. You accomplish this through multiple-field indexes. When you sort a table by a multiple-field index, Access sorts first by the first field listed in the Indexes window. If there are records with duplicate values in the field, Access sorts next by the second field listed, and so on.

To create a multiple field index, you must open the Table window in Design view, and open the Indexes window by choosing Indexes from the View menu. As you can see in Figure M4.9, you can define multiple indexes on a single table. You create multiple-field indexes for a table by including a row for each field in the index and entering the name of the index only on the first row. Access treats all rows as part of the same index until it comes to a row with another index name in the first field. Access creates each index when you save the table or index definition.

- Open the *Prem* database.
- Open the *Part* table.
- Change to the Table Design View.
- Select Indexes from the View menu.
- Click in the first available (or empty) row in the Index Name column.
- Enter WREHSENM, ITEMCLSS, and then TAB to move to the Field Name column.

Figure M4.9

Multiple field
index

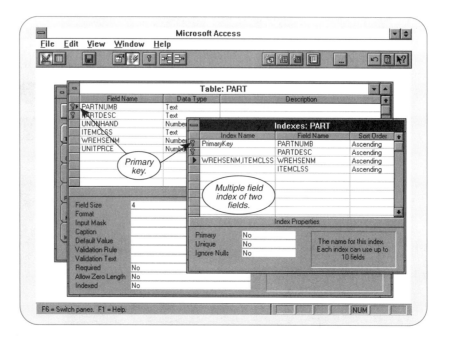

■ Enter WREHSENM for the name of the index field.
(You could also press Alt and the Down arrow to select the field from a list.)
■ Move to the Field Name of the second row of the new index.
■ Enter ITEMCLSS for the name of the second index field.

Your screen should now look like Figure M4.9. To save your index, switch to the
datasheet view and save your changes when prompted.

■ Click on the Datasheet view button. (Or select Datasheet from the View menu.)
■ Click on OK to save the changes before switching views.

**Removing
Unwanted
Indexes**

Primary indexes are important, and each table should have one. If you no longer plan
to use a secondary index, however, you should remove it. There are two reasons for
doing so. First, the index entries occupy space on your disk. Second, when you update
records in your table, the index has to be updated as well, which takes extra time. If
an index is necessary, you are usually willing to put up with these problems; if it is
not, it would be foolish to retain the index.

To remove a secondary index, select the table for which you want to delete the
index and then select Indexes from the View menu. Then, for each index that you wish
to delete, simply select and delete the index line. For example, let's remove the warehouse number and item class index that you created in the last section.

■ Click on the Table Design view button, or select it from the View menu.
■ Click on the Indexes button, or select Indexes from the View menu.
■ Highlight the item class and warehouse number rows.
■ Press Delete to remove the two rows from the indexes list.
■ Select Datasheet from the View menu, or click on the Datasheet view button.
■ Click on OK to save the changes before switching views.

M4.4   SEARCHING

When we refer to *searching* in the database environment, we mean looking for records
that satisfy some condition. Looking for a customer whose number is 522, looking for

a customer whose name is Tom Daniels, and looking for all customers who are represented by sales rep 6 are all examples of searching. Access provides some powerful facilities for searching. Some of them were introduced earlier with query. We will now examine searching facilities using the Find command.

Using Find to Search for records

You should still be in the *Parts* table. If you are not, open the *Prem* database and the *Parts* table. The first search we want to do is to find the part with a description of "Toaster".

- Select Find from the Edit menu.

The Find dialog box shown in Figure M4.10 is displayed. Enter the data that you want to search for, and set any conditions in the option and check boxes.

- Enter *Toaster* in the Find What text box.
- Press Enter or click on Find Next.

Access will jump to the next occurrence of the text you searched for.

Figure M4.10

Using FIND to search for records

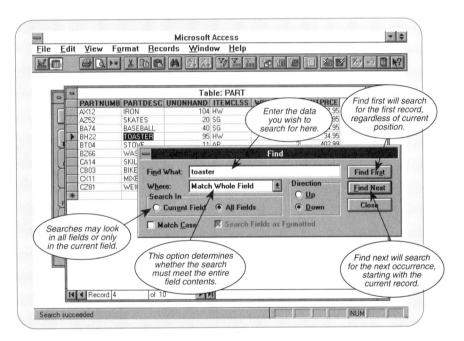

Repeating a Search

Sometimes, you don't want to find the first record that satisfies the condition. Maybe you want to find the second such record or perhaps the fifth. You may want to examine all such records, one after another. In other words, you need to be able to repeat a search. In Access, you can do so by selecting Alt + F to repeat the search process.

Using Wildcards

When you use query forms for searching, you can use the same wildcard symbols as in your queries in Module 3. You can use "*" to represent any combination of characters or "?" to represent any single character. Figure M4.11, for example, illustrates the search for records with an Item Class with a "S" in the first position, and any other character in the second position. In addition, with this search, the process has been limited to searching only in the *Itemclss* field. Clicking on Find next will jump to the next occurrence.

- Click in the *Itemclss* field.
- Select Find from the Edit menu.
- Enter "S?" in the Find What field.

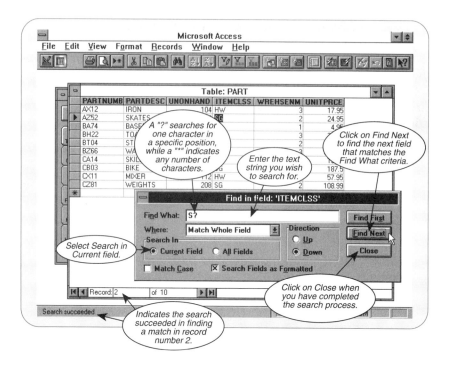

- Click on Current Field option.
- Click on Find next.

As in Figure M4.11, Access should have moved the current record indicator to the first row with a "S" in the first position of *Itemclss* field. If you want to continue the search, you only need to press Alt + F. When you have finished searching, close the Find dialog box and the table.

- Click on Close when you are finished with the Find process.
- Select Close from the File menu to close the table.

M4.5   CHANGING THE STRUCTURE OF A TABLE

Changing the structure of a table is a simple task in Access. We will illustrate various changes by modifying the *Customer* table. First we will add a field called *Custtype* and then fill in values for it. Next we will change the characteristics of *Custtype* and change the values in this field using query forms. Finally, we will delete the field.

- Select the *Customer* table.
- Select Table Design from the View menu.

You will then see the current database design. Now you are ready to make changes to the design.

**Adding
Fields**

To add a field, you first switch to the design view and then move your cursor to the location where you want to add the field. In our example, you will place a field immediately prior to Slsrnumb, so you will move the cursor to Slsrnumb. (If the field is to be the last field, you would move past all the existing fields.) To make room for the new field, you insert a row.

- Move the cursor to the Slsrnumb row.
- Select Insert Row from the Edit menu.
- Enter Custtype in the field name.

- Enter `Text` for the Data Type.
- Change the Field Size to 1 in the Field Properties.

Your screen should now look like the one shown in Figure M4.12. Your database now contains the new field. When you switch to the Datasheet view, the modified table should be on your screen.

- Click on the Datasheet view button.
- Click on OK to save the changes before switching views.

Your screen should now look like the one shown in Figure M4.13. (The window has been maximized in the figure.) Notice the new column.

Figure M4.12

Adding a field to a table

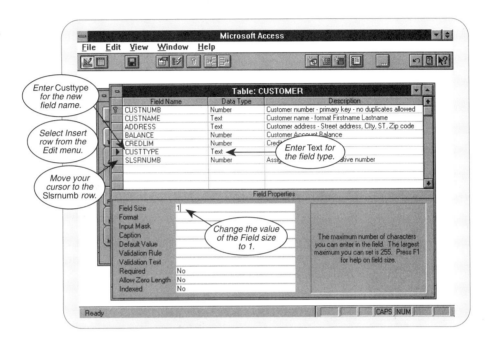

Figure M4.13

Field has been added

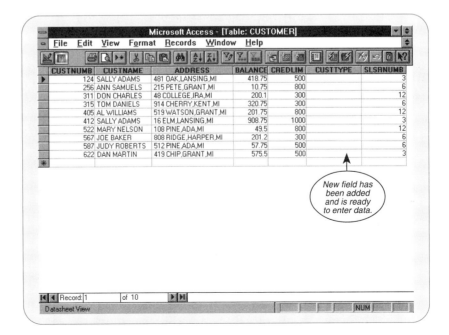

Figure M4.14

Filling in values for new field

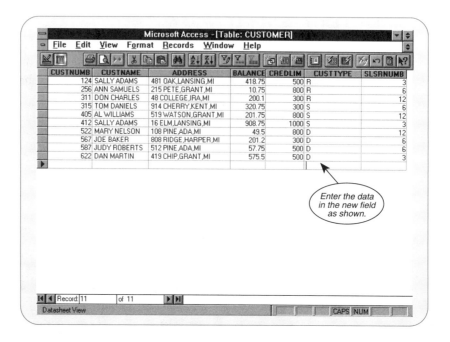

Enter the data in the new field as shown.

■ Enter the customer type values, as Figure M4.14 shows.

Your database is now updated.

Changing Characteristics of Existing Fields

Let's suppose we need two-character codes for customer type. We need to change the width of the *Custtype* field to 2. The process is similar to the one for adding a new field. Instead of adding a new field, we make changes to the entries for an existing field.

■ Click on the Table Design view button.
■ Move the cursor to the *Custtype* field.
■ Change the field size from 1 to 2 and press the Enter key.
■ Click on the Datasheet view button.
■ Click on OK to save the changes before switching views.

The width of the field has now been changed.

Making Entries for New Fields

To change the entries for `Custtype,` you simply proceed through each record. Whenever you encounter a record on which the value for Custtype is R, change it to RE; if the value is S, change it to SP; and so on. Does this approach seem cumbersome to you? Even with only ten records, it probably seems like a lot of busy work. What if there were several thousand records? It would take a long time to make these changes, with many chances to make errors. Fortunately, there is an alternative. Access provides the Replace command for such tasks (Figure M4.15).

■ Move the cursor to the first row in the *Custtype* field.
■ Select Replace from the Edit menu.
■ Enter R in the Find What box.
■ Enter RE in the Replace With box.
■ Click on Replace All.

Answer No to the request to start the search process from the beginning of the file again. Then answer YES to update the records without being able to undo the procedure. The results of the first replace are shown in Figure M4.16. This will change three records. Use the same technique to change the value for *Custtype* to SP for all records on which it is S and DI for all records on which it is D.

Figure M4.15

Using the Replace command

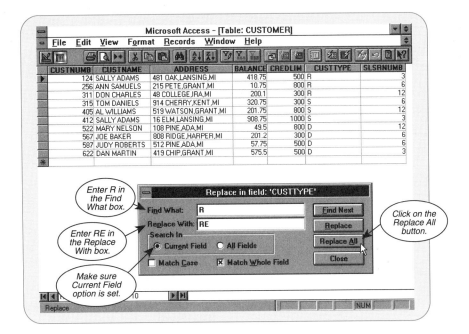

Figure M4.16

Results of the Replace command

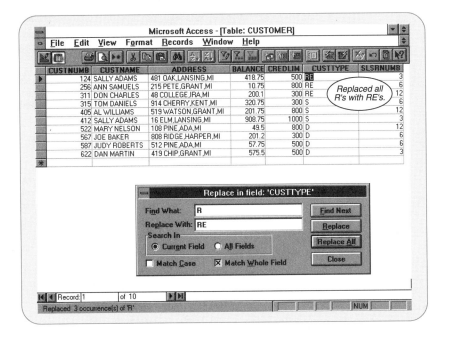

- Repeat the above process for S and D.
- Click on the Close button when finished with the replace operation.

The changes are now complete and all records now contain an appropriate value in the *Custtype* field. Your screen should now look like the one shown in Figure M4.17.

Deleting Fields

Suppose you decide at some later point that you no longer need the *Custtype* field. Deleting a field is similar to adding or changing one.

- Click on the Table Design view button.
- Click on the *Custtype* field, as Figure M4.18 shows.

Figure M4.17

A completed
Replace command

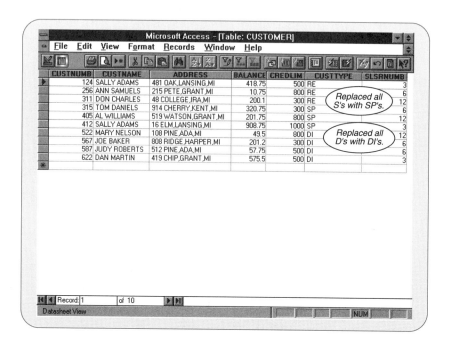

Figure M4.18

Selecting a field
for deletion

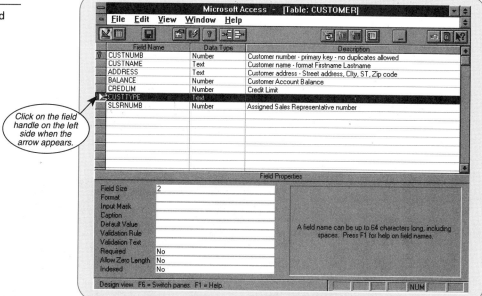

- Pressing the Delete key deletes the field that is selected. (You could also select Delete Row from the Edit menu.)
- Click on OK to verify the deletion (Figure M4.19).
- Click on the Datasheet view button.
- Click on OK to save the changes before switching views.
- Select Close from the File menu.

M4.6   VIEWS (SAVED QUERIES)

Often, there is no need to save a query. You enter conditions, specify the fields you want, see the results, and you are done. It's a one-time operation. Sometimes, however,

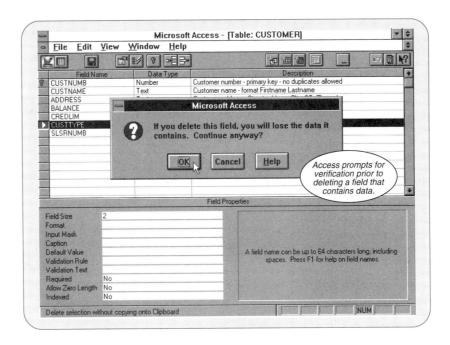

you anticipate the need to ask for the same conditions and the same set of fields at a later date. Rather than go through all the work of respecifying the conditions and re-selecting the fields, it would be easier to save your work. Then you can simply use the same query design at any time, which is easy to do.

A saved query is called a **view.** The data is not saved in the form specified in the view. Rather, Access saves the conditions and list of fields that make up the query and applies this query whenever you ask to see the data.

Creating a View

The closest thing to a view in Access is a saved query. Thus, to create the view, you must first create the corresponding query; that is, we must make the necessary entries in appropriate query forms. Let's use as an example the following query: "For each order, list the sales rep number, the sales rep name, the customer number, the customer name, the order number, and the order date."

To create a view, simply create the new query and then save it for later use.

- Click on the Query object button.
- Click on the New button.
- Click on the New Query button to bypass the Query Wizard.

At this point, because you were not currently using a table, Access prompts the tables to be added to the query.

Note: The position of the table boxes and the Add dialog box has been relocated to show the tables. Your screen may hide a table when it has been added to the query.

- Double click on the *Slsrep* table.
- Double click on the *Customer* table.
- Double click on the *Orders* table.

Your screen should look similar to Figure M4.20 now.

- Click on Close to complete adding tables.

Next, we need to add the fields that we want in the query.

- Drag the Slsrnumb from the *Slsrep* table, to the first column.
- Drag the Slsrname from the *Slsrep* table, to the second column.

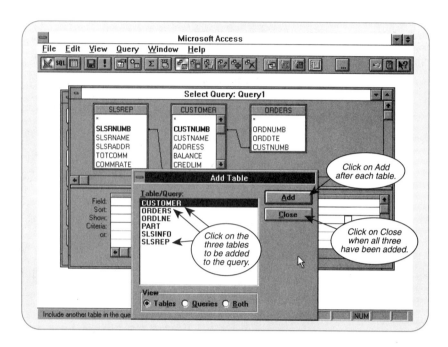

- Drag the Custnumb from the *Customer* table, to the third column.
- Drag the Custname from the *Customer* table, to the fourth column.
- Drag the Ordnumb from the *Orders* table, to the fifth column.
- Drag the Orddte from the *Orders* table, to the sixth column.

Note: You will need to use the scroll bars to move the columns left or right to view all six of them.

At this point, we wish to view the results of the query sorted in ascending order by the sales representative number. To do this, click on the *Sort* field for the Slsrnumb column. Press Alt and the Down arrow and then select Ascending from the list.

- Select Ascending from the *Sort* field in the Slsrnumb column.

The *Slsrep* and the *Customer* tables are linked by the *Slsrnumb* field. Similarly, the *Customer* and *Orders* tables are linked by the values in the Custnumb columns in both tables. As you would expect, we create these links by dragging the *Slsrnumb* field in the *Slsrep* table to the *Slsrnumb* field in the *Customer* table and the *Custnumb* field in the *Customer* table to the *Orders* table. When you do this, Access indicates the join link by drawing a line connecting the two tables (Figure M4.21).

- Drag the *Slsrnumb* field name from the *Slsrep* table to the *Slsrnumb* field in the *Customer* table.
- Drag the *Custnumb* field name from the *Customer* table to the *Custnumb* field name in the *Orders* table.

Figure M4.21 shows that the *Slsrep, Customer,* and *Orders* tables have been added, the six fields added to the field row, and relationships between the tables established with joins. Although the first and second columns have scrolled off the left side of the screen, you may use the scroll bar arrows to view the entire form.

Now that you have created the query, you need to save it. Let's call it Sales Orders.

- Select Save As from the File menu.
- Enter `Sales Orders` (Figure M4.22).
- Click on OK to save.

Figure M4.21

A completed query

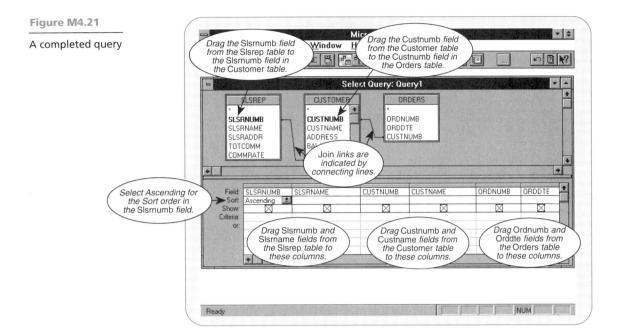

Figure M4.22

Saving a query

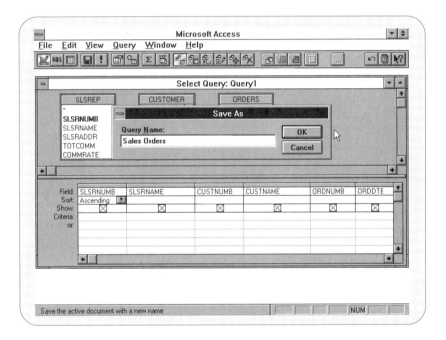

The query is now saved, and you have created a view.

- Select Close from the File menu.

Using a View   To use a view, you need to re-execute the saved query. In Access jargon, the view has been saved as a query. By opening a saved query, you recreate the view of the data (Figure M4.23).

- Click on the Query object icon.
- Click on the Sales Orders query.
- Click on Open to view the query.

Figure M4.24 shows the result of the saved query.

Figure M4.23

Selecting a saved query

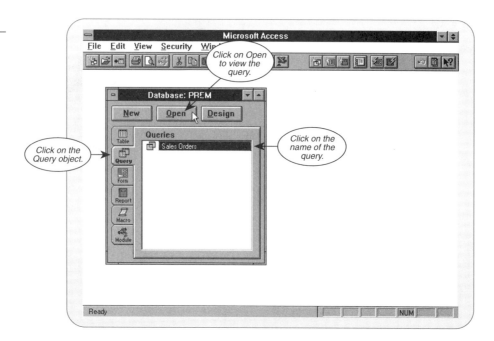

Figure M4.24

A completed query

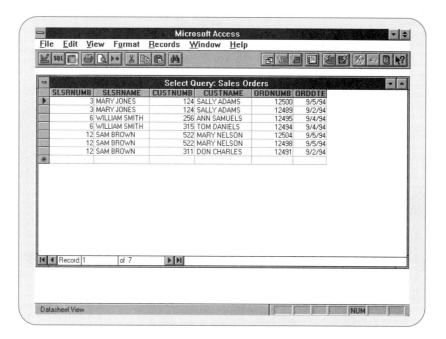

M4.7   DEFINING RELATIONSHIPS

Once you have defined two or more related tables, you should tell Access how the tables are related. If you do this, Access will know how to link all of your tables when you need to use them later in queries and reports.

For example, for each customer added to the *Customer* table, there should be a matching sales representative in the *Slsrep* table. Likewise, you refer to this as a one-to-many relationship because for each single sales representative, there can be many customers but each customer may have only one sales representative.

To define relationships, begin from the database window. Close all tables or queries windows by selecting Close from the respective File menus. Then choose Relationships from the Edit menu to open the dialog box shown in Figure M4.25.

Figure M4.25

Selecting tables in relationships

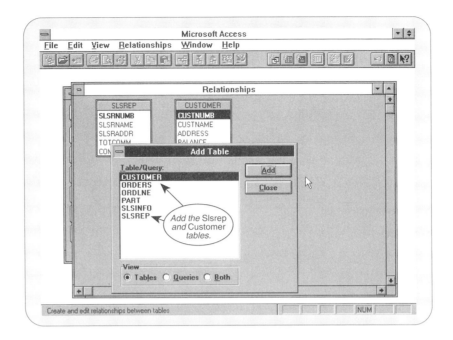

- Close any open tables or query windows.
- Make sure that the *Prem* database window is open.
- Select Relationships from the Edit menu.
- Choose Add Table from the Relationships menu.
- Click on the *Slsrep* table and then the Add button.
- Click on the *Customer* table and then the Add button.
- Click on the Close button when you are finished selecting tables.

To create the relationships between the tables, simply point to the field in the primary table, hold down the mouse button, and drag the field to the related table. In this example, we want to create a relationship between the *Slsrep* and *Customer* tables with a link through the *Slsrnumb* field (Figure M4.26).

Figure M4.26

Defining relationships between tables

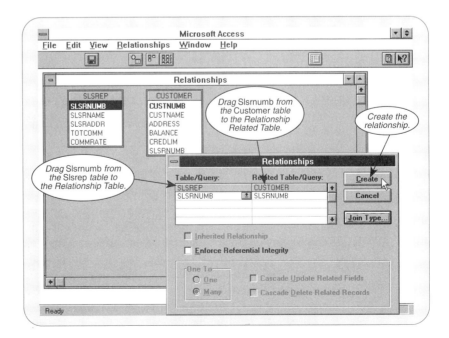

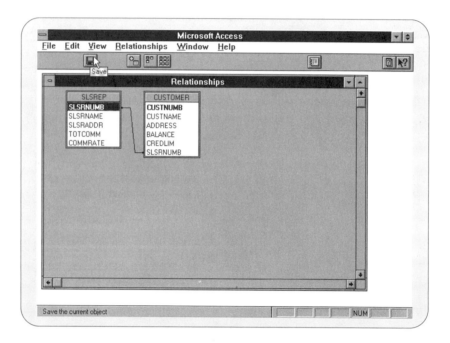

- Drag the Sfield icon from the *Slsrep* table to *Customer* table.
- Click on the Create button to make the relationship.
- Click on the Save button (Figure M4.27) to save the relationship in the database.
- Select Close from the File menu.

If you want Access to always validate the foreign key values on the "many" side of a relationship, check the Enforce Referential Integrity check box. This option will not allow you to create a record in the "many" table that doesn't have a matching value in the "one" table. For example, you can't add a customer with a sales representative number of 4 because there is not a match in *Slsrnumb* in the *Slsrep* table.

SUMMARY

The following list summarizes the material covered in this module:

1. To list the tables, open a database and click on the Tables object icon.
2. To list the structure of a table, open a table and then select Table Design from the View menu.
3. You can use indexes to provide rapid access to records and as an efficient alternative to sorting.
4. To create a primary index, click on the Primary key icon while in the proper field in the Design view.
5. To create a secondary index, select Indexes from View menu while in the Design view and enter the field names for each index.
6. To delete a secondary index that is no longer required, press Delete on the field name in the Indexes dialog box.
7. To add a field to a table, switch to the Design view, move the cursor to the desired location, and insert a row. Type the field name and characteristics.
8. To change the characteristics of a field, move to the field while in the Design view and simply type the new value.
9. To delete a field from a table, move the cursor to the field to be deleted, and press Delete. Select OK.
10. To create a view, create a query for the view. Then save the query.

11. To set relationships between tables, open a database, and select Relationships from the Edit menu. Select the proper table for the primary table and a related table. Then select the matching field between the two tables.

EXERCISES

1. How can you obtain a list of the tables in the current directory?
2. List two benefits to using indexes.
3. How can you create a primary index? Are there any restrictions on the key fields for the primary index?
4. How can you create a secondary index? How can you delete one?
5. How do you add a new field to a table?
6. How can you change the characteristics of an existing field in a table?
7. How do you delete a field from a table?
8. How do you create a view? How do you use one?
9. What is a relationship? How do you create one? What does a relationship do once you have created one?

COMPUTER ASSIGNMENTS

In each of the following problems, accomplish the task and also explain the steps required:

1. List the tables in the *Books* database. Print the design structure of each of the tables.
2. Create a query on the *Book* table to list the titles of all books.
3. Use the Find command to search the *Book* table for the first book of type ART. Find the next such book.
4. Add a field called *Csttype* to the *Bookcust* table. This field should be of Text type with a width of 1 and not be indexed. Fill in data for this field. Set the value to the letter A for the first five records and the letter B for the rest. Print the current contents of this table.
5. Increase the width of *Csttype* to 2. Change the value to AX on all records on which it is currently A. Change the value to BY on all records on which it is currently B. Print the contents of the table.
6. Delete the *Csttype* field from the *Bookcust* table. Again, print the contents of the table.
7. Create a view called Authbook. The fields in the view should be *Author Number, Author Name, Book Code, Book Title,* and *Price.* You will need to include the *Book, Author,* and *Wrote* tables in this view.

5 *Data-Entry Support*

OBJECTIVES

1. Explain how to create custom forms.
2. Discuss how custom forms can be used during data entry.
3. Investigate the use of validity checks in controlling data entry.
4. Discuss the creation and use of multi-table forms.

M5.1 INTRODUCTION

In this module, we turn our attention to how you can use Access to enhance the data-entry process. Section M5.2 begins by examining the way you can create custom forms. We will also see how to use the forms and how to enhance them. Next, section M5.3 investigates the use of validity checks, which allows us to make sure users enter only valid data. We look at the creation and use of multi-table forms (forms that contain data from several tables) in section M5.4.

M5.2 CUSTOM FORMS

You have already used the Data Entry mode to add records to a table, using a simple form on the screen. Although the form was helpful, it was not particularly pleasing. The fields were merely lined up next to each other. Only the nondescriptive names of the fields displayed on the screen.

You can use Access forms to enter, change, view, or print data. In this section, you will use Form Wizard to create your own custom forms to use in place of the simple Data Entry form Access supplies. For example, you will create a custom form for the *Part* table.

Form Creation

You can create Access forms with or without the assistance of Form Wizard. However, Form Wizard makes the process much faster and easier. A form must be connected to a table or query. You also may connect forms to data in one table or to more than one table if you base the form on a query.

This discussion provides an opportunity to introduce Access Wizards. Access uses Wizards to assist the user in creating tables, queries, forms, and reports. In the previous modules, we bypassed using Wizards and created tables and queries from scratch. Here, we will use Form Wizard to create a New Form. Using Form Wizard is an easy, fast way to create a new form. You can either use the form as is (the default creation of Form Wizard), or you can modify the form to meet your needs. To create a form:

- You should be in the *Prem* database window.
- Click on the Form object button and then the New button, or select New and then Form from the File menu.

Access Form Wizard first must attach the form to a table or a query (Figure M5.1). Since you want to create a form of the *Part* table, you could simply type in the name of the table in the dialog box. A faster method would be to click on the drop-down arrow to the right of the table name text box and select *Part* from the list of tables and queries, as Figure M5.2 shows. Then you have the option of letting the Form Wizard prompt you through the various options or starting from a blank form.

- Click on the drop-down arrow and select *Part* table.
- Click on the Form Wizards button.

Figure M5.1

Creating a new form

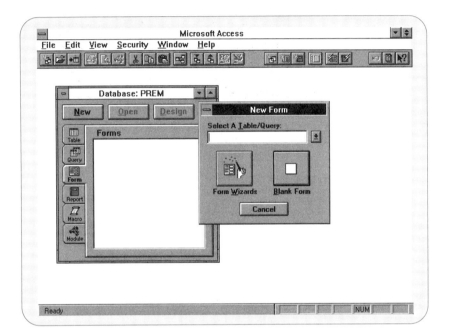

Figure M5.2

Selecting a table for the form

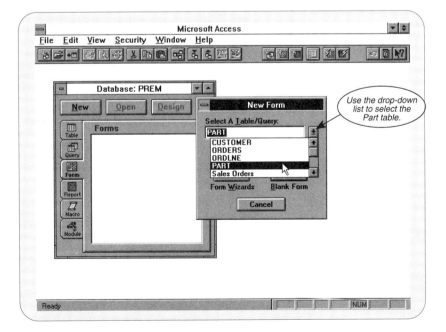

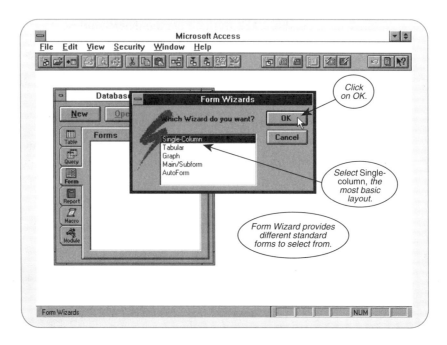

Access will prompt you through a few screens of data about the form. Figure M5.3 shows the first screen, which provides options for four standard form designs. We will begin with the most basic form, *Single-column,* and then modify it. The first step is to tell Form Wizard the type of form to create.

■ Click on the Single-column form.

Selecting Fields

A series of screens prompt you for the data needed to create a form. The first screen (Figure M5.4) requests the fields to include and the order in which to list them. You select the first field and then click on the ">" button to copy each field from the available fields from the left box to the right box for fields to be printed. However, the ">>" button will copy all fields in the order listed in the Available fields box.

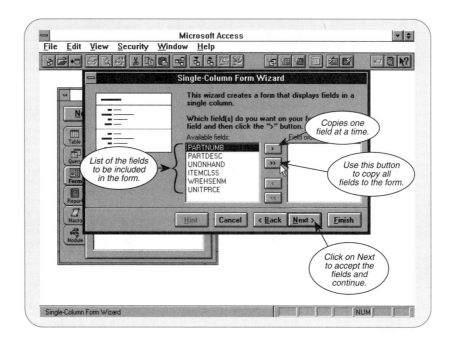

■ Click on the "≫" button.
■ Click on the "Next >" button to move to the next screen.

The next screens prompt for the type of appearance you want and the title that you want printed on the top of the form. A single-column form offers a standard, chiseled, shadowed, boxed, or embossed style of appearance. We will use the default, embossed.

■ Click on the Next > button to accept the embossed appearance and move to the next screen.

The following screen (Figure M5.5) asks for the title to be printed on each form. The default is the name of the table. However, we want to list the company's name in the title. Once you have entered this, you can open the form and view data in the new form or select design to modify the default form that was created.

■ Enter `Premiere Products Parts` for the form title.
■ Click on Open to view the table in the New form view.

Figure M5.5

Adding a title to a form

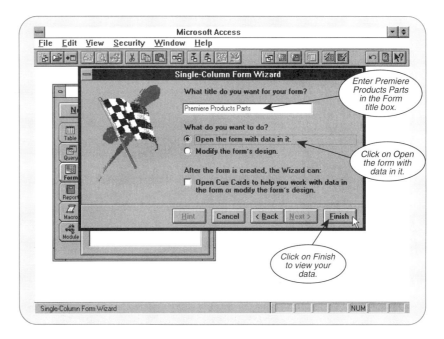

Form Wizard will create the standard form, including the fields that you copied, and display the field labels in the appearance that you selected. When you open a new form, the first record in the table is displayed in the new form's format, as in Figure M5.6.

The single-column design simply displays all fields straight down the left side of the screen. It helps to modify the form so that it is easier to read. To modify the form design, click on the Design button. Figure M5.7 displays the fields and their attached labels.

■ Click the Design button.

Moving Fields on a Form

The first action you can take to improve the form is to move the fields into a different arrangement. We will move the Part description to the right of the part number and then the warehouse data into a group.

To move a field's text box, you must first click on the field—either the attached

Figure M5.6

Default single-
column form

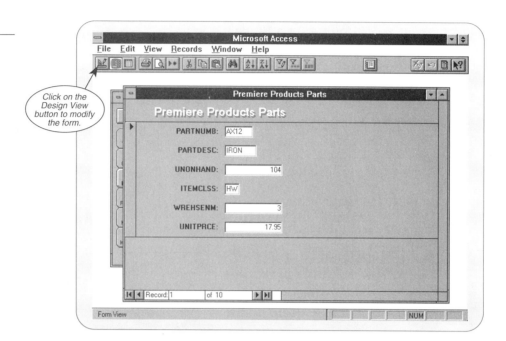

Figure M5.7

Form Design
screen

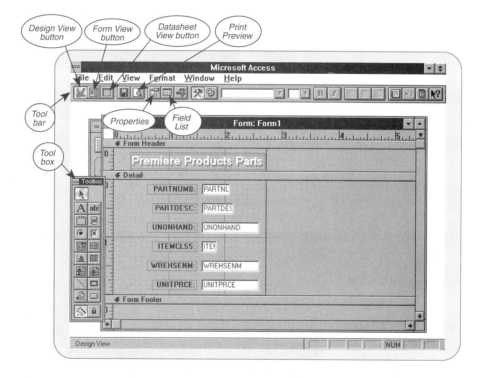

field label (the box on the left side), or the field control box (the box on the right). Figure M5.8 shows that the field now has handles on the corners of each box. In this example, the field's attached label is selected. Notice that the box on the left (the attached label) has a larger handle on the top-left corner and smaller handles on each side and corner. These handles are used to move and size the fields. The handle in the upper-left corner is used to move the selected box, while the smaller handles are used to change the size of the boxes. When you hold down the left mouse button on the move handle, the selected box, either the field control box or the attached label, moves independently of the other. In addition, when you move the pointer on top of one of the box edges, the

Figure M5.8

Modifying a form

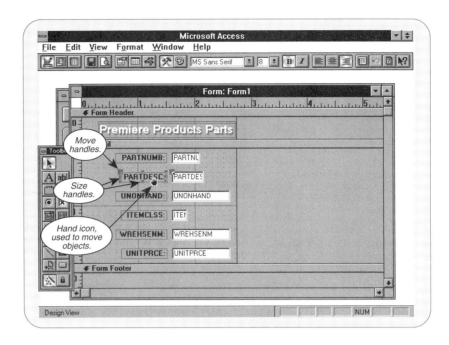

arrow changes to an open hand icon. The hand is used to drag both the field control box and the attached label around the form. To move the *Description* field, first click on the field and then move the pointer until the arrow changes to a hand. Hold down the left mouse button (with the hand pointer) and drag the field up and to the right of the *Part number* field.

- Click on the *Partdesc* field.
- Move the pointer to edge of the field until it changes to a hand.
- Drag the field to the new location, as Figure M5.9 shows, and release the mouse button.

Figure M5.9

Moving a field

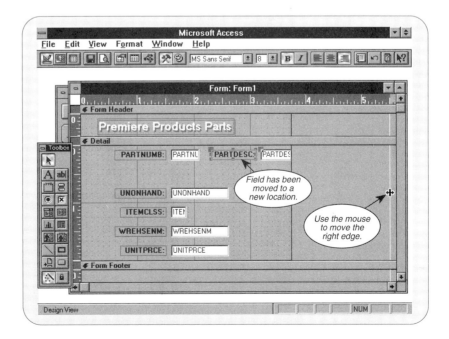

Once you release the mouse button, the field will have been moved to the new location. That is all there is to moving fields on a form. Now let's finish moving the rest of the fields.

■ Relocate each field to match Figure M5.10.

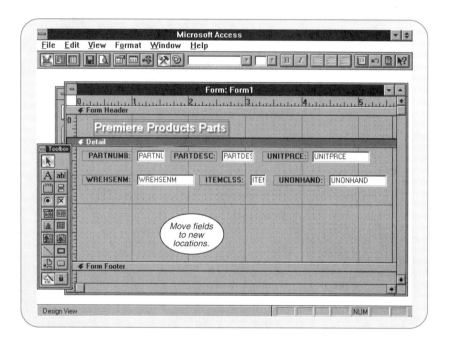

Access uses the field names for the labels for each field. The words preceding each field box, *Partnumb, Partdesc, Unonhand,* and so on are called **prompts.** They prompt the user to enter data and indicate the type of data that is to be entered. When the cursor is next to the word *Partdesc,* for example, the user knows it is time to enter a part description. We certainly want prompts on our forms, we also want them to be more descriptive.

You will now change your prompts (*Partnumb, Partdesc, Unonhand,* and so forth) and replace them with those shown in Figure M5.11. To change the existing prompt, click on the prompt just to the left of the colon at the end of each prompt. When you click on the prompt, the pointer changes to an insertion bar. To erase the current prompt, repeatedly press Backspace. To insert the new prompt, simply type it.

■ Change the prompts to the ones shown in Figure M5.11.

Changing the Size of Field Boxes

The size of the field also may not match the actual length of the data in each field. Because Access uses the field names as the label for each field box, some of the boxes are much larger than they need to be. For example, the length of each warehouse number is only one character, while the box on the form is much longer. Therefore, we can change the size of each field box to match the needs of the data in that field.

To change the size of each field, click on the field box that you want to edit. The field will have handles on each corner, as they did when you moved the field location. Handles on the center of one of the edges of a box may be used to change the size of the box in one direction only. For example, you may use the handle in the center of the top or bottom of a box to change the height of a box, whereas you would use the handle in the center of the left or right side to change the width of a box. You may use handles on the corners of a text box to change the size of the box in any direction. As you move the pointer over the lower-right corner of each field, the arrow changes to a resize

Figure M5.11

Changing field
size, location, and
prompts

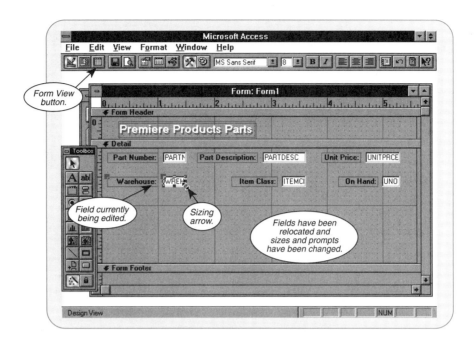

arrow like that one shown on the *Wrehsenm* field in Figure M5.11. With the double
arrow showing, drag the size handle to the left to shrink the field or to the right to
enlarge it.

■ Change the size of each field box to match Figure M5.11.

Note: After you have changed the prompt, location, and size of each field, you
may need to move each field again in order to get the final version to match Figure
M5.11.

Saving the Form

After you have completed the changes, you want to save the form as follows:

■ Select Save As from the File menu.
■ Enter `Parts` for the form name.
■ Click on OK to save the form.

Viewing Data Using a Form

To view data in the new form, click on the Form View button in the tool bar. Your
screen should now look similar to Figure M5.12. To scroll through the table, press the
PageDown key, to view each record one at a time in the form you just created. Pressing
PageUp scrolls backward through the table.

Adding an Additional Field

We will now examine the process of adding an additional field to a form by adding
the on-hand value to the form you created for the *Part* table. (On-hand value is the
product of the units on hand and the unit price.) We will place this new field directly
below the unit price.

■ Click on Design button on the tool bar.
■ Click on Text box icon in the tool box.
■ Move the text box icon below the unit price.

Your screen will now be similar to Figure M5.13.

■ Click the left mouse button to place the new field.

Your screen will now be similar to Figure M5.14. Before you describe the calculated
value of this field, you can change the field prompt. Follow the process previously used
to change the field prompt to On-hand Value.

Figure M5.12

Form View screen

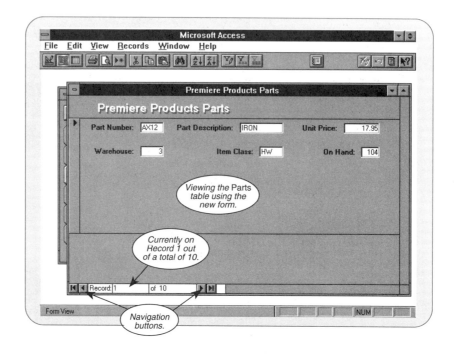

Figure M5.13

Adding a field to a form

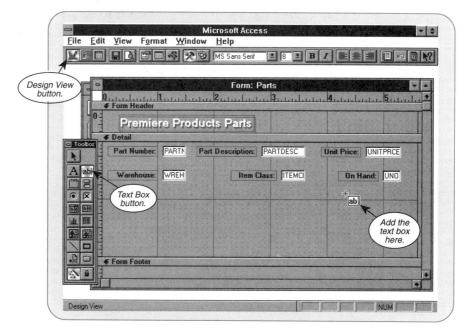

■ Change the new field's prompt to On-hand Value.

To describe the calculated value of the new field, as Figure M5.15 shows, click on the new text box and delete the current entry. Then type in the new formula:

■ Click in the new text box.
■ Enter the new formula: =[Unitprce]*[Unonhand].
■ Press the Enter key.

One more change necessary before you view our new form is the format to display the data. Without describing the desired format, Access will use a general numeric format. Therefore, to use the first record for an example, the units on hand currently

Figure M5.14

Adding a field to a form

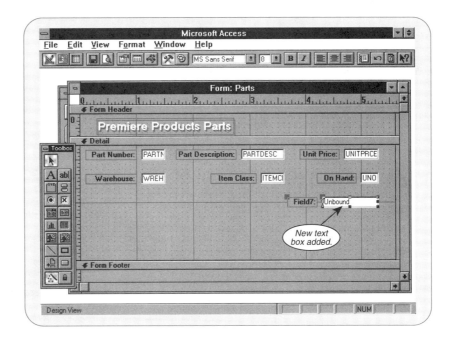

Figure M5.15

Describing a calculated field

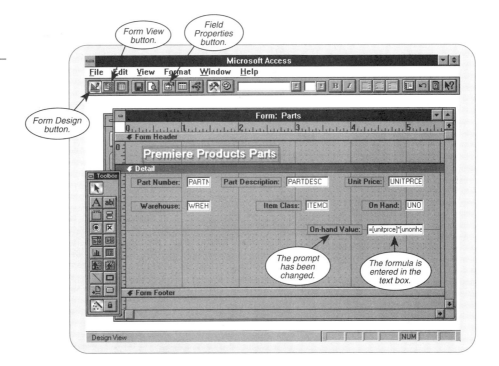

equals 104, and the unit price equals $17.95. The product of these two values would produce a value of 1866.8 unformatted. Let's change the format of both the on-hand value and unit price fields. Refer to Figure M5.16 to make the following changes:

- Click on the On-hand Value box.
- Click on the Field Properties button.
- Click on the Format box.
- Click on the drop-down arrow.
- Select the Currency format.

Figure M5.16

Changing field
properties

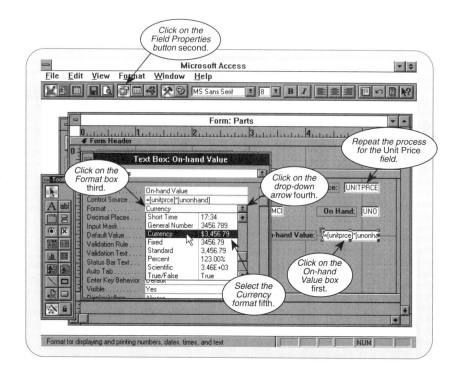

Figure M5.16

Changing field
properties

■ Repeat the process for the *Unit Price* field.
■ Click on the Form View button to see the completed form.

Your screen should now look like Figure M5.17.

**Adding Boxes
to a Form**

You can improve the look of a form by adding boxes to set off portions of the form. In
this form, you will place a box around (actually under) all of the fields. Begin by
selecting the rectangle box tool and placing a box on the form (Figure M5.18).

■ Click on the Rectangle Box tool.
■ Click to place the box on the form.

Figure M5.17

Using calculated
fields

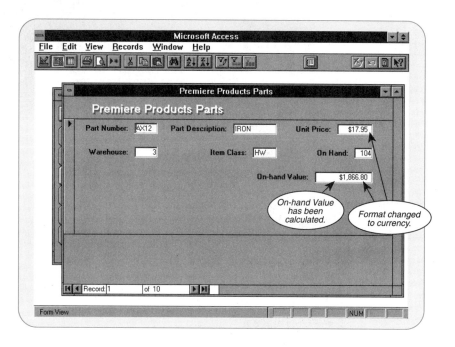

Figure M5.18

Adding a box to forms

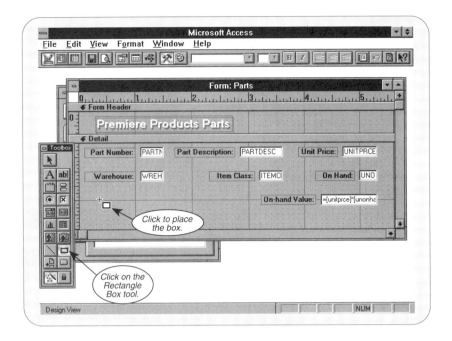

The default rectangle box is placed on the form similar to Figure M5.19. You can use this box to start with and modify it to meet your needs. First relocate the box to the upper-left corner of the screen by moving the pointer over the box. As you do, the pointer changes to a hand grabber. Press the mouse button and drag it so that the top-left corner of the box is just above and left of the *Part number* prompt. When you release the mouse button, the field is placed at the point of release.

■ Move the pointer over the box so that it changes into a hand.
■ Drag the box to the upper-left corner of the form.

Next, change the size of the box to include all of the fields. Use the size arrows to change the size of the box to match Figure M5.20.

Figure M5.19

Adding a box to forms

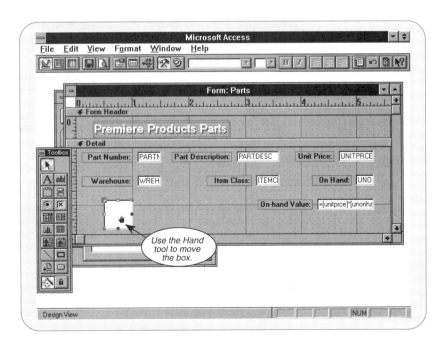

Figure M5.20

Resizing a box

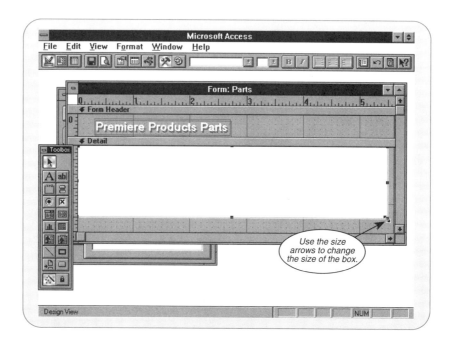

■ Move the pointer over the lower-right corner so that it changes into a double arrow.

■ Press the mouse button and drag the lower-right corner to the location indicated in Figure M5.20.

The box as it is created by default is a solid white color and is placed in the top or front layer of the form. By moving the box to the back, it places a frame for the fields.

■ Select Send to Back from the Format menu. (This places the box behind the fields.)

Figure M5.21

Using a box as a background

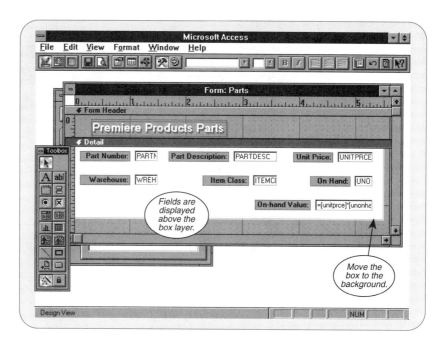

Changing Fonts

The box improved the look of the form. You can enhance the form further by using special attributes such as color and font. We will illustrate the process by changing the form's font. The easiest way to select each of the fields is to select the *Select All* fields from the Edit menu.

- Select the *Select All* option from the Edit menu.
- Select the Times New Roman font from the drop-down font list in the tool bar.

Note: If your fonts are not the same as those shown, you could select the *MS Serif* font or some similar font.

Your screen should now look like the one shown in Figure M5.22.

Figure M5.22

Changing fonts in a form

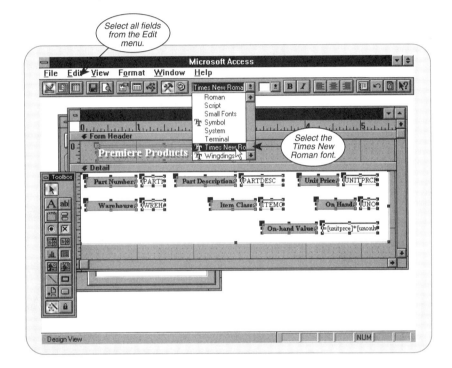

Changing Colors

In addition to changing the form's font, we want to add another dimension to the form by changing the color of the box we added. The color option allows you to change the color scheme for an area of the screen, for the box, or for individual fields. You could even make the background one color, the prompts another, and the data in the fields a third color. We will change the white background box that we added to a dark shade with a border.

Access the color options by clicking on the Palette button. The palette box is used to change colors and appearance of objects on the screen. The Palette is shown in Figure M5.23.

- Click on the white background box to select it for modification.
- Click on the Palette button.
- Select a dark shade for the Fill color.
- Select a black border.
- Select a single line border.
- Close the palette by clicking on the control-menu box and then Close.
- Click on the Form View Button.

Figure M5.24 shows the final form product.

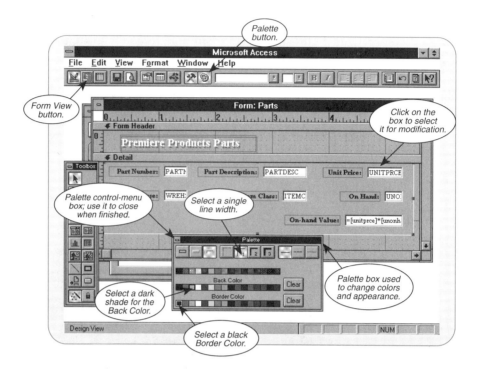

Figure M5.23

Changing colors in a form

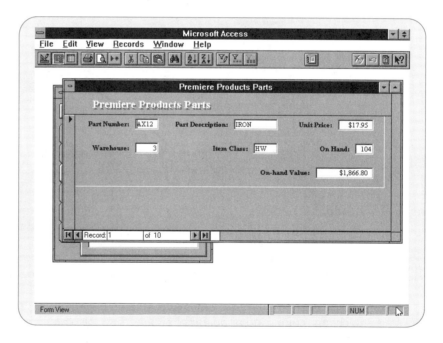

Figure M5.24

Final form design

Closing a Form

When you have finished designing a form, be sure to save it. You should have saved the form previously with the name *Parts*. If the form has already been saved, use the File menu's Save command. To save a form for the first time, use the Save As command. Assuming that you are saving for the second time:

- Select Save from the File menu.
- Select Close from the File menu.

Using a Form

To recall and use a form at a later time when you're in a database window, simply click on the Form object button and select the form.

- Click on the Form object.
- Select the Parts form.
- Click on the Open button.

The first record of the *Part* table is now visible and available for searching, editing, adding records, or printing.

M5.3   VALIDATION RULES

The work we have completed so far makes data entry more pleasing. It does nothing, however, to ensure that users only enter valid data. In this section, we will create **validation rules** that limit data entered in a record to values that meet certain requirements. You create validation rules by setting an expression that is evaluated when data in a field is added or changed. Access checks the data to ensure it follows the expression rule before saving the record in the table. As you will see, Access will prevent users from entering data that violates rules. Validation rules are set in a field's properties box and are entered as expressions that may include such rules as minimum or maximum values, a range of values, or specific values in a list.

Note: Access automatically ensures that a table's primary key fields and other fields with unique indexes contain values, so you don't need to set a validation rule to require data entry in those fields.

Specifying a Required Field

Suppose that you want to ensure that when a record is added or changed, a value is given for the number of units currently on hand. To do so in Access, you set the Required option of the Field Properties to Y for yes.

Specifying Ranges

To specify a range into which entries for a field must fall, use an expression for the values. Let's assume that the value in the *Unonhand* field cannot be lower than 0 or greater than 999.

The *Parts* form should be in the active window. If it is not, select the Form object, and then open Parts and change to the Design View.

- Click on the Design View button.
- Click on the *Unonhand* field.
- Click on the Field Properties button.
- Click on the Validation Rule box.
- Enter >=0 And <=999 in the Validation rule.
- Enter Must be between 0 and 999 in the Validation Text box.

Your screen should now look similar to Figure M5.25. The validation rule entry checks the entered data against the expression. If the two do not match, the validation text is displayed in an error box.

- Repeat the above process to create a validation rule and text for Wrehsenm field.
- Click on the *Warehouse* field box. (You may have to drag the properties box to a different location to get to the field.)
- Enter the validation rule as a range between 1 and 3.
- Close the Properties dialog box.

Specifying Default Values

If you want to enter a default value for a field, simply select the field on which you want to set the value, click on the properties button, and enter the default value in the appropriate box. For example, if most parts were assigned to warehouse number 1, you could set the default value in the field properties. (However, we do not have any such fields at this time.)

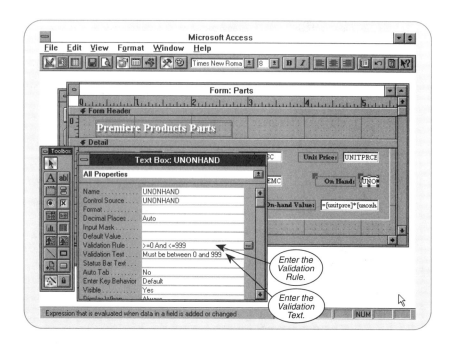

Specifying a Collection of Legal Values

Access provides a variety of methods to enhance your form to select a value from a list rather than enter the data. A list of choices is presented for the user to choose from. Access has two types of controls that provide a scrollable list of choices: list boxes and combo boxes. List boxes present a box with a list of valid alternatives that are displayed at all times. A combo box does not display the list until you click on the drop-down arrow to display it. The advantage of this type is that it takes up less room on the form and you only see the selected value.

You can choose to create your list or combo box on your own, or you can use the List Box or Combo Box Wizard to prompt you through the options. If you know the list of data and the format that you want it displayed in, it is often easier to create the field without the assistance of a Wizard. To demonstrate this feature, we will create a combo box for the *Itemclss* field. First, you must delete the current *Item Class* field so that you can add another one with the same name.

- Click on the *Itemclss* field.
- Press the Delete key to get rid of the old field.
- Click on the Control Wizards tool so that it is *Not* highlighted.
- Click on the Combo box tool.
- Place the new combo box in place of the old field.

Use the same procedure to place the new combo box as you did in placing a new text box. (Refer back to Figure M5.13.) The new combo box may not fit in the given space without changing the field's size and location. Use the size and move handles of the new box to resize and place the field, as Figure M5.26 indicates. Then use the field properties to enter the list of options.

- Resize and locate the new box, as Figure M5.26 shows.
- Change the label of the field to `Item Class:`
- Click on the Properties button.
- Enter `ITEMCLSS` in the Name and Control Source lines.
- Enter `Value List` in the *Row Source Type* field.
  (You could also click on the drop-down arrow and select Value List.)
- Enter `HW;SG;AP` in the Row Source field.

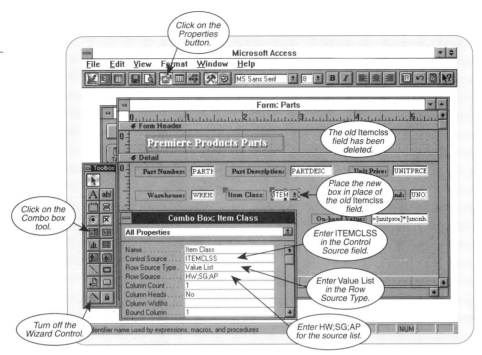

Your screen should now resemble Figure M5.26. To view the form (Figure M5.27) and see how the combo box works, enter the Form View.

- Click on the Form View button.
- Click on the *Itemclss* drop-down list.

When you create a new form, Access automatically assigns a tab order that controls which field the tab moves to between entering data in each field. To set the tab order yourself:

- Click on the Form Design button.
- Select Tab Order from the Edit menu (Figure M5.28).

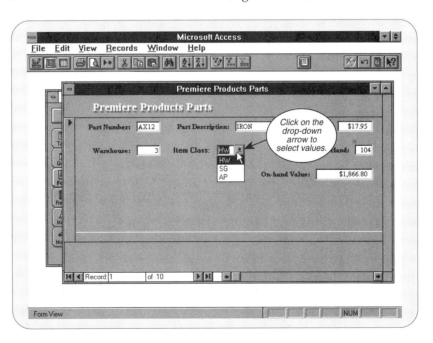

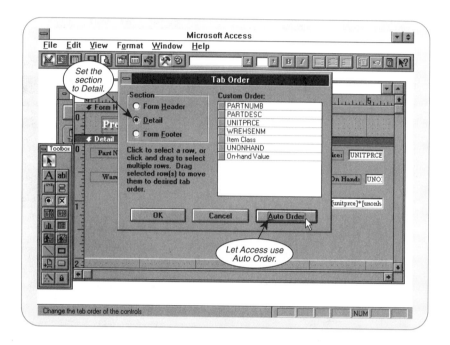

- Click on the Detail option.
- Click on Auto Order to let Access automatically assign the order as Left to Right, Top to Bottom.
- Click on OK.
- Change to the Form View.

When you change back to the Form View and press Tab through each field, the order will flow logically across each row.

- Select Close from the File menu.
- Respond Yes to save and close the form.

Table Lookup Validation

Another type of validation is to use a Combo box to look up valid data from another table or query. This lookup is used to ensure that data in one table matches data in another. This type of validation is similar to the previous Combo box we created for *Itemclss*, but we will select Table/Query for the source instead of a Value List. For example, we will use this feature to make sure that the sales rep number for any record in the *Customer* table actually matches the sales rep number on some row in the *Slsrep* table. (It wouldn't make sense to have a customer represented by sales rep 4 if there were no sales rep 4.) We will now create this validity check.

(Open the *Prem* database if it is not already open.)

- Click on the Form object.
- Click on New to create a new form.
- Select Customer from the drop-down list of table/queries.
- Click on FormWizard button.
- Select the Single-column format and then OK.
- Click on the >> button to copy all available fields to the field order list.
- Click on the Finish button to move to the end of the FormWizard.
- Click on the Design icon to modify the form.

Your screen should now look similar to Figure M5.29. (You may need to resize the form to match the form in Figure M5.29.) We next want to replace the *Slsrnumb* field with a combo box that uses a table to validate the sales rep number.

Figure M5.29

Creating a
Customer lookup
form

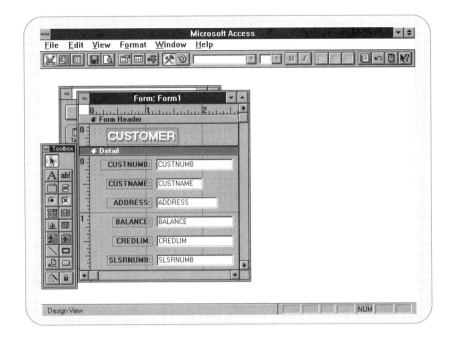

Note: Before continuing, make sure that the Control Wizards button (the lower-left button on the Toolbox) is turned OFF.

- Delete the original *Slsrnumb* field by clicking on it and pressing the Delete key.
- Click on the Combo field button.

As you move the mouse pointer into the form window, the arrow pointer changes to a combo box icon.

- Click the mouse button to place the new combo box in the original space, as Figure M5.30 indicates.

Figure M5.30

Creating a
Customer lookup
form

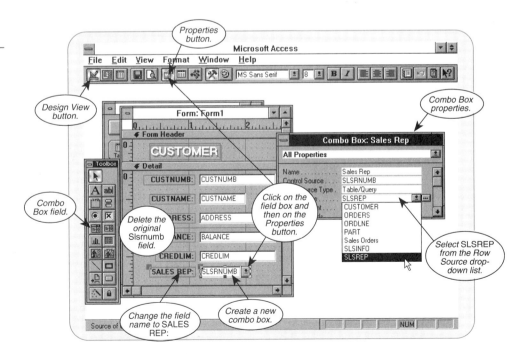

- Change the field's attached label to `Sales Rep.`
- Click on the field box and then on the Properties button on the tool bar.
- Select *Slsrnumb* from the Control Source drop-down list in the Combo Box properties.
- Select *Slsrep* from the Row Source drop-down list in the Combo Box properties.
- Click on the Form View button to see how the combo box works.

Your completed lookup query should look similar to Figure M5.31. Click on the drop-down arrow for the Sales rep number. Only the three valid sales representatives are listed. Click on number 3 to select the number. Since we are finished with this form and do not need to keep it, close it without saving.

- Select Close from the File menu.
- Answer `No` to skip saving the form.

Figure M5.31

Using a lookup

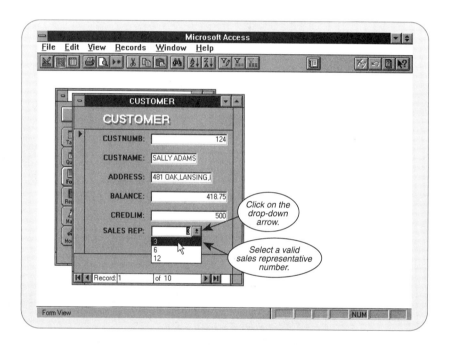

M5.4   MULTI-TABLE FORMS

This section investigates the creation and use of forms that involve more than one table. Figure M5.32 gives a sample of such a form. The top of the form contains data from the *Orders* table (order number, date, and customer number). The box near the top shows data from the *Customer* table (name, address, balance, and so on). The box in the lower portion of the screen contains data from the *Ordlne* table (part number, number of units ordered, and quoted price).

The main part of the form is called the **main form,** and the table it updates is referred to as the **main table.** In our example, the *Orders* table is the master table and the portion of the form containing the order number, order date, and customer number is the master form.

Forms included from other tables are called **subforms.** Although you can include an subform from a table that is not related to the master table, it is not particularly useful. Thus, there should be some relationship between the tables. In addition, the relationship must use the first field in the tables for the embedded forms. Pro-

Figure M5.32

Multi-table form

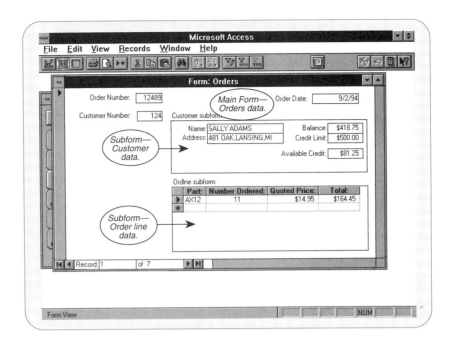

cessing will be most efficient if the field is keyed. As long as the table is related to the master table in this way, it is called a **linked table.** Subforms are especially effective when you want to show data from tables or queries with a one-to-many relationship. The main form represents the *one* side of the relationship. The subform represents the *many* side of the relationship.

There are two main possibilities for the relationship:

1. The field from the master table matches the unique key for the linked table. In this case, each master table record will be related to exactly one detail record. In our example, the master table (*Orders*) will be linked to the *Customer* table by using the customer number (*Custnumb*) fields from both tables. Since the *Custnumb* field is the unique key for the *Customer* table, there will be only one customer for each order.

2. The field from the master table matches a portion of the unique key for the linked table. In this case, each master table record can be related to several detail records. In our example, the master table (*Orders*) will be linked to the *Ordlne* table by using the order number (*Ordnumb*) fields from both tables. Since the key for the *Ordlne* table is the combination of the *Ordnumb* and *Partnumb* fields, there can be several order lines for each order.

The subform for the *Customer* table only needs to display a single customer, the one customer who placed the order that is currently on the screen. This is called a *single-record subform.* It would be helpful, however, if the subform for the *Ordlne* table could display several order lines at a time. Since this means display *multiple* records, this type of subform is called a *multi-record subform.*

The process to combine all of the above sections requires the creation of three forms. First we will create two subforms (Customer, and Ordlne) and then create a master form (Orders) that will include the two subforms.

**Creating a
Single-Record
Subform**

A subform (especially a single-record subform) is very much like any other form. There are only two special issues we need to address in such a subform:

1. Since the subform must fit *inside* another form, we must make it small enough so that it can do so.
2. We won't include the field that will be used to link the tables together, since it will be included as a field from the master table. In our example, we will not include the *Custnumb* field in this subform.

With these in mind, let's create the Customer subform.

- Click on the Form object button.
- Click on the New button.
- Select Customer from the Table/Query list. (Use the drop-down list.)
- Click on Blank Form.

A blank white box appears on the screen titled Form1. We will drag the four necessary fields from the *Custnumb* table's field list, and add a calculated field.

- Select Field List from the View menu.

A list of fields from the *Customer* table displays, as Figure M5.33 shows.

- From the field list, drag the Custname, Address, Balance, and Credlim fields from the list onto the blank form.
- Change the position and size of the boxes to match those in Figure M5.34.

You are finished adding fields from the Field List, so you can close that box by double clicking on its control menu box or by clicking once and then selecting Close.

- Close the Field List box. (Double click on the control-menu box.)

To change the size or position of a field, click anywhere on the field. You will notice that the field box has handles on the corners. The handle in the upper-left corner of the field and the field name boxes are larger than the others. These larger handles are used to move the prompt or field box independently. When you move the pointer to the large handle on the top-left corner, the arrow changes to a hand with the finger pointing to the area to select. Use this handle to change the beginning (top-left) position of either the prompt or the field box.

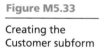

Figure M5.33

Creating the
Customer subform

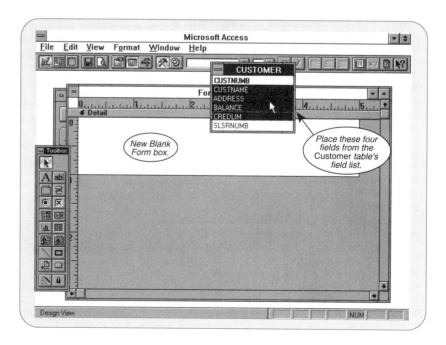

Figure M5.34

Creating the
Customer subform

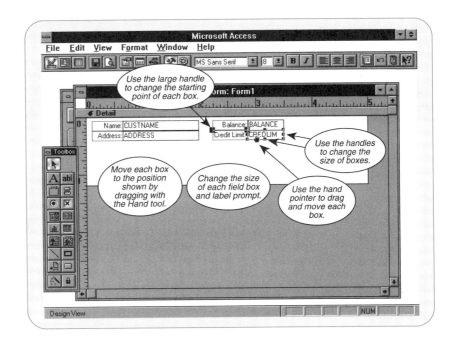

You can move the box by pressing the left mouse button with the hand pointer on the handle in the upper-left corner and dragging to the new location.

When you move the pointer to one of the edges of the box, the pointer changes to a hand. Use this hand pointer to drag the box to change its position or size. Change the position of each of the fields and their prompts to match Figure M5.34.

Let's replace the field names for new prompts on the form now. Change the field names by clicking on the prompt to select the prompt box and then click again at the point in the prompt text that you want to edit and click again. Use the Backspace key to erase the old text and type in the new prompt.

- Rename the prompts: `Name, Balance, Address, Credit Limit.`

You also need to resize the background box to make room for another field. To resize the box, place the pointer on the right edge of the white background box. You will notice that the pointer changes to a left-right arrow that is used to move the edge of the box. Use this pointer to change the size of the box (as Figure M5.35 shows) so that the right side of the box is on the 3 1/2-inch mark.

- Change the size of the box as Figure M5.35 shows.

Next, we will add a field for available credit (Available credit = credit limit − balance).

- Click on the Text Box icon in the tool box.
- Place the box below the *Credit Limit* field.
- Rename the prompt: `Available Credit:`
- Change the "unbound" in the field box to: `=[credlim]-[balance]`
- Use the hand pointers to change the position and size to match Figure M5.36.

Next let's design the subform to look like we want it to appear in the main form. You can change the appearance of fields by changing the settings for each field's properties. Click on the field prompt to change the prompt alignment, and click on the field content on the right side to change the format and alignment of the data. Start with the *Available Credit* field.

- "Click on the field you want to modify (starting with *Available Credit*).

Figure M5.35

Modifying the size
of the background
box

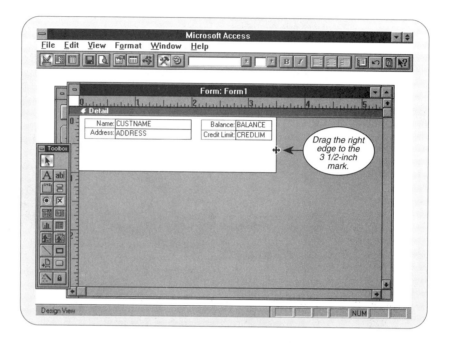

Figure M5.36

Adding a field to
the Customer
subform

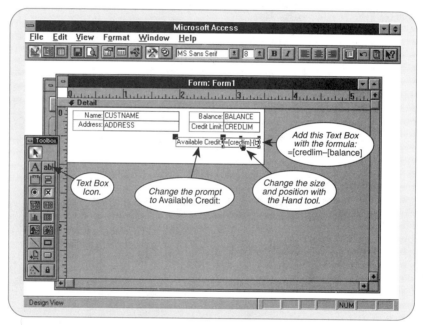

- Select Properties from the View menu. (See the box in Figure M5.37.)
- Change the Format to `Currency`.
- Scroll down and change the Text align to `Right`.
- Repeat the process for the *Limit* and *Balance* fields.
- Click on the label prompts and change each Text align to `Right`.
- Close the Properties box.

Your final Customer subform should now look like the one in Figure M5.38. If it does not, modify it to be as similar as possible.

- Click on the Form View button.
- Select Save As from the File menu, and save it as `Customer subform`.
- Double click the control-menu box to close the Customer subform.

Figure M5.37

Modifying field
properties

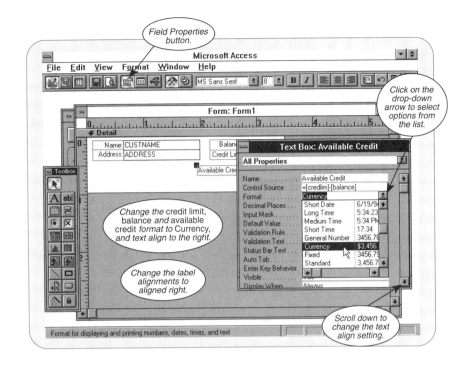

Figure M5.38

Completed
Customer subform

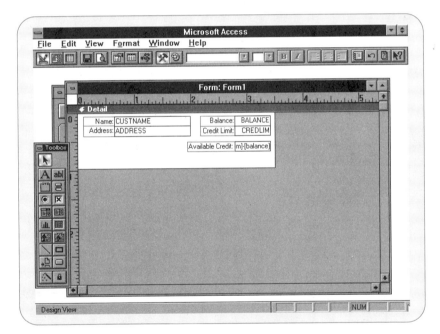

Next we repeat the above process to create the Ordlne subform.

■ Click on the Form object button and then on New.
■ Select the *Ordlne* table.
■ Click on the Blank Form button.
■ Select Field List from the View menu to get the fields *Partnumb, Numbord,* and
    Quotprce.

Place each of the fields in the background box, as shown in Figure M5.39. Placement is not as important in this form because we will only be viewing the subform in
the datasheet view.

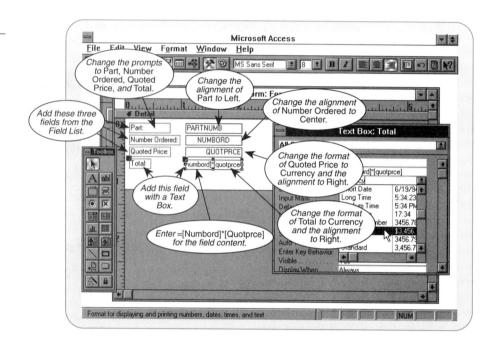

- Change the prompts to `Part`, `Number Ordered`, and `Quoted Price`.
- Change the Number Ordered, properties alignment to `Center`.
- Change the Quoted Price properties format to `Currency` and alignment to `Right`.
- Add a Text box for Total.
- Change the prompt to *Total*.
- Change the Unbound value in the field to `=[Numbord]*[Quotprce]`.
- Change the Total properties format to `Currency` and alignment to `Right`.

We want this form to be the same width as the Customer subform. Change the width of the white background box by moving the arrow pointer to the right edge and moving the right side to the 3 1/2-inch mark and the bottom edge to the 1/14 inch mark.

- Change the size of the white background box so that the right side is at 3 1/2 inches and the bottom is at 1/14 inch.

When you are finished, change to the Datasheet view and save the form.

- Click on the Datasheet View button.
- Save the form as `Ordlne subform`.

The last step we need to take in creating the two subforms is to set the appearance and views allowed in the form properties. To access the Form Properties, select the form from the Form window, click on the Design button, and then click on the Properties button (Figure M5.40).

- Click on the Design button.
- Click on the Properties button.
- Change the Default View and Views Allowed to `Datasheet` using the drop-down list.
- Change the Scroll Bars to `Neither`.
- Save and close the Ordlne subform window.

Now repeat the above process for the Customer subform.

- Click on the Design button.
- Click on the Properties button.

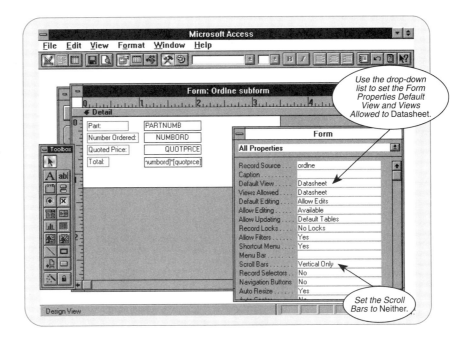

- ■ Change Default View to `Single Form`.
- ■ Change Views Allowed to `Form`.
- ■ Change Scroll Bars to `Neither`.
- ■ Change Record Selectors to `No`.
- ■ Change Navigation Buttons to `No`.
- ■ Save and close the Customer subform window.

Creating a Multi-table Form

Begin creating the multi-table form just like any other form. The only difference is that, before you are done, you will place forms from other tables within this form.

- ■ Click on the Form object button and then New.
- ■ Select the *Orders* table.
- ■ Click on the Blank Form.

Next, add the fields and prompts. Select the Field List and add the three fields, as shown in Figure M5.41.

- ■ Select Field List from the View menu.
- ■ Drag the Ordnumb, Orddte, and Custnumb to the form as shown.
- ■ Close Field List.
- ■ Change the labels and field box size, as Figure M5.41 displays.
- ■ Select Save from the File menu.
- ■ Save the form with the name Orders.

Once you have the three main form fields added, you can add the two subforms to the main form. Accomplish this by dragging the icon for the two subforms from the Database window to the main form (Figure M5.42).

- ■ Select Database: *PREM* from the Window menu.
- ■ Click on the Form object icon.
- ■ Drag the icon for the Customer subform to the main form.
- ■ Select Database: *PREM* from the Window menu.
- ■ Drag the icon for the Ordlne subform to the main form.

Next, realign the two subforms on the main form. Click on the Customer subform and use the box handles and the hand pointer to move the subform outline, as

Figure M5.41

Creating the Multi-
table Orders Main
Form

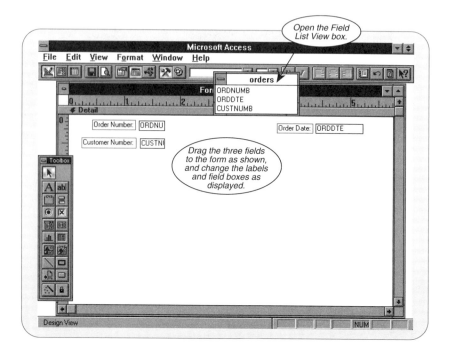

Figure M5.41

Creating the Multi-
table Orders Main
Form

Figure M5.42

Adding the
Customer and
Order line
subforms

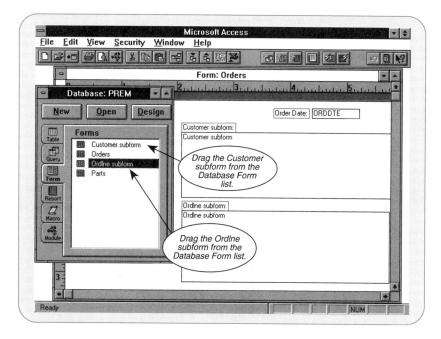

shown in Figure M5.43. Then click on the Ordlne subform and place it directly under
the Customer subform.

■ Drag the Customer and Ordlne subforms to the positions shown in Figure M5.43.

Once you have moved the subform outlines to the correct locations, change the
size of the white background box. We want the size of the background box to be
5 1/2 inches by 3 inches. Use the double arrow pointer when the arrow is moved to
the edge of the box to drag the box edge.

■ Resize the background box so that the background box measures 5 1/2 inches
by 3 inches.

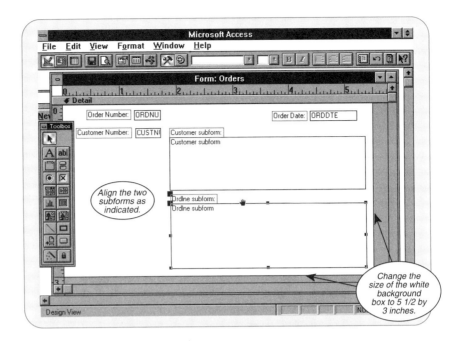

Once you have added the subforms to the main form, you need to link them with the main form. When you display a form that contains subforms, you want the subform to show the records that are related to the main form. For example, if orders are displayed on the main form, the subforms should display records related to the current main record.

After you have dragged a subform to the main form, you can check to see if Access has automatically created a link for you. If the link is not set, you will have to link the forms. To check or set the link between a subform and the main form, open the properties for the subform. The *LinkMasterFields* and *LinkChildFields* should have the linking fields entered. For example, the properties settings for the `Customer` subform should have the *Custnumb* field as the link.

With the current screen in the Design view for the Orders main form, click on the Customer subform and click on the Properties button. As Figure M5.44 shows, enter the *Custnumb* as the *Master* and *Child Link* fields.

- Click on the Customer subform.
- Click on the Properties button.
- Enter `Custnumb` for the LinkMasterFields and LinkChildFields.
- Close the Properties box.

Then repeat the settings for the Ordlne subform.

- Click on the Ordlne subform.
- Click on the Properties button.
- Enter `Ordnumb` for the *LinkMasterFields* and *LinkChildFields*.
- Close the Properties box.

Your form should now be complete. Switch to the Form view and save the main form. The links between the subforms and the main form will be saved with the main form.

- Select Save from the File menu.

Your screen should now look like the one shown in Figure M5.45. Notice that order 12489 is on the screen. Data about the customer who placed the order (Sally

Figure M5.44

Linking the
Customer subform
with the main
form

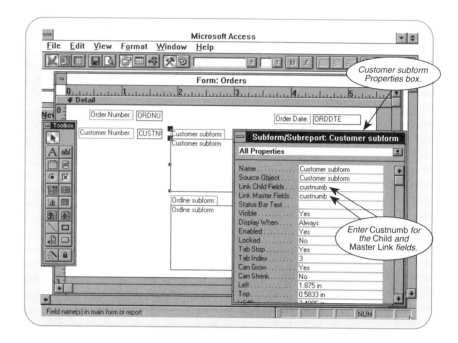

Figure M5.45

Orders form in the
Form view

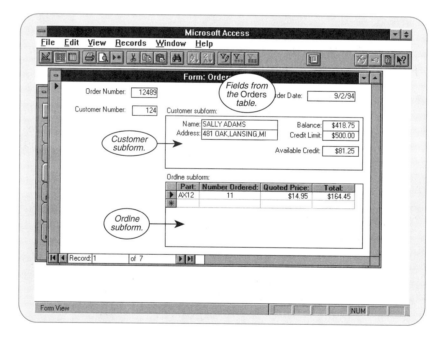

Adams) is shown in the upper box. The lower box contains only the order line for this
order.

**Using the
Multi-table
Form**

You can use the multi-table form by activating it just as you did other forms. For
example, to search for specific records using a form, open a form and click on the Find
icon (the binocular icon on the tool bar.) To demonstrate, we will search for order
number 12494.

- Click in the *Order number* field.
- Click on the Find (binocular) button.
- Enter 12494 in the Find What field (Figure M5.46).
- Click on Find First.

Figure M5.46

Searching for
order number
12494

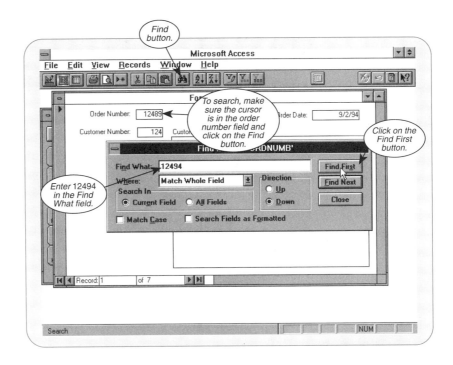

Figure M5.46

Searching for
order number
12494

When you have found the record, click on the Close button to quit the search and
the desired record is displayed (Figure M5.47).

Let's use this form to add a record. (Refer to Figure M5.48 for prompts.)

Note: You could also use the DataEntry option to add records with this form. The
process of actually filling in the form would be the same in either case.

To add records in the form view, move to the last record and click on the next
record button. This will insert a new blank record. Enter the order data, including the
customer number. When the customer number is added, the customer data will automatically be filled in. You can then click in the first *Part Number* field and add the first
order line.

Figure M5.47

Completed order
number search

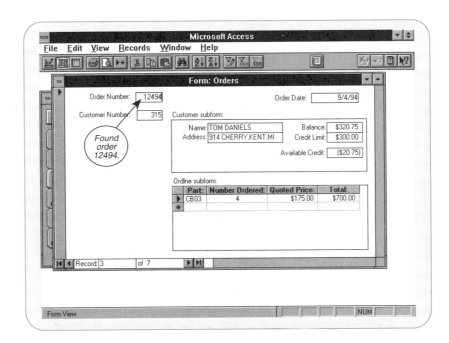

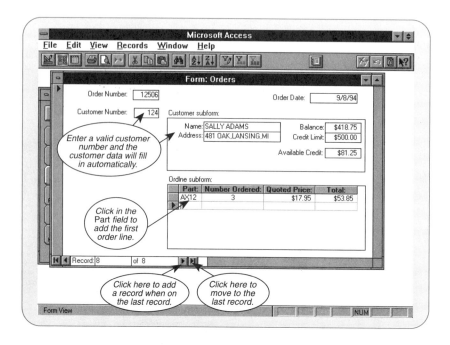

- Click on the Last record button to move to the end of the table.
- Click on the Next record button to insert a new record.
- Enter 12506 as the order number and press Enter.
- Enter 9/08/94 as the order date and press Enter.
- Enter 124 as the customer number and press Enter.

The customer data will be filled in for the Customer subform. You can now move to the Order line subform to add the first order line.

- Click on the first part number field.
- Enter AX12 as the part number and press Enter.
- Enter 3 as the number ordered and press Enter.
- Enter 17.95 as the quoted price and press Enter.

The Total value will be calculated. Press Enter; you are now ready to add a second order line if you had any to add. You are now finished working with the forms. Double click on the control-menu box to close the form.

SUMMARY

The following list summarizes the material covered in this module:

1. To create a custom form, click on the Form object button and then on Design. Select the table for which you are creating the form, and then click on Form Wizards. Form Wizard will create an initial form using the prompts that you provide. You can then modify this initial form. To modify a form, use the same procedure: This time, when you click on design, you are automatically in the form and may modify as necessary.
2. When creating a form, use the Field List to place fields.
3. To use a form you have created, click on the Form object and select the form you want to use. The appropriate table(s) will automatically be selected.
4. To construct a multi-table form, first construct the subforms with a single or multi-record structure. Then save the subform as you would any other form.

5. Create a subform just like any other form. The form must be small enough to fit at the appropriate location within the master form.
6. To include a subform within the multi-table form, create the main form as you would any other form. Then drag the icon for the subform from the database form list to the main form and modify the size and position as desired.
7. Create links between the subforms and the main form by selecting the subform in the Design view. From the Form Properties, set the LinkChildFields and Link-MasterFields to the same linking field.
8. Use a multi-table form like any other form. Any data in a subform that is linked to the main form will be filled in automatically.
9. One-to-many records are created in the Relationships of the tables. When you use these with multi-table forms, these relationships are automatically recognized.

EXERCISES

1. What is the purpose of a custom form? How do you create one?
2. How can you use the standard FormWizards form as a starting point in designing a form?
3. How do you move fields on a form? How do you add fields? How do you delete them?
4. How can you make Access automatically use a custom form instead of the standard form?
5. How can you add a box to a form? How can you change the visual characteristics of a form?
6. What is a multi-record form? How do you construct one?
7. What is a multi-table form? What is the master form? What are subforms?
8. Describe the process of including a subform in a multi-table form.
9. How do you move between the subforms within a multi-table form? When you are using a multi-table form, how do you move to another record in the master table?

COMPUTER ASSIGNMENTS

1. Create a Form Wizards form for the *Publishr, Branch,* and *Author* tables in the *Books* database. In each case, your new forms should be the forms that appear automatically when a user switches to Form View. Describe the steps involved.
2. Create a multi-table form for the *Book* table and describe the steps involved. The multi-table form should contain three subforms:
   a. A form Listing the publisher's name and city. This form should be display only.
   b. A multi-record form giving the Author number and the sequence number. There should be room in the subform for at least two authors.
   c. A multi-record form giving the branch number and number of units on hand for each branch that has copies of the book. There should be room in the subform for at least three branches.
3. Use your form to add a book of your choice. The book should have at least two authors. There should be units of the book in at least three branches. Describe the steps involved.
4. Use your form to modify the book you just added. Change the publisher. Delete one of the authors and add another. Change the number of units on hand at one of the branches. Describe the steps involved.

6

Advanced Reports

OBJECTIVES

1. Explain how to create reports with ReportWizard.
2. Discuss how to use subtotals in a report.
3. Explain the use of embedded subreports.
4. Explain how to include multiple levels of grouping in a report.

M6.1 INTRODUCTION

In this module, we continue to investigate the report-creation process that we began in Module 2. Section M6.2 begins by examining the use of ReportWizard. Next, in section M6.3, we use the ReportWizard to create a Tabular Report listing records in a list format. Section M6.4 focuses on the use of groups in reports to include such things as subtotals. We look at embedding subreports in section M6.5 using multiple tables in the report. Finally, in section M6.6 we design a multi-table report with group headers and footers based on a multi-table query.

Note: In case you find you don't want to complete a report design during a single session, do the following:

1. Save the design.
2. When you are ready to begin again, click on the Report object button and then select the report and click on Design.
3. Continue designing the report from the point at which you left off.

M6.2 USING REPORTWIZARD

To assist you in creating reports, Access provides you with ReportWizards. A ReportWizard prompts you through a series of questions about what you want printed and then creates the report based on your answers. Even the experienced report designer may want to use ReportWizard to quickly generate a report. Once you have created these reports, you can modify and refine them. ReportWizards can create Single-Column, Group/Totals, Summary, Tabular, and Mailing Labels reports. In addition, Access provides AutoReport, a Wizard that creates an immediate Single-Column report without any prompts. This is the fastest way to get a report of any table with all the fields displayed.

You used the ReportWizard earlier to produce a simple report. However, you can create many other types of reports with this feature. In this section, you will create an AutoReport to see how the report feature works and then modify that report. In the next section, you will use the ReportWizard to produce reports provided with the ReportWizard using the tables in *Premiere Products*.

ReportWizard Options

As just mentioned, Report Wizards can create Single-Column, Group/Totals, Summary, Tabular, and Mailing Labels reports. Figure M6.1 shows the initial Wizard selection screen with the different types of reports. Figure M6.2 illustrates a single-column report created with the AutoReport option. Notice that all fields in each record are displayed in a single-column without any special formatting. Each record is listed one after the other.

A group/total report (Figure M6.3) groups records by one or more fields—Sales

Figure M6.1

ReportWizard options

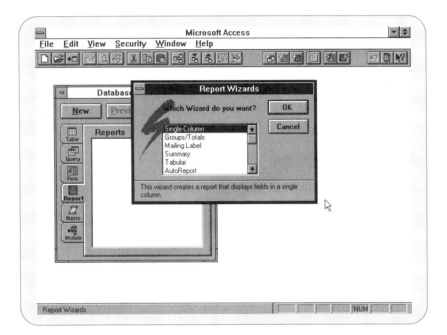

Figure M6.2

AutoReport generated Single-Column report

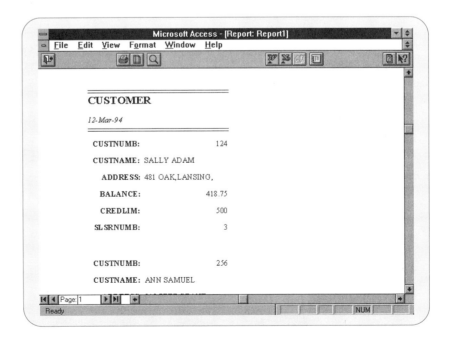

Figure M6.3

ReportWizard
generated Group/
Total report

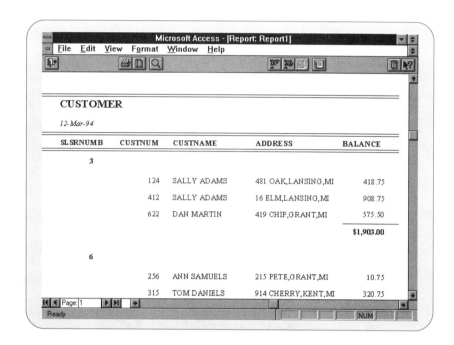

Rep in this example—sorted by customer within each group. In addition, totals and subtotals are generated for each sorted group.

Summary reports (Figure M6.4) are similar to group/total reports, but only the totals and percentages are printed for the sorted groups.

Perhaps the most common report is the Tabular report, displayed in what might be referred to as a multi-column format. Figure M6.5 shows individual records printed on a separate row with each field arranged in fixed columns. This report is the easiest way to generate a list of records in a table.

The last type of report generated with a Wizard is the Mailing Labels format. Figure M6.6 shows that this format displays records in a two-column format with fields arranged both side-by-side and on multiple rows to fit on standard mailing labels. You

Figure M6.4

ReportWizard
generated
Summary report

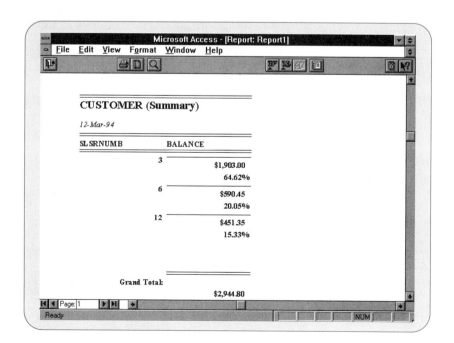

Figure M6.5

ReportWizard
generated Tabular
report

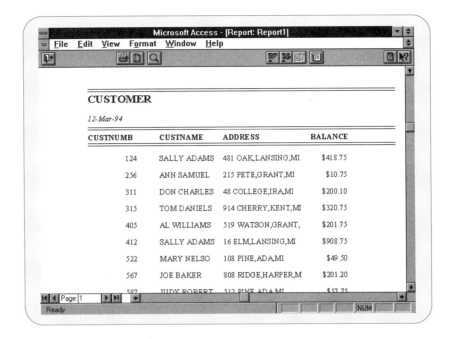

Figure M6.6

ReportWizard
generated Mailing
Labels report

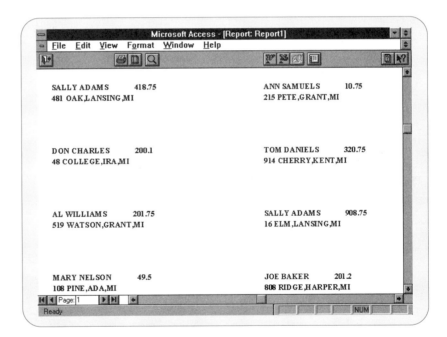

could also use this report to generate a single-column report with fields arranged side-by-side, something that the AutoReport single-column report doesn't do.

**Creating an
Initial Report** Let's illustrate the process by creating an AutoReport for the *Customer* table (Figure M6.7).

- Open the *Prem* database.
- Click on the Report object button.
- Click on the New button to create a new report.
- Select the *Customer* table from the drop-down list.
- Click on Report Wizards to have Access generate the report for you.

Figure M6.7

New Report dialog
box

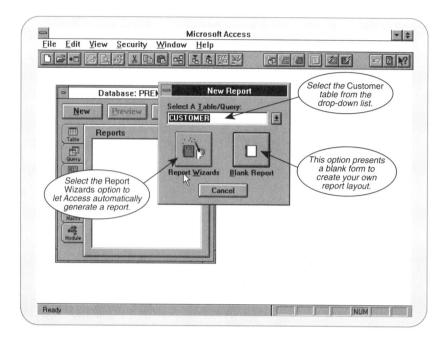

The next step (Figure M6.8) is to tell ReportWizard which type of report style to create. In this section, we will choose the AutoReport format, which creates the default Single-Column report.

- Select AutoReport.
- Click on OK.

Access will work for a short while creating the report and then it will display the initial Single-Column report (Figure M6.9). Notice that all the fields are included. They are arranged vertically. The field name is used for labels. We will now rearrange these fields and add two fields that are currently not included. Also, we will change the labels.

Figure M6.8

ReportWizard
options

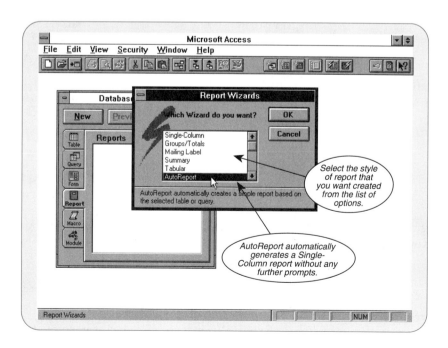

Figure M6.9

AutoReport generated Single-Column report

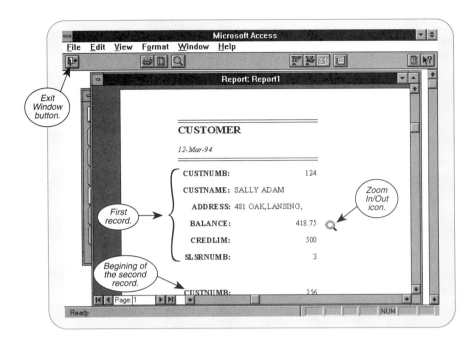

Changing Fields and Labels

First, we will rearrange the layout, moving the *Balance* and *Credlim* fields to a different position. Switch to the Design view (Figure M6.10) by clicking on the Exit Window button in the top-left corner of the window.

■ Click on the Exit Window button to switch to the Design View.

The first action you wish to take is to move the *Balance* and *Credlim* fields to the right of the *Customer Number* and *Customer Name* fields, as shown in Figure M6.11.

■ Click on the *Balance* field and drag to the right of the *Custnumb*.
■ Drag the *Credlim* field to the right of *Custname*.

Figure M6.10

First report in the Design view

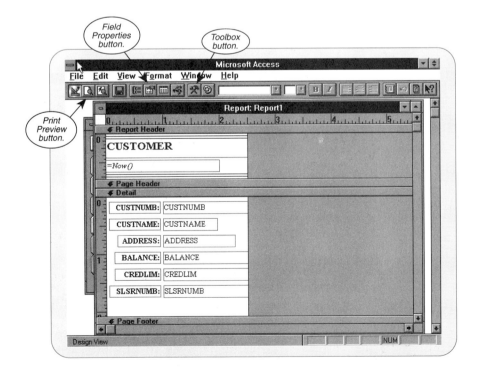

Figure M6.11

Moving fields in a
Report Design

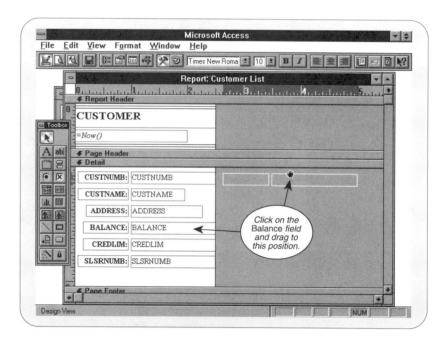

For this report, you do not need the *Sales Representative* field. To delete a field, simply click on a field to select it and press the Delete key.

■ Delete the *Slsrnumb* field.

In addition to moving fields to different locations, we also want to change the size of the fields and the labels. When the fields were moved from the lower part of the form to the upper-right side, the space at the lower half of the form was left blank. To reduce the space reserved for each record, you can move the mouse pointer to the field footer edge. As you do this, the arrow changes to a double-pointing arrow that you may use to move the footer up or down. At the top of the report is a header with the table name used as the default title. We want to expand the title from the simple table name to Customer List. We want to change the report to match that shown in Figure M6.12.

Figure M6.12

Removing fields
and changing
labels

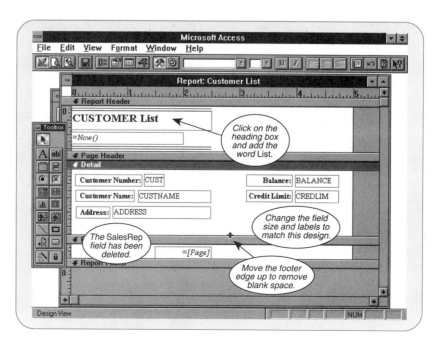

■ Change the size and labels of the fields to match Figure M6.12.
■ Add the word `List` to the title.
■ Move the footer edge up to remove the blank space.

Adding Computed Fields

You will now add the computed field for the available credit at the position indicated in Figure M6.13.

■ Click on the Text Box button.
■ Move the Text Box symbol to the location shown in Figure M6.13 and click.
■ Change the label of the text box to `Available Credit`.
■ Click in the Field Contents box and delete the "Unbound" entry.
■ Enter the computed field formula: `=[credlim]-[balance]`.

Figure M6.13

Adding a computed field

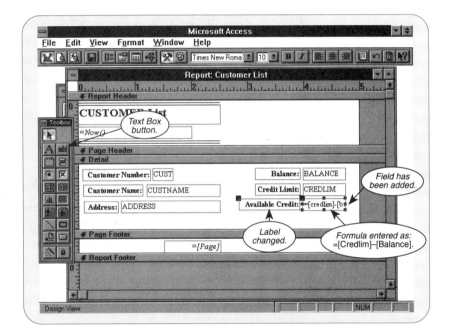

Changing Field Properties

Our report is now complete except for the format of the three numerical fields. The *Balance, Credit Limit,* and *Available Credit* fields currently will display values in the simplest, default format that displays numbers without any set number of decimal places. We want to set the format to `Currency` with two fixed decimal places. To set the format characteristics of a field, select the Properties dialog box and select the format style you wish (Figure M6.14).

■ Click on the *Available Credit* field.
■ Click on the Properties button.
■ In the *Format* field, select Currency from the drop-down list.
■ In the Decimal Places field, enter `2.`
■ Close the Properties dialog box.

Repeat the above process for the *Balance* and *Credit Limit* fields.

Viewing the Completed Report

To see what the report looks like,

■ Click on the Print Preview button.

The report is now complete and should look like the report shown in Figure M6.15.

■ Select Save from the File Menu and enter Customer List for the file name and then press Enter.
■ Select Close from the File Menu.

Figure M6.14

Changing Field
formats

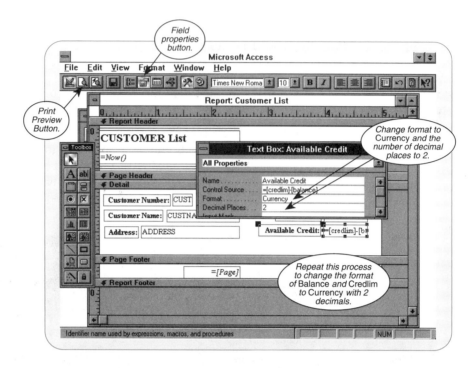

Figure M6.15

Completed Report

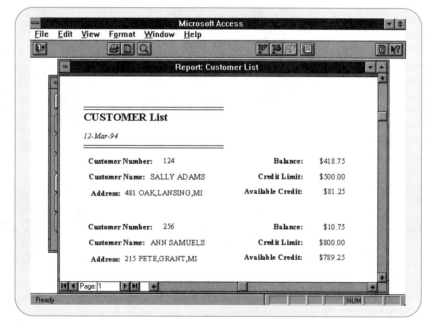

M6.3 TABULAR REPORTS

You have already constructed a report with the assistance of the ReportWizard. The
previous report listed each record one at a time in a single column. We then modified
the report to list the fields in a different order than the one designed by the Wizard.
We will now use the ReportWizard to create a report listing each record on individual
lines, with each field listed in vertically aligned columns.

**Creating a
Tabular
Report**

We begin by again using the ReportWizard to create a report of the Customer table.
The steps to create a report are similar to the previous report. First select the Report
object and click on a New report button. You will create a report on the *Customer* table
and select the fields to be listed in columns.

- Click on the Report object.
- Click on New to generate a new report.
- Select the *Customer* table from the drop-down list.
- Click on the ReportWizard button.

The first option that the ReportWizard presents is the type of report you want.

- Select the Tabular report type.

You will design a simple report with the *Customer Name, Credit Limit,* and *Balance* fields in three columns. Then you will change the format of these columns and add a column for the available credit. The next screen prompts you for the fields that you want to have included in your report. Figure M6.16 shows the Field Selection screen.

- Click on the *Custname* field in the leftmost box of available fields.
- Click on the Copy button.
- Repeat the process to copy the *Credlim* and *Balance* fields.
- Click on the Next button to move to the next screen.

Figure M6.16

Copying fields to a
Tabular Report in
ReportWizard

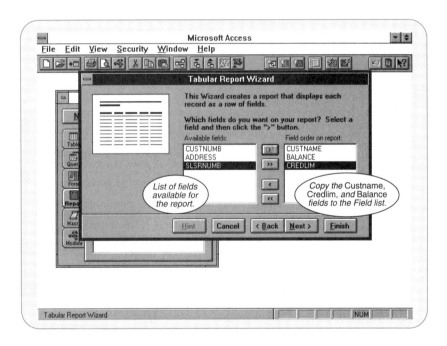

Once you have selected the three fields, ReportWizard needs to know what field to sort the report on (Figure M6.17).

- Click on the *Custname* field.
- Click on the Copy button.
- Click on the Next button.

The next screen offers different styles of appearances for the report (Figure M6.18).

- Click on the Next button to accept the default screen.

The final screen prompts for a report title (Figure M.19) and different options for opening the report. Enter Customer Listing for the title and check the See all fields on one page option.

- Enter `Customer Listing` in the Title box.
- Check the *See all fields on one page* option.
- Click on the Finish button to see the report.

Figure M6.17

Sort order in a
Tabular Report

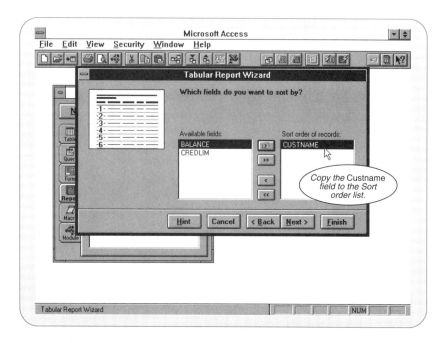

Figure M6.18

Setting the Report
style

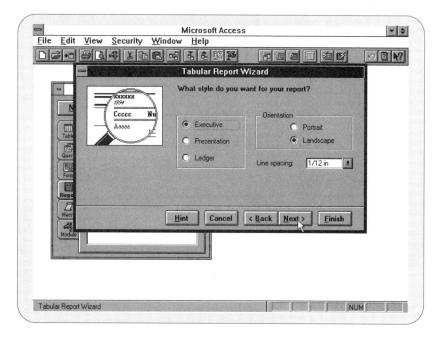

ReportWizard will take a short while to create the report and then will display a preview of a printed page (Figure M6.20).

Formatting Fields in a Report

You will notice that the columns are not formatted and that the width of the columns cause the right-aligned values to print farther to the right than the headings. You want to format the Credit limit and Balance columns to a currency format with two decimal places and align them to the right. Then you will add a calculated column for each customer's available credit. Before you make any of these changes, however, you should save the report.

■ Select Save from the File menu.
■ Enter `Customer Tabular List` for the file name, and press Enter.

Figure M6.19

Entering the
Report title

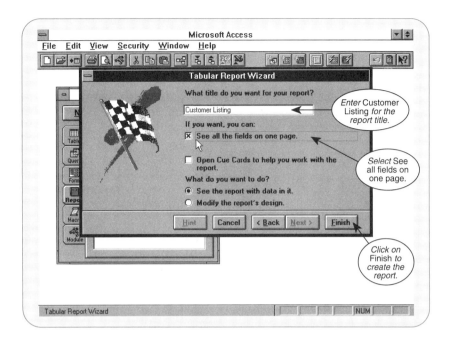

Figure M6.20

Previewing the
default Wizard
generated Tabular
report

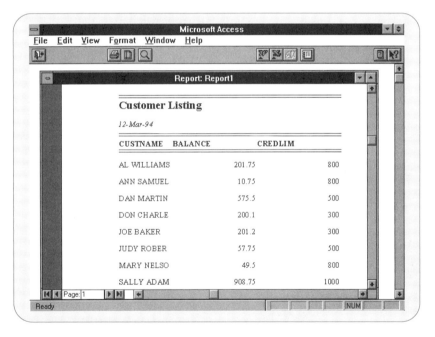

To make these changes, you must switch to the Design view. First, close the current window by clicking on the Close Window button on the left side of the tool bar. When you are returned to the Reports list of the Database window, the current report should be highlighted. When this occurs, click on the Design button.

■ Click on the Close Window button.
■ Click on the Design button with *Customer Tabular Listing* highlighted.

The Design view of the tabular report is separated into five sections. There is a Report Heading that prints only at the beginning of the report. The Page Heading prints at the top of each page, normally including page numbers, dates, and column headings. The Detail section includes the fields of each record to be printed on individual lines. The Page and Report Footers typically include totals of the columns of data.

Figure M6.21

Design view of a
Tabular report

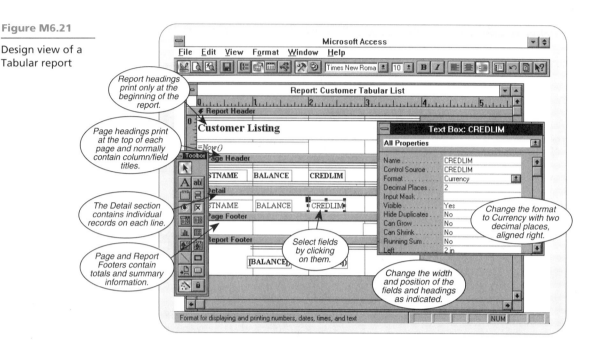

In the next section, we want to change the format and alignment of the *Balance* and *Credlim* fields. To change the format of a field, you select (click on) the field to modify, and then click on the Field Properties button. As Figure M6.21 shows, the *Credlim* field is selected and the Properties dialog box is open. To change one of the options, either tab to the field, or click on it with the mouse, and then choose the format from the drop-down list. (Alternatively, you could press the first letter of the option to select it. For example, in the *Format* field, simply pressing a C will bring up Currency.) Click on the *Credlim* field in the Detail section, and change the format to Currency, and Decimal places to 2. Then scroll down the Properties box and change the Alignment field to Right. Repeat this process for the *Balance* field in the Detail section, and the two *Sum* fields in the Report footer section. You will also need to change the location of the Balance and Credlim headings by dragging them to the position indicated in Figure M6.21.

- Click on the *Credlim* field.
- Click on the Field properties box.
- Change the Format to `Currency`.
- Change the Decimal places to `2`.
- Change the Alignment to `Right`. (You will need to scroll down in the dialog box.
- Repeat this process for Balance, and the two *Sum* fields.
- Change the field width and placement to match Figure M6.21.
- Change the position of the column headings as indicated.

Adding Calculated Fields to a Report

In addition to changing the position and format of the current fields, add a calculated field. The credit remaining is the difference between the customer's balance and the credit limit. To calculate that field, we must add a text box in a new column and enter the formula in its field properties.

Begin by clicking on the text box button in the Toolbox. The pointing cursor changes to a small icon of a text box. Move the cursor to the location in the Report design where you want the field and click to place the field. You can then use the field handles to resize and/or move the field.

- Click on the Text box button in the Toolbox.
- Place the field to the right of the *Credlim* field in the Detail section.

The formula used to calculate the available credit is entered in a field's properties box. Once you have placed a new text box in the Detail section (with the new field still selected), click on the Field Properties button in the toolbar. Normally, the *Control Source* field in the properties box is where the field name to be printed is entered. Formulas for calculated fields are also entered in the control source. Enter the formula as =[Credlim]-[Balance] in the Control Source and then change the format and alignment to Currency, 2 decimal places, and right-aligned. Figure M6.22 shows the Properties box as it should look when finished.

- Select the new text box.
- Click on the Field Properties button.
- Enter the formula =[Credlim]-[Balance] in the *Control Source* field.
- Change the format to Currency.
- Change the Decimal Places to 2.
- Scroll down in the Properties dialog box and change the Alignment to *Right*.

As Figure M6.22 indicates, you must also add a column heading in the Headings section. Enter headings as labels. Use the Label button (to the left of the Text Box button) on the Toolbox. When you select the label button, your pointing cursor changes to the label icon, an uppercase A. Move the cursor to the Page Header section and click to place it to the right of the Credlim heading. Then enter the label as Avail. Credit. You will notice that the other labels in the header are formatted bold. To set the label to bold, simply click on the bold text icon shown as a highlighted B on the toolbar.

- Click on the Label button in the Toolbox.
- Place the label to the right of the Credlim heading in the Page Header section.
- Type Avail. Credit.
- Click on the Bold button on the toolbar.
- Position the field as indicated in Figure M6.22.

Once you have completed the report, you will want to save it. Select Save from the File menu. To view the report, click on the Print Preview button on the toolbar.

- Select Save from the File menu.
- Click on the Print Preview button to view the completed report.

Figure M6.22

Adding a calculated field to a report

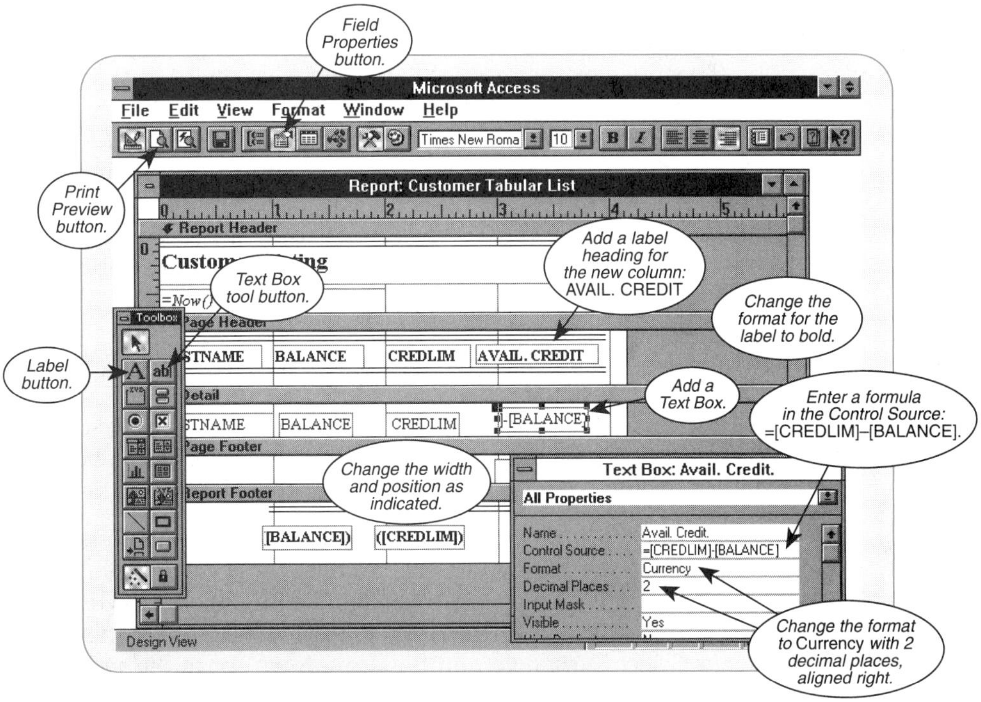

Figure M6.23

Completed Tabular report

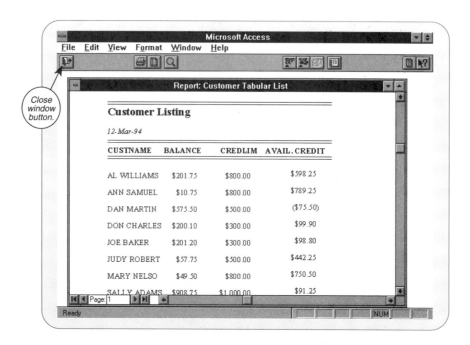

Figure M6.23 shows the completed report. To close the Print Preview window, click on the Close button in the upper-left corner of the preview screen.

M6.4   SORTING AND GROUPING REPORTS

You have already constructed a report that contained totals at the end. Now you will construct a report containing subtotals (intermediate totals). In particular, you will construct a report that groups customers by sales rep. At the end of the group of customers for a particular sales rep, we include subtotals for the *Balance* field. To create this report, we will modify the report designed in the previous section. The Design view of the tabular report should be displayed on your screen.

Sorting Records in a Report

In order to calculate the subtotal of groups of records, you first must sort the records by the values in one or more fields. For this report, we will sort first on the sales rep number and then sort the customer names alphabetically within each sales rep. You set the sort order for a report in the Sorting and Grouping box. With the report in the Design view, click on the Sorting and Grouping button on the toolbar. The top portion of the dialog box is used to set the sort order on up to ten fields. The lower portion of the dialog box is used to set the group properties of each sorted field. To switch between the two parts of the dialog box, use the F6 function key or click in the desired box with the mouse.

- Open the Customer Tabular List in the Design view, if necessary.
- Click on the Sorting and Grouping button on the toolbar.

Select *Slsrnumb* from the drop-down list for the first sort field/expression, and then move to the Properties section and select **Yes** for both the Group Header and Group Footer. Next, select *Custnumb* for the second sort field/expression.

- Select *Slsrnumb* from the drop-down list in the first Field/Expression field.
- Press the F6 key to switch to the Properties section.
- Select *Yes* from the drop-down list in the Group Header and Footer fields.
- Press the F6 key to switch to the Field/Expression section.
- Select *Custnumb* from the drop-down list in the second Field/Expression field.

Figure M6.24

Sorting and
Grouping dialog
box

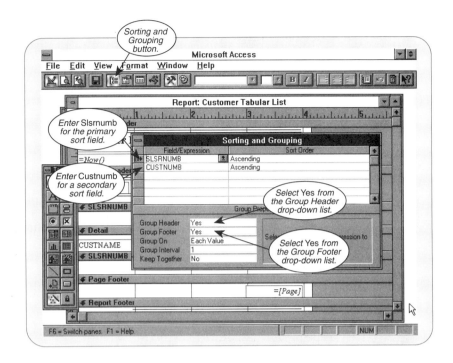

Your screen should now look similar to Figure M6.24. Press Alt-F4 to close the Sorting and Grouping dialog box, or select Close from the dialog's control box.

■ Press Alt-F4 to close the Sorting and Grouping dialog box.

Adding Group Headers and Footers

As Figure M6.25 shows, new sections have been added for a SLSRNUMB Header and Footer. Labels or text box fields that are added in the header will print prior to each new group of the sorted records. After each group of records are printed, and prior to the next header, data in the footer will be printed. In this example, we will print the *Slsrnumb* in the header, and a total of the Balance column for each sales representative. As you did when you added the *Available Credit* field in the detail section, we will add

Figure M6.25

Header and Footer
sections

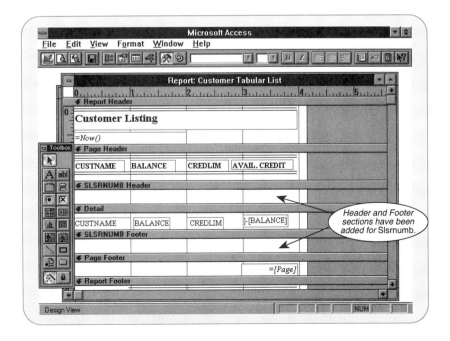

a label to describe the data and then use the text box to print the sales representative number and the calculated sum of balances.

- ■ Click on the Label button in the Toolbox.
- ■ Place the label at the left of the SLSRNUMB Header section.
- ■ Type `Sales Representative:`
- ■ Click on the Bold format button on the toolbar.
- ■ Position the label so that it begins printing at the far-left edge of the header.
- ■ Click on the Text Box button in the Toolbox.
- ■ Place the text box to the right of the label.
- ■ Type `=[Slsrnumb]` in the text box.

Next, add a label and text box in the footer to print totals for each sales rep section.

- ■ Place a label at the left edge of the SLSRNUMB Footer.
- ■ Type `Total:`
- ■ Place a text box to the right of the label, under the Balance detail column.
- ■ Type `=sum([Balance])`
- ■ Place another text box under the Credlim detail column.
- ■ Type `=sum([Credlim])`
- ■ Place another text box under the Credlim detail column.
- ■ Type `=sum([Credlim]-[Balance])`
- ■ Position and resize the text boxes so that they match Figure M6.26.

You also need to set the properties for each of the new fields. For the Total label, the only change we need to make is to set it to bold. Do this by clicking on the label and then clicking on the Bold button on the toolbar.

- ■ Click on the Total label.
- ■ Click on the Bold button on the toolbar.

Next, for each of the three *Sum* fields, we want to change the format to Currency with two decimal places, bold text, and aligned to the right (Figure M6.27).

- ■ Click on the *=Sum([Balance])*field.
- ■ Click on the Bold button on the toolbar.
- ■ Click on the Right Align button on the toolbar.

Figure M6.26

Sorting and
Grouping

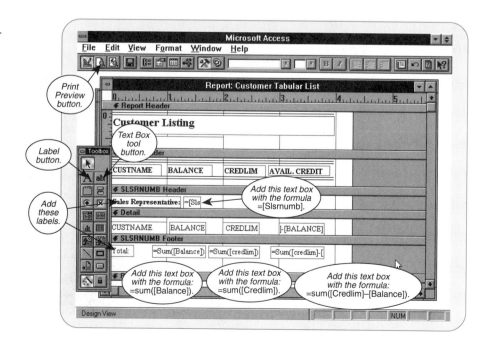

Figure M6.27

Setting Field
properties

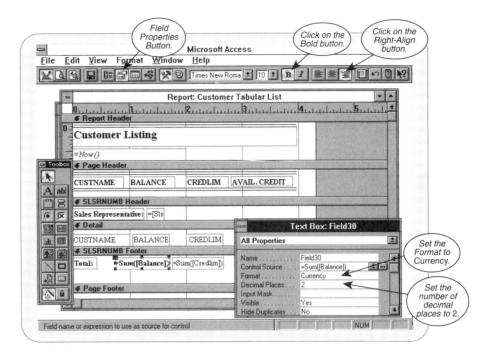

- Click on the Field Properties button.
- Select Currency from the Format drop-down list.
- Enter 2 for the number of decimal places.
- Repeat the above process for the *=Sum([Credlim])* and *=Sum([Avail.Credit])* entries.

When you have set the properties for the three fields, the report is complete. Save the file and then click on the Print Preview button to view the report.

- Choose Save from the File menu.
- Click on the Print Preview button on the toolbar to view the report.

Figure M6.28 shows the completed report. To close the report:

- Click on the Close Window button.

Figure M6.28

Completed Sorting
and Grouping
report

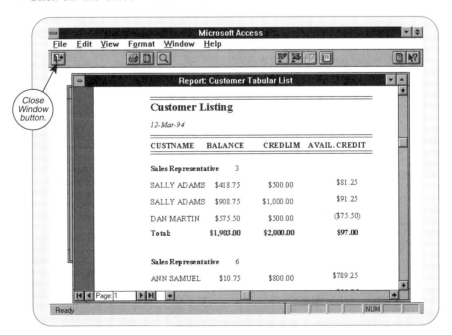

M6.5 EMBEDDING SUBREPORTS IN A MAIN REPORT

You can include data from multiple tables in a single report. Do this by designing a subreport that is included in the main report. For example, in the previous report, the records are sorted and grouped on the sales representative number. However, the sales representative name is not part of the *Customer table* and therefore is not part of the report. You can design a report consisting of only the sales representative name and then embed that short report next to the salesrep number.

Creating a Subreport

Begin by returning to the *Prem* database window and creating a new report and then dragging the new report onto the main report. You should have the Customer Tabular List report open in the Design view.

- Select Database: *Prem* from the Windows menu.
- Click on the Reports object icon.
- Click on the New button.
- Select the *Slsrep* table from the drop-down list of tables.
- Click on the Blank Report button in the ReportWizard dialog box.
- Click on the Text box button on the toolbox.
- Place a new text box in the Detail section.

The subreport should include only the text box of the salesrep name—and not the label. Therefore, you need to select the label and delete it.

- Click on the "Field1" label.

When this label is highlighted (as in Figure M6.29), delete the label. We also need to identify *Slsrname* as the control source of the field.

- Press Delete.
- Enter `=[slsrname]` in the unbound text box.

Next, we want to resize both the field and the Detail box to fit in the main tabular report.

- Drag the text box to the upper-left corner of the Detail section.
- Drag the right-side handle of the text box to just less than 2 inches wide.

Figure M6.29

Designing a subreport

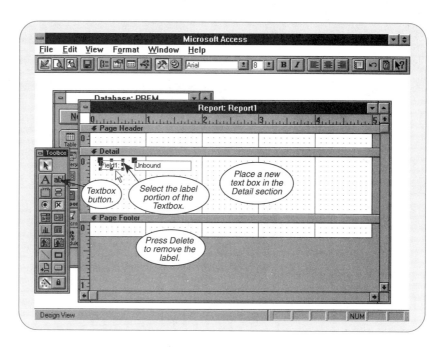

Move the pointer to the lower-right corner of the Detail section. The arrow changes to the resize icon. Change the size of the Detail section to match that shown in Figure M6.30.

- Change the size of the Detail section to 1/4 inch by 2 inches.
- Click on the Slsrname box.
- Select *Times New Roman* from the Font box.
- Select a point size of *10* from the Font Size box.
- Select Save As from the File menu and save as `Sales rep name.`
- Close the Report window.

Figure M6.30

Designing a subreport

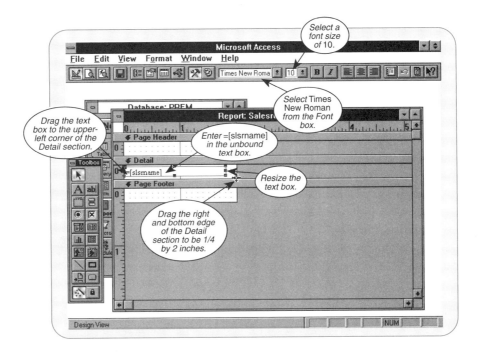

Embedding a Subreport

The next step is to embed the subreport in the main tabular report. You do this by dragging the icon and/or name of the subreport from the Database window to the desired location in the main report. With the Tabular report open, select the Database: *PREM* window and drag the Salesrep name subreport to the SLSRNUMB Header section.

- Select Database: *Prem* from the Windows menu.
- Drag the Salesrep name to the SLSRNUMB Header section.

You will notice that the subreport is larger than necessary and that a label is attached (Figure M6.31). We will delete the report label and resize the box.

- Click on the label *Salesrep name:*
- Press Delete.
- Drag the Subreport box to the right of the Slsrnumb text box.
- Change the size of the Subreport box and the SLSRNUMB Header to match Figure M6.32.

To view the report, click on the Print Preview button.

- Click on the Print Preview button to view the report (Figure M6.33).

This completes the Customer Tabular report with sales representative information. Save the report and close the window.

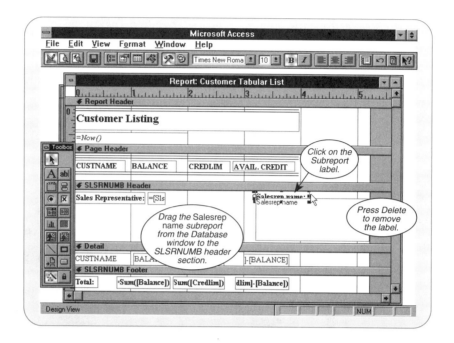

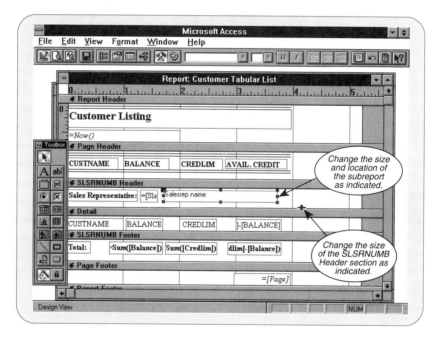

- Select Save from the File menu.
- Close the Report window.

M6.6   SORTING AND GROUPING MULTIPLE TABLE REPORTS

As the previous example demonstrated, reports tend to bring together data from different tables. In this next section, you will construct a report of customers, their orders, and the order lines within these orders. The records are grouped by customer. Within customers, the records are grouped by orders. This report will include fields from the *Customer, Orders, Ordline,* and *Part* tables. One of the easiest methods to begin designing

Figure M6.33

A main report with
an embedded
subreport

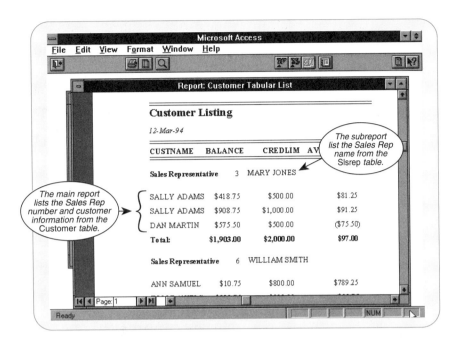

such a report is to first create a query that combines all of the fields from the different
tables. You can then use the query as the source of the report.

The steps to creating this report are:

1. Design a multiple table query.
2. Design a report using the new query.
3. Define sorting and grouping criteria.
4. Define the Headers and Footers.

Creating the Report Query

You will recall that queries can be designed with multiple tables, linked by common
fields. We will begin by designing a query with the *Custnumb* and *Custname* fields from
the *Customer* table; *Ordnumb* and *Orddte* from the *Orders* table; *Partnumb, Numbord,* and
Quotprce from the *Ordlne* table; and *Partdesc* from the *Part* table. In addition, we will
include a calculated field for an Extension of the number ordered and the quoted price.

Create a new query in the *Prem* database using the new query button, bypassing
the QueryWizard.

- Open the *Prem* database, if necessary.
- Click on the Query object button.
- Click on the New button.
- Click on the New Query button to bypass the QueryWizard.

Access next prompts for the tables to be included in the query. Click on each of
the four tables that will be used (*Customer, Orders, Ordlne,* and *Part*), clicking on the
Add button after each selection. When you are finished, your screen should be similar
to Figure M6.34. Click on the Close button when you have finished.

- Click on each of the four tables, one at a time, choosing Add after each table.

Note: If any of the relationship lines between tables do not display, you can drag
the key field of one table to the related field in another table.

Note: If you wish to add another table after you have closed the Add Table dialog
box, you can click on the Add Table button on the tool bar to display the dialog box.

The next step in creating the query is to select the fields to be included. When
adding fields to a query, you may use one of two methods. The first method of adding

Figure M6.34

Adding tables to a
multi-table query

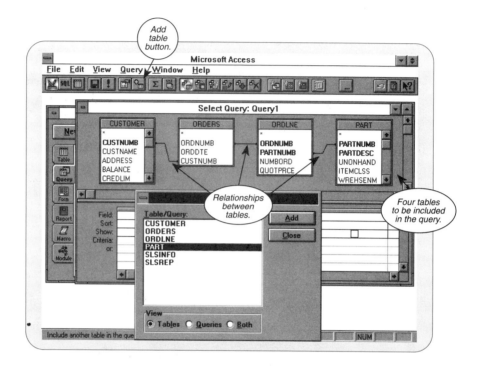

Figure M6.34

Adding tables to a
multi-table query

fields is to simply drag the field name from a table to one of the columns in the query. A second method is to use the drop-down list in Field row. We will use this method to add the fields to the query. The fields that we want included in the query are: *Custnumb, Custname, Ordnumb, Orddte, Partnumb, Partdesc, Numbord,* and *Quotprce.* In addition, we will add a calculated field for an Extension of the number ordered times the quoted price. Refer to Figure M6.35 when adding fields. Figure M6.35 shows the first three fields added with the drop-down list highlighting the *Orddte* field from the *Orders* table. In addition, the *Custnumb* and *Ordnumb* fields are set to sort in ascending order.

- Click on the drop-down list in the Field row of the first column.
- Select the *Custnumb* field.

Figure M6.35

Adding fields to a
multi-table query

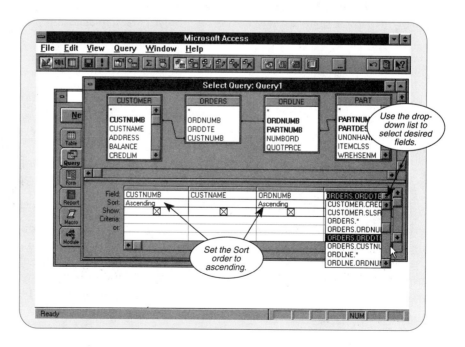

- Repeat the process for each of the above fields.
- (For the *Custnumb* and *Ordnumb* fields, set the Sort to ascending.)

Once each of the fields are added, we need to add the calculated field for the price extension. As Figure M6.36 shows, you enter the expression in the Field row of the desired column. Enter the expression as: *=[Numbord]*[Quotprce].*

- Enter the expression: `=[Numbord]*[Quotprce].`
- Press the Enter key.

Figure M6.36

Adding a calculated field to a query

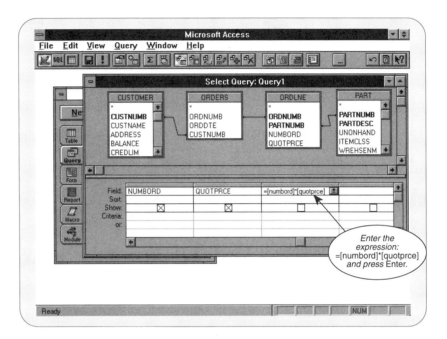

To change the label, you only need to replace the Expr1 label with EXTENSION. In addition, we want to set the format to Currency for the quoted price and the price extension.

- Change the Expr1 label to `EXTENSION`.
- Click on the Properties button on the tool bar.
- Enter `Currency` in the *Format* field.
  (Actually, when you type a c, the entire word Currency appears.)
- Repeat the properties format setting for the Quotprce column.

Since you will be using the query to generate our report, you need to save the query prior to designing the report.

- Use the `Save as` option to save the file as Orders by Customer.

To test the query, choose the Datasheet option from the View menu. You will need to change the column width of columns to see all of the fields on the same screen.

- Choose Datasheet from the `View` menu.

Your screen should now be similar to Figure M6.37. Close the query and return to the *Prem* database window.

- Close the query by selecting Close from the File menu.

Creating a Multi-Table Report

You will now construct the report using the query. Begin by clicking on the Report object button and then New. Again, we will bypass the ReportWizard to create the report.

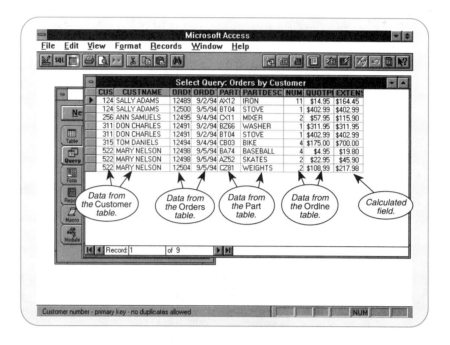

■ Click on the Report object icon, or select Report from the View menu.
■ Click on the New button.

Access needs to know which table or query to use for the data in the report. At this point, you can only select one table or query to use. If you want to select multiple tables, you must use one table for the main report and then embed subreports as we did in the previous example. Alternatively, you can use the query that combines multiple tables.

■ Select the *Orders by Customer* query from the drop-down list.

Your screen should have a list similar to Figure M6.38, although the location and window size may be different than yours.

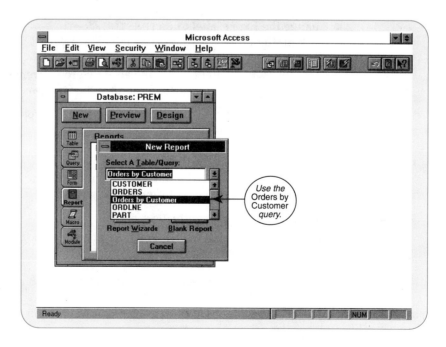

■ Click on the Blank Report button to bypass the ReportWizard.
■ Select Field List from the View menu to display the list of available fields.

Figure M6.39 shows the first screen for a blank report, prior to adding fields or headers and footers. First you will add each of the fields to print in the various sections. Then you will determine the order for grouping and sorting, followed by designing the headers and footers for groups, pages, and the entire report.

To add the fields in the report, you drag each of the fields in the field list to the Detail section of the report. As you drag each field to the Detail section, the field label is attached. To remove the label, drag the field to a blank area, click on the label (the left box), and press Delete. Figure M6.40 shows the first field, *Custnumb*, placed in the Detail section and the label selected.

Figure M6.39

Creating the multi-table report

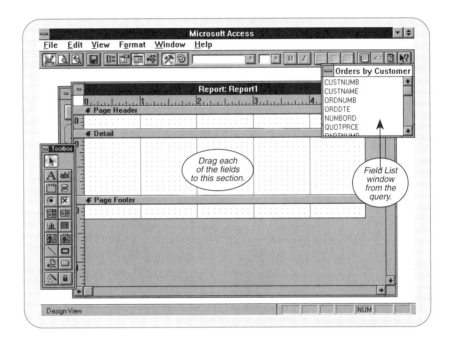

Figure M6.40

Creating the multi-table report

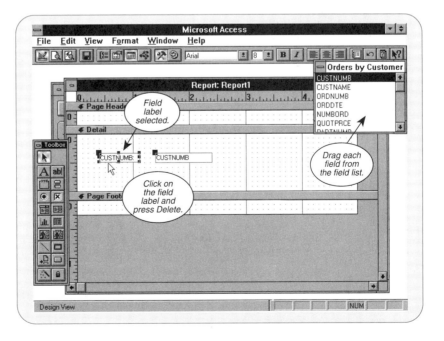

■ Drag the *Custnumb* field to the Detail section.
■ Click on the label box.
■ Press the Delete key.

Next you need to place the field in its print position in the report and resize it as necessary. To move the field, place the pointer on one of the box edges so that the arrow changes to the hand symbol and then drag the box to the desired location. To resize the field's text box, move the cursor to the lower-right handle and resize the box to an adequate size.

■ Drag the Custnumb text box to the upper-left corner of the Detail section.
■ Resize the text box to match the field as Figure M6.41 shows.

Figure M6.41

Moving and
resizing fields

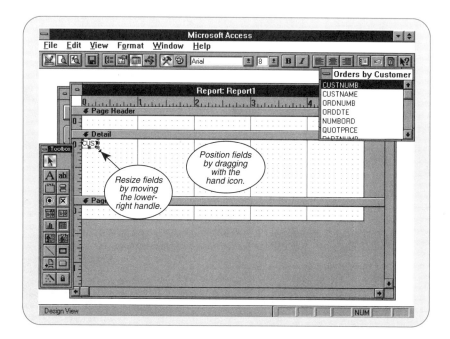

One at a time, drag the fields to the Detail section in the following order: *Custnumb, Custname, Ordnumb, Orddte, Partnumb, Partdesc, Numbord, Quotprce,* and *Extension*. As you add each field, delete the label, and then move and resize the text box so that the field layout is similar to Figure M6.42. You will notice that the lower edge of the Detail section has been moved so that it is just under the field text boxes. This allows records to print right after each other.

■ Add each of the fields, delete the labels, resize the text boxes, and place them as shown in Figure M6.42.
■ Move the lower edge of the Detail section so that it is just under the field boxes.

The last change we need to make to the Detail section is to set the format of the quoted price and extended price to Currency with two decimal places, aligned right.

■ Click on the *Quotprce* field in the Detail section.
■ Click on the Properties button on the toolbar.
■ Set the format to Currency and the decimal places to 2.
■ Scroll down the Properties dialog box to set the text alignment to Right.
■ Repeat these format settings for the *Extension* field.

Before continuing, we should save the report design. Select Save As from the File menu and save the report as Orders by Customer.

■ Select Save As from the File menu and save the report as `Orders by Customer.`

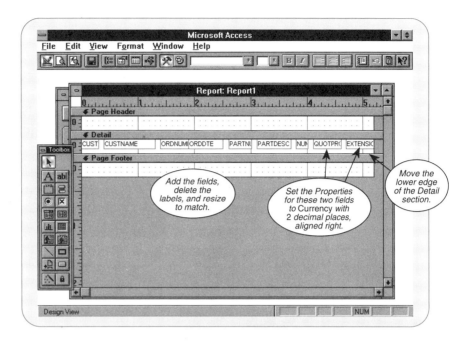

Adding Report and Page Headings

You also want to add report and page headings. By default, the page heading is already in place without any labels or text boxes in it. However, you need to add the Report heading. For the report heading, you want to print Orders by Customer in a large font size in the center of the header.

- Choose Report Header/Footer from the Format menu.
- Add a label `Orders by Customer` in the center of the Report Header.
- Change the label's font size to `12` or a font size larger than the report default size.

For the page header, add column headings for each of the fields and the page number.

- Add a text box for each column heading.
- Enter the text for each column heading. (Refer to Figure M6.43 for each of the column headings.) Hold down the Shift key while pressing Enter to move to the second line of a heading without creating a new label.
- Click on Line tool.
- Add a solid line under the column headings.
- Add a text box for the page number. (Delete the label for the box.)
- Enter the page number expression as: `= "Page " &[Page] & " of " &[Pages]`

When you are finished, the report design should match Figure M6.43. To preview the report, click on the Print Preview button. The result is shown in Figure M6.44.

- Click on the Print Preview button on the tool bar.

Setting Group Headers and Footers

Notice in the report that there are three customers who have repeating lines in the report. In addition, two of the customers have repeating orders. There also are no totals for each customer's order. You can fix these problems by appropriately using sorting and grouping.

Use the Sorting and Grouping option to determine how the report should break for different groups of data, depending on the sorted order. You are prompted for the fields to use as the sorted breaks. Sequence the report first on customer number and then on order number within each customer.

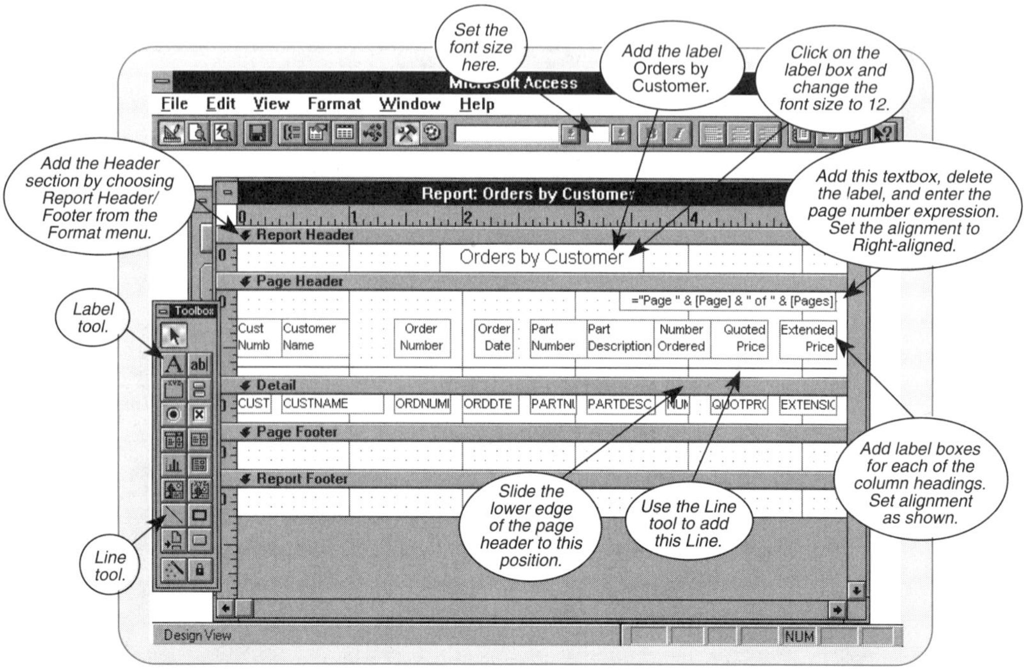

Figure M6.43

Defining page and
report headers

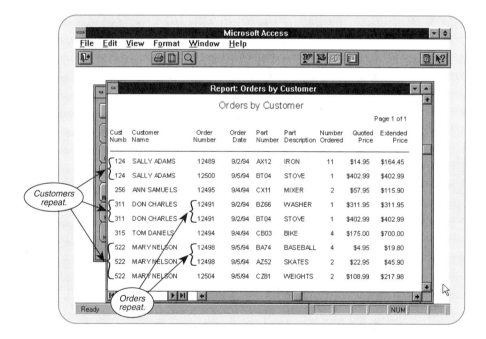

Figure M6.44

Multi-table report

- Click on the Close Window button on the tool bar.
- Select Sorting and Grouping from the View menu.
- Enter Custnumb in the first Field/Expression row.
- Enter Yes in the Group Header and Group Footer.
- Enter Ordnumb in the second Field/Expression row.
- Enter Yes in the Group Header and Group Footer.

Figure M6.45 shows that headers and footers have been added for both groups. Close the Sorting and Grouping dialog box by either selecting Close from the dialog

Figure M6.45

Adding group
headers and
footers

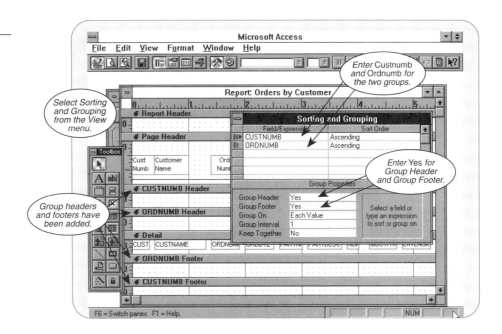

box's Control Box menu or by selecting Sorting and Grouping from the View menu again to close the box.

■ Select Sorting and Grouping from the View menu to close the box.

Next we want to move the customer number and name from the Detail section to the Custnumb header and the order number and date to the Ordnumb header. This action, which will cause the customer and order information to print only once per group, is the way that we indicate that we want to group orders *within* customers. That is, within a group of records with the same customer number, we will create subgroups consisting of records with the same order number.

■ Modify your group headers to match those shown in Figure M6.46.
■ Drag the Custnumb and Custname textboxes from the Detail section up to the Custnumb header. Leave the fields in the same relative position under the column headings but drag them up to the header.
■ Drag the Ordnumb and Orddte text boxes from the Detail section up to the Ordnumb header.
■ Move the lower edge of the headers and Detail sections so that they are just under the text boxes by moving the pointer to the lower edge. As the arrow changes to a resize symbol, hold down the left mouse button and drag the edge to the desired location.

You also want to print totals for each order and each customer in the appropriate footers. This is done by adding a Text Box and changing the label to *Order Total:* and *Customer Total:* in the respective footers. Then enter an expression in the Text Box as: = *Sum([Extension])*.

■ Modify your group footers to match those shown in Figure M6.47. Click on the Text Box tool and place a box in the footers, placing them in the positions as indicated.
■ Change the label to `Order Total:` and `Customer Total:` in the respective footers.
■ Enter an expression in the Text Boxes as: `=Sum([Extension])`.
■ Set the properties of the two sum boxes to: `Currency` with 2 decimal places, aligned right.

Figure M6.46

Adding group
headers and
footers

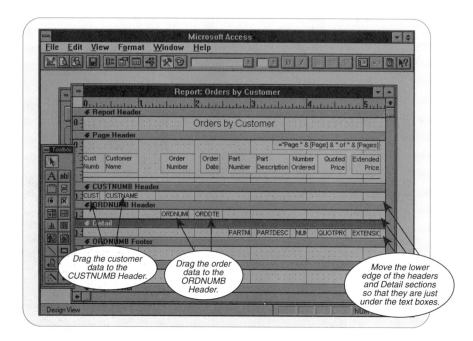

Figure M6.47

Adding totals to
group footers

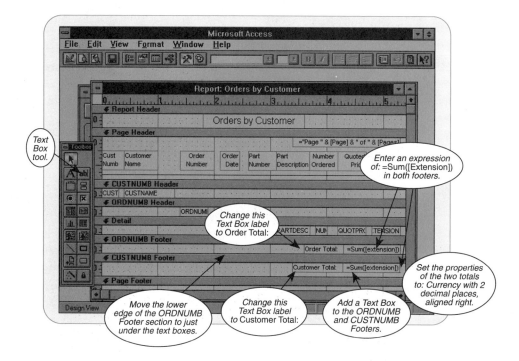

■ Move the lower edge of the footer sections so that they are just under the text boxes.

Check your work by viewing the report to see if it matches the one shown in Figure M6.48. You may need to use the scroll bars to move the viewing window to see the entire report. To print the report to the printer, simply click on the Printer icon on the tool bar or select Print from the File menu.

■ Click on the Print Preview to check your work.
■ Save your report.

```
                      Orders  by  Customer


                                             Page 1 of 1

Cust  Customer      Order  Order  Part   Part          Number  Quoted  Extended
Numb  Name          Number Date   Number DescriptionOrdered    Price   Price

 124  SALLY ADAMS
                    12489  9/2/94
                                  AX12    IRON             11   $14.95   $164.45
                                                Order Total:            $164.45
                    12500  9/5/94
                                  BT04    STOVE             1  $402.99   $402.99
                                                Order Total:            $402.99
                                             Customer Total:            $567.44

 256  ANN SAMUELS
                    12495  9/4/94
                                  CX11    MIXER             2   $57.95   $115.90
                                                Order Total:            $115.90
                                             Customer Total:            $115.90

 311  DON CHARLES
                    12491  9/2/94
                                  BZ66    WASHER            1  $311.95   $311.95
                                  BT04    STOVE             1  $402.99   $402.99
                                                Order Total:            $714.94
                                             Customer Total:            $714.94

 315  TOM DANIELS
                    12494  9/4/94
                                  CB03    BIKE              4  $175.00   $700.00
                                                Order Total:            $700.00
                                             Customer Total:            $700.00

 522  MARY NELSON
                    12498  9/5/94
                                  BA74    BASEBALL          4    $4.95    $19.80
                                  AZ52    SKATES            2   $22.95    $45.90
                                                Order Total:             $65.70
                    12504  9/5/94
                                  CZ81    WEIGHTS           2  $108.99   $217.98
                                                Order Total:            $217.98
                                             Customer Total:            $283.68
```

SUMMARY

The following list summarizes the material covered in this module:

1. To create an AutoReport, select the Report object, choose New, and identify the table to use for the data. Then use the ReportWizard's AutoReport to generate a default single-column report.
2. To place a field from the table on a free-form report, select Field List from the toolbar, and drag the field to the form.
3. To place a calculated field on a report, select the Text box icon from the Toolbox and click to place the file in the desired position. Then enter the expression for the calculation in the *Control* field.
4. To change the format and behavior of a field, select the Properties button on the toolbar where the format, alignment, and other characteristics may be set.
5. A tabular report is one with fields listed in a columnar style. You may modify this style, however, to create a free-form design. The easiest method is to use the ReportWizard and select the Tabular report type. ReportWizard will then prompt you for the necessary information to create a default tabular report.

6. To sort or group records in a report, use the Sorting and Grouping option on the toolbar. The records will be sorted as specified and group header and footer sections will be added. The contents of the group header print before a group and the contents of the group footer display after the group. Group footers typically contain subtotals.

7. To construct a report using data from several tables, construct a subreport containing all the required fields from an individual table or query. If you are going to merge numerous tables, it is easier to create a query containing multiple tables. In that case, use the new table or query as a section of the report.

8. You can have more than one level of grouping in a report by using Sorting and Grouping to set the sort order and groups/subgroups for a report. This gives you the option of printing headers and footers for each group level.

EXERCISES

1. How do you create an AutoReport? What is the difference between a single-column report and a tabular report?
2. How do you add a regular field to a report? How do you add a calculated field?
3. How do you add a field from another table to a report? Are there any restrictions on the other table?
4. How do you change the format of a field in a report?
5. How do you change the location and size of a control field or an attached label in a report?
6. How do you group records in a report? How do you sort in a report?
7. How do you construct a report that involves data from more than one table?
8. How do you add summary information for a group subtotal? Where would you be most likely to add one?
9. Describe the process of adding multiple sets of subgroups to a report.

COMPUTER ASSIGNMENTS

1. Create a single-column report for data in the *Part* table. The report should include all fields from the part record as well as an additional field called *Ohvalue*. This field, which represents on-hand value, is defined to be the product of units on hand and unit price (Unonhand * Unitprce).

2. Create a tabular report for data in the *Part* table. The report should include all fields from the part record as well as *Ohvalue* (see Assignment 1). The records should be grouped by item class. The group header should have the words *ITEM CLASS:* followed by the item class. The group footer should contain a total of *Ohvalue* for all parts in the item class. The report footer should contain a grand total of *Ohvalue*. The page header should follow the style of the page headers in the reports we produced within this module.

3. Use the *Book* database to create a report of publishers and books. The report should include the publisher's code, name, and city as well as the book code, title, book type, and price. The records should be grouped by publisher code. The group header should have the publisher's code, name, and city. The group footer should contain a total of the price. The report footer should contain a grand total of the prices of the books.

4. Create a report of parts, orders, and customers. The report should include the part's number, description, units on hand, and warehouse number. In addition, for each order on which the part is present, the report should include the number of units ordered, the order number and date, and the number and name of the customer who placed the order. The report should be grouped by warehouse number with

the warehouse number displaying in the group header. The group footer will be a single blank line. Within each warehouse, the records should be grouped by part number with the part number, description, and units on hand appearing in the group header. The detail lines should contain the order number, date, customer number, name, and number of units ordered. The group footer should contain the total number of units of the part that were ordered on all orders. There will be no report footer.

5. Create a report of publishers, books, and authors. The report should be grouped by publisher number with the publisher code, name, and city displaying in the group header. The group footer will list the number of books published by the publisher. Within the books of each publisher, the records should be grouped by book type, with the book type displaying in the group header and the number of books of the given type appearing in the group footer. Within the books of a given publisher and a given type, the records should be grouped by book number with the book number, title, and price appearing in the header. The footer will be a single blank line. The detail lines should contain the author's number and name.

7 Access and the Functions of a Database Management System (DBMS)

OBJECTIVES

Compare the features Access furnishes with those that a database management system (DBMS) should provide in the following areas:

1. Data storage, retrieval, and update.
2. A user-accessible catalog.
3. Support for shared update.
4. Security services.
5. Integrity services.
6. Services to promote data independence.
7. Backup and recovery services.
8. Utility services.

M7.1 INTRODUCTION

In this module, we investigate the manner in which Access furnishes the functions that are typical of most microcomputer DBMS discussed in Chapter 7. We discuss storage and retrieval in section M7.2 and the catalog in section M7.3. Section M7.4 investigates the facilities Access provides to support shared update. Section M7.7 presents the security features of Access. Section M7.6 covers integrity support; section M7.7 covers data independence. The backup and recovery facilities are examined in section M7.8. Finally, section M7.9 discusses some of the utility services Access provides.

M7.2 STORAGE AND RETRIEVAL

A DBMS must furnish users with the ability to store, retrieve, and update data in the database.

Like any other good DBMS, Access furnishes users with such capabilities. Without them, the product couldn't even be considered a legitimate DBMS.

M7.3 CATALOG

A DBMS must furnish a catalog in which descriptions of data items are stored and which is accessible to users.

Some aspects of the catalog described in Chapter 7 are furnished by Access through its graphical interface. Access shows a hierarchical structure of database files in the directory; then, by choosing the object type (table, query, form, or report, and so forth), a list of available files is displayed. For example, in Figure M7.1, a dialog box lists the

Figure M7.1

Hierarchical
structure of
database files

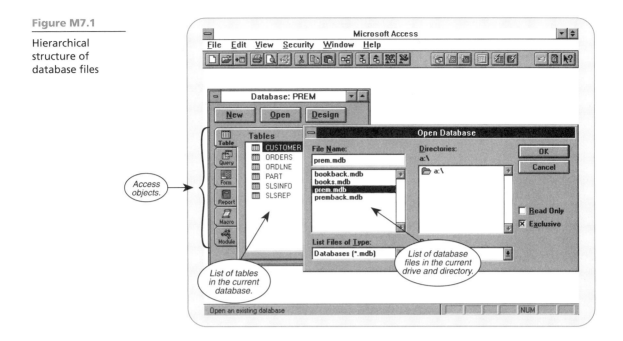

database files on the current drive, whereas a list of tables for the *Prem* database are listed in the current database window.

Remember that you can display the structure of an individual table by selecting the table whose structure you wish to examine and then displaying the table in the Design view (Figure M7.2). The details concerning the structure of the table are displayed in the top half of the window showing the data type and description of each field. In the lower half of the window, the properties of each field are displayed showing aspects of the field such as the data type, format, default value, and validation rules. To view properties, simply move to the table and field that you want to see and select the Design view.

Figure M7.2

Table structure
information

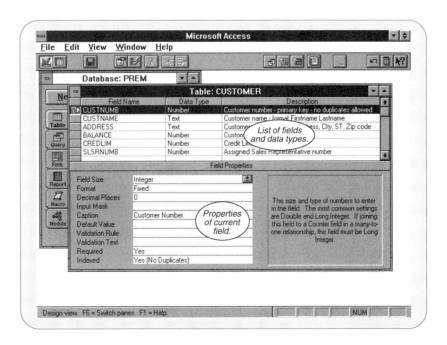

M7.4   SHARED UPDATE

A DBMS must furnish a mechanism to ensure accuracy when several users are updating the database simultaneously.

When Access is installed on a network, two or more users can access the same database at the same time. If your computer is connected to a network, you can use Access to view and edit data that others are also using. Access works the same whether you are using a database exclusively or sharing it with others. When you open a database, you specify whether you want exclusive or multiuser access to the database by checking the Exclusive option in the Open Database dialog box. (Refer to Figure M7.1.) To prevent incorrect data entry, Access has a comprehensive locking scheme.

Types of Locks

Various objects (records, tables, queries, forms, and reports) can be locked in a variety of ways. The types of locks a user can place on an object are controlled by the Read Only and Exclusive options displayed in Figure M7.1.

1. *Exclusive = Yes, Read Only = No.* You can open and edit the database. No other user can access the object.
2. *Exclusive = No, Read Only = Yes.* The user with this lock can view the object but not modify it. Other users can both view the object and modify it.
3. *Exclusive = No, Read Only = No.* You can open and modify the object while other users are also using the database.
4. *Exclusive = Yes, Read Only = Yes.* Every user can open or modify the object.

Choosing a Locking Strategy

For the most part, Access handles the placement and removal of these locks automatically. While you edit a record in a multiuser environment, Access automatically locks that record to prevent other users from changing it before you are finished. You set this control under the Options settings under the View menu shown in Figure M7.3.

The settings for Options are outlined below.

No Locks: With this setting, the current record isn't locked when you are editing it. (This is the default setting.) The No Locks setting offers the greatest flexibility, however, its use can result in the following problems:

Another user can overwrite your changes.

Another user can lock a record that you are currently using.

Figure M7.3

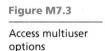

Access multiuser options

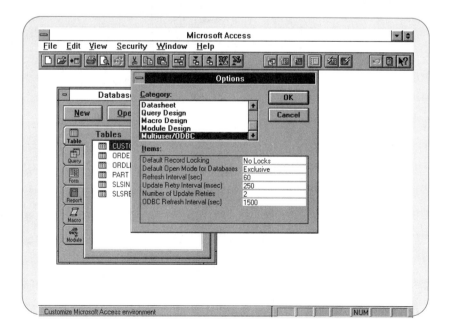

Edited Record Lock: This setting locks every record being edited so that no other user can change it. Although this setting may delay other users being able to access a record while you are using the data, it guarantees that all changes are made and conflicts with other users' work is avoided.

All Records Lock: This setting locks the entire object (and underlying tables) when you open it. When you attempt to open a form or datasheet in the All Records locked setting, you must be the only user using all underlying tables. Because this setting is very restrictive, it should be used in situations where only user is going to be using the database.

Exclusive Lock: Opens a database for your exclusive use.

Shared: Opens a database for shared use in a network environment.

Note: Regardless of the Record Locking option selected, Access always locks a record when you are reading or saving it in a file. If a record is locked by another user when you are trying to access it, you are notified with a message.

Refreshing the Screen

At any point in time, you might have several records displayed on your screen. If another user updates one of these records, the data on your screen would no longer be accurate. To overcome this problem, Access uses a process called *refreshing* your screen. **Refreshing** simply means updating the data on your screen with the current data in the table. The control for refreshing your screen is also set in the Options menu. (Refer to Figure M7.3.)

Refresh Interval: Sets the time interval to update your screen with the current data. This is useful in a multiuser database with other users having editing privileges.

Update Retry Interval: Waits for the selected interval period before trying to access or save a locked record.

Number of Update Retries: Tries the given number of times to save or access a locked record.

M7.5 SECURITY

A DBMS must furnish a mechanism that restricts access to the database to only authorized users.

When Access is installed on a network, a Security utility, **Permissions,** is available to assign passwords to users. (Refer to Figure M7.4.) There are several types of privileges (permitted actions) that may be associated with these passwords. Permissions are given in two categories for each of the various database objects. These permissions are broken down between the access and design of the database and the level of access to the data itself. The various possible permissions are:

1. **Read Design.** User can view, but not modify, an object's design.
2. **Modify Design.** User can view and modify an object's design.
3. **Administer.** The administrator has control over the design of the database objects and also all permissions.
4. **Read Data.** User can retrieve data from the table but cannot make any changes.
5. **Update Data.** User can change existing data but cannot add new records or delete existing records.
6. **Insert Data.** User can change existing data as well as insert new records but cannot delete records.
7. **Delete Data.** User can change existing data, insert new records, or delete existing records.

The actual process of assigning these privileges is as follows: Someone, typically the database administrator (DBA), uses the Permissions option to assign owner passwords to each table or object. Only the DBA can use these owner passwords. When

Figure M7.4

Access Permissions
control

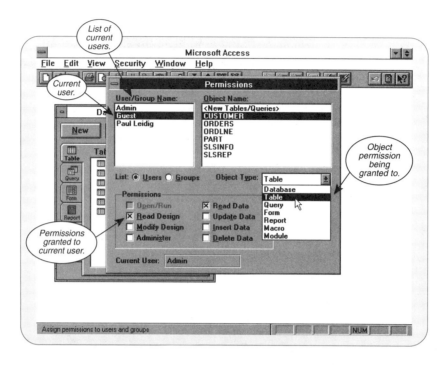

Access is installed in a network environment, a new user is given an account and
password under the Permissions option, which, in turn, determines the level of access
for that user.

M7.6 INTEGRITY

**A DBMS must furnish a mechanism to ensure that both the data in the database and
changes in the data follow certain rules.**

Access provides comprehensive integrity support in the Design view by changing
the Properties of a table or other object (Figure M7.5). All the following various types
of integrity constraints can be enforced:

Figure M7.5

Integrity support
in field properties

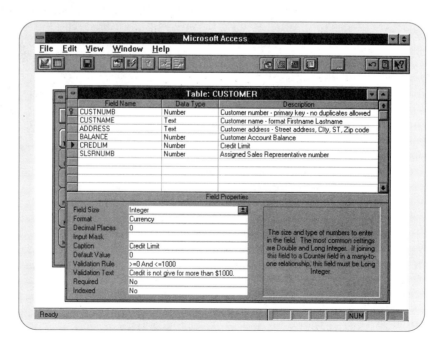

1. **Data type.** Like most DBMSs, Access will ensure that the data that is entered for a given field must agree with the field type; for example, a field that has been specified as numeric will not be allowed to contain nonnumeric data.
2. **Format.** We can specify a wide variety of formats by selecting one of the predefined format options.
3. **Legal values.** You can allow a field to contain a range of values by creating a validity check using Validation Rules. You can create a validity check with the Input Mask option to allow a field to contain only certain specific values .
4. **Key constraints.** You can enforce primary key constraints simply by designating the field or fields that compose the primary key as key fields. A primary key is designated as being required or not and also whether an index should be created and used for the key.

M7.7   DATA INDEPENDENCE

A DBMS must include facilities that provide programs with independence in terms of their relationship to the structure of the database.

You can easily change the structure of a table in Access by modifying the object in the Design view. New fields can be added, existing fields can be deleted, and the characteristics of existing fields can be altered. Although many types of changes in existing fields are possible, by far the most common is a change of length.

These changes have some impact on users. If a field is changed or deleted, the field is automatically changed in reports and forms. Thus, you should examine each such report or form and make any necessary adjustments.

The last type of change involves changes to relationships. If you are not using views (stored queries), implementation of any relationships must be built directly into a table. If you are using views, however, this might not be the case. Once you have changed the query definition to handle the change in the relationship, you can potentially use the query as you did before.

M7.8   RECOVERY

A DBMS must furnish a mechanism for recovering the database in the event that the database is damaged in any way.

Recovery facilities furnished by most microcomputer DBMSs typically consist of a facility for making backup copies of live databases and a facility for copying a backup copy over the live database. They do *not* usually include facilities for maintaining and using a journal that contains a record of any change made in the database.

Access is no exception, although it contains two special features that can be helpful. The first concerns the process of making a backup copy. To backup a database in Access, all you need to do is make a copy of the .MDB file to another drive or directory or copy it with a new name. For example, to copy the *PREM*.MDB database, you would need to copy the file to another disk or copy it with another name such as *PREM-BACK*.MDB. Remember, Access stores all design and data for tables, queries, forms, and so forth in one file.

The second special feature that relates to recovery concerns the ability to undo changes by selecting Undo from the Edit menu. As you have seen, you can undo a wide variety of changes, provided you discover your mistake in time. Many DBMSs do not provide this flexibility. Of course, once you leave Access or begin working on another table, you will no longer be able to undo your changes in this manner. The only way to undo the changes at that point would be to use your most recent backup copy.

M7.9   UTILITIES

A DBMS should provide a set of utility services.

You can change the structure of a table easily by using the Design view. You can create new indexes with the same option. The File menu allows access to macros and add-in services as well as to the transfer of data between Access and other products.

Access includes a number of fourth-generation tools. Among these are the easy edit facilities furnished through the Table View and Form View and the query capabilities furnished by Query-by-Example. Access contains a full-function procedural language, Access Basic. Access Basic uses the same syntax structure as Microsoft's Visual Basic, providing a powerful application generation platform. This topic is covered in more detail in the next chapter.

Finally, the menu system provides a menu-driven interface that is both highly functional and easy to use.

SUMMARY

The following list summarizes the material covered in this module:

1. Access provides support for data storage, retrieval, and update.
2. Access provides a user-accessible catalog through the hierarchical display of a database including tables, queries, forms, reports, and so forth to facilitate finding appropriate objects.
3. Access provides support for shared update through automatic locking and multiuser controls. It also can refresh a user's screen when the data displayed on the screen is changed by another user. Although automatic locking handles most situations, explicit locking is also available, both from the menu system and within programs.
4. Access provides security services through a sophisticated password scheme by which a variety of types of privileges can be associated with user passwords.
5. Through the use of primary indexes and validity checks, Access provides services to support many different types of integrity constraints.
6. Access provides services that make it easy to change the structure of a table as well as to create or remove indexes. These services help promote data independence.
7. Access provides backup and recovery services by including all tables and associated objects in one file. Another feature related to recovery is the ability to undo various types of changes.
8. Access provides a large number of utility services.

EXERCISES

1. Describe the Access features that furnish the user-accessible catalog. What benefits do we obtain from using it?
2. How do the Access features compare with the catalog described in Chapter 7? What, if anything, is missing?
3. How does Access compare in the area of support for shared update with the typical microcomputer system described in Chapter 7? Indicate whether Access provides each of the facilities listed there and, if so, in what manner the facility is provided.
4. Of the facilities that Access provides, which ones relate to backup and recovery?
5. What types of integrity constraints can Access enforce? How can they be enforced?
6. Discuss the various kinds of changes that you can make in the structure of a database in Access. How are they accomplished? In what way do such changes affect programs that currently access the database?

7. Of the utility services listed in Chapter 7, which ones does Access provide? Which Access facilities are used to provide these services?

COMPUTER ASSIGNMENTS

1. Use Access facilities to list as much information as you can about the tables, indexes, forms, reports, and views in the *Premiere Products* database; then do the same with respect to the tables, indexes, and views in the *Books* database.

2. Create a new subdirectory on your data disk called Students. (Make sure the drive containing your data disk is the current drive and then type `md\students`.) Make this the current subdirectory (type `cd\students`) and then start Access in the usual manner. Create the tables that are necessary for the database you designed in exercise 8 of Chapter 6. Make up some sample data and enter it. Create custom forms for data entry and custom reports to display the data in the various tables in your database.

8

Advanced Topics

OBJECTIVES

1. Present the creation and use of *Memo* and *OLE* fields.
2. Explain the use of a one-to-many relationship.
3. Describe the use of radio buttons, check boxes, and option buttons.
4. Present the capabilities of macros to automate processing.

M8.1 INTRODUCTION

At this point, you know how to create tables and how to update the data in your tables. You can query your tables in various ways. You can display attractive data entry forms on the screen. Finally, you can produce professional-looking reports that present data from your files.

This module introduces some of the more advanced features of Access for Windows. Section M8.2 modifies the *Slsrep* table to include *Memo* and *OLE* fields. Section M8.3 looks at some advance form techniques. Finally, section M8.4 introduces Macros to show you how to automate your applications.

M8.2 MEMO AND OLE FIELDS

The tables in the *Premiere Products* and *Books* databases include fields with numeric, currency, date, and text formats. In addition, to be able to store text with a longer variable length, we will use *memo* fields and graphic images created in another application called *OLE* (Object Linking and Embedding) fields.

As examples of these two types of fields, we will add two fields to the *Slsrep* table: a *Comment* field to store long notes about each sales representative and a *Signature* field to store a graphic picture of each salesrep's signature.

Object Linking and Embedding Fields

Access allows you to store *objects* such as an Excel worksheet, a Microsoft Graph chart, or a picture created with Paintbrush that were created in another Windows application. OLE is a Windows feature that allows you to not only include object from other applications but also makes it easy to create and edit those objects directly from within Access. You can insert an object into a field by jumping to the application to create the object.

The difference between linking and embedding relates to where the object is physically stored. When an object is linked, it is kept in a separate file where it was created. The form or report only holds a link to the location of the object file and displays the object. An embedded object, on the other hand, is created in another application but becomes part of the database as a field in one of the tables.

To add an OLE object to a form or report, you must first add a field to a table or query where the linked or embedded object is attached. The field's data type is set to OLE. Then, while in a form or report, you can insert an object into an object frame. An object frame is either *unbound* or *bound* to a specific record in a table. An example of an unbound object could be a picture of your company logo that would be displayed on all forms or reports. We will create a sample bound object, a signature picture that will be bound to each sales representative.

Adding OLE and Memo Fields to the Table

We begin by opening the *Slsrep* table in the *Prem* database in the design view. At this point, we want to add a field for the signature as an OLE type and a *Memo* field for a note on each sales representative.

- Open the *Prem* database.
- Select the *Slsrep* table.
- Click on the Design button.

To add a field, move to the first available Field Name row and enter the name of the new field. In this example, move to the first empty row and enter Signature for the name of the field and then use the pull-down list of data types to select OLE. Then move to the next row and enter Note with a data type of Memo.

- Enter SIGNATURE in the first empty field.
- Select *OLE* from the pull-down list of data types.
- Enter NOTE in the next empty field.
- Select *Memo* from the pull-down list of data types.

You screen should now match Figure M8.1. These are the only additions we need to make to the table structure, so we can save it and create a form to display the new data.

- Select Save from the File menu.
- Exit and close the table.

Figure M8.1

Modifying the *Slsrep* table

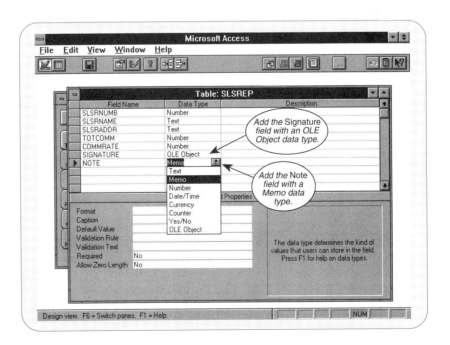

Creating a Form with OLE and Memo Fields

We are now ready to create a new form with the two new fields.

- Click on the Form object button.
- Click on the New button.
- Select on the *Slsrep* table.
- Click on the *Blank Form* button.

A blank screen is displayed where you place the fields. To begin, display the field list, and one field at a time, place the sales rep number, name, address, commission, commission rate, and note fields on the form. At this point, do not drag the *Signature* field to the form because we will add a bound object frame to display the signature picture. You will need to resize the text boxes and change the attached labels, as Figure M8.2 shows.

- Display the field list by selecting Field List from the View menu.
- Drag all of the fields (except Signature) to the form, as Figure M8.2 shows.
- Change the labels as indicated.
- Resize and position the fields.
- Close the Field List window.
- Open the Properties window by clicking on the Properties icon or by selecting Properties from the View menu.

Figure M8.2

Creating a new form

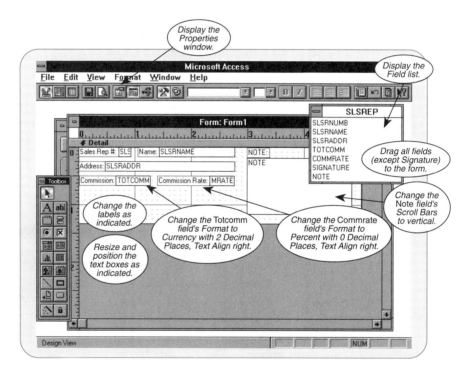

Note: You will need to scroll down the properties list to change the following property settings.

- Change the Format of the *Totcomm* field to Currency with 2 Decimal Places, Text Align `right`.
- Change the Format of the *Commrate* field to Percent with 0 Decimal Places, Text Align `right`.
- Change the Scroll Bars of the *Note* field to Vertical.

We now need to add a bound object frame for the *Signature* field (Figure M8.3). When you place an object frame in the space provided, the form automatically enlarges

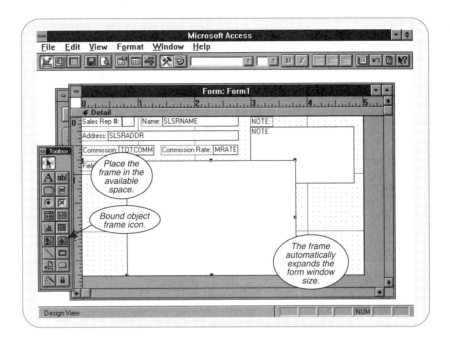

to make space for the default frame. You can move the label and change the size of the object frame window.

- Click on the Bound Object Frame button.
- Click in the space below the commission amount to place the new frame.

Access places a large bound frame in the Form window and expands it to fit. You need to change the size and position of the *Signature* field.

- Change the label to Signature: and position it as shown in Figure M8.4.
- Change the size and position of the bound object frame as indicated.
- Drag the lower edge of the form to match Figure M8.4.

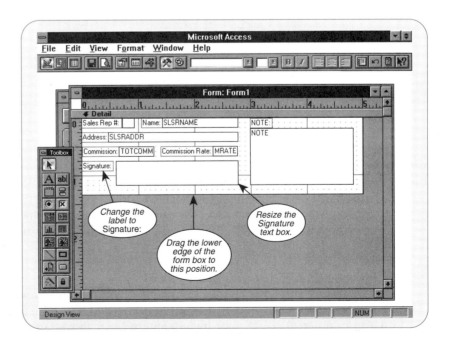

You also need to set the property of the Signature object frame so that it will be bound to the *Signature* field in the *Slsrep* table. This allows you to insert an object from another Windows application into an object frame in a form or report and have the object become part of a table in an Access database.

- Click on the Signature object frame box.
- Display the Properties box by clicking on the Properties tool bar button.
- Move to the Control Source entry box.
- Select *Signature* from the drop-down list of available fields to be bound to.

Note: If the Control Source is not set, you will not be able to embed an object into a field and have it bound to a specific record in the table.

Once your form matches Figure M8.4 and you have set the bound property, you should save the form as Sales Reps before proceeding. To check your layout, change to the Form view and maximize the window. Your screen should now resemble Figure M8.5.

- Select Save As from the File menu and save with the name Sales Reps.
- Change to the Form view by clicking on the Form View button or selecting Form from the View menu.
- Click on the Maximize button or select Maximize from the Control-box menu.

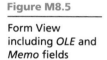

Figure M8.5

Form View including *OLE* and *Memo* fields

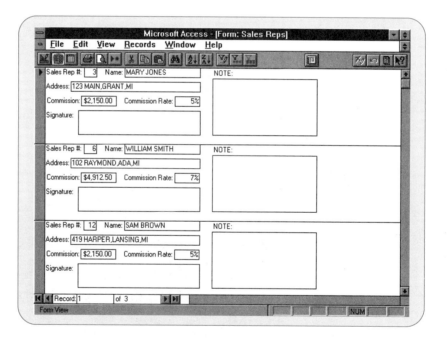

Embedding OLE Objects

To embed *OLE* fields, you could copy an object to the clipboard and then return to the Access form and paste the contents of the clipboard into the field. You could also use the Insert Object method to transfer to another application such as Windows Paintbrush, load or create a picture, and then update the table automatically. We will use the last method and create signatures for the *Slsrep* table with Paintbrush.

- Click in the Signature object frame.
- Select Insert Object from the Edit Menu.

The Insert Object Dialog box displays with a list of available Windows applications. You can scroll down the list of applications and select the one that you want to switch to in order to create your image. Figure M8.6 shows an example of such a list. Your list of applications will most likely include different applications; however, the default

Figure M8.6

Inserting an object
in an *OLE* field

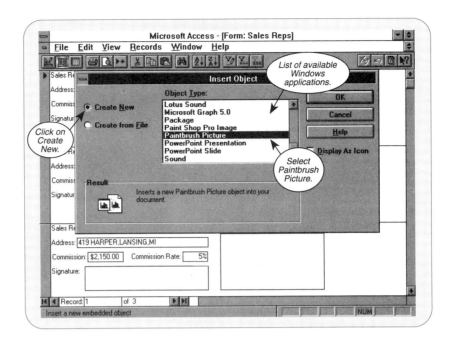

applications that come with Windows should be in the list. You will want to use one
of these applications, Paintbrush Picture. You will also create a new object, as opposed
to loading it from a file.

- Scroll down through the list of applications to select Paintbrush Picture.
- Click on Create New.
- Click on OK.

The selected application, Paintbrush, opens in a window under the control of
Access (Figure M8.7). You can create an image and then update the form. Click on the
Brush tool to change the pointer icon to a small dot that you can use to draw the

Figure M8.7

Using the
Paintbrush
application within
Access

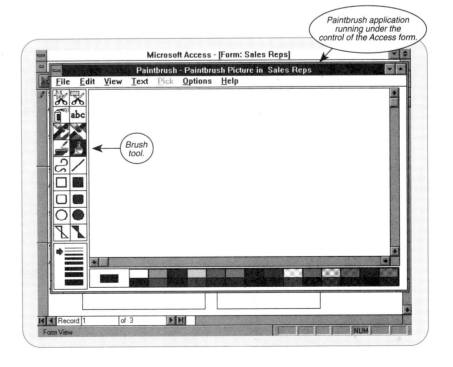

signature. To draw, hold down the left button on the mouse as you move the mouse around. You can release the button to move the mouse without drawing.

- ■ Click on the Brush tool.
- ■ Use the brush tool to draw a signature for Mary Jones as it displays in Figure M8.8.

Figure M8.8

Using the Paintbrush application within Access

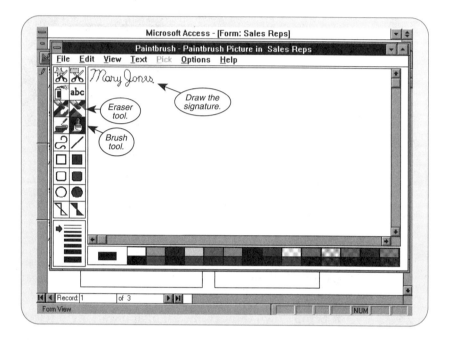

Note: You can use the Eraser tool to erase and redraw any part of the image.

Once you are satisfied with your signature, update the Access record by selecting Update from the File menu or Exit & Return to Sales Reps form. The image in the Paintbrush application is entered automatically into the object frame of the Access form.

- ■ Select Exit & Return to Sales Reps from the File menu (Figure M8.9).

Figure M8.9

Updating an OLE object

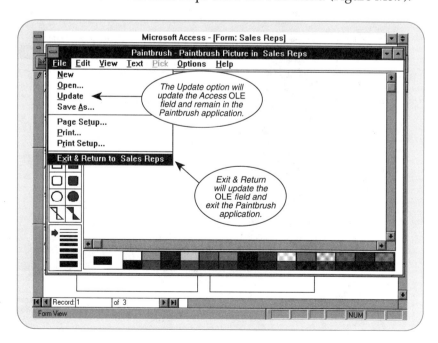

Before closing the Paintbrush application, you are prompted to verify that you do want to exit and pass the image to Access (Figure M8.10).

- Click on OK to verify the update.

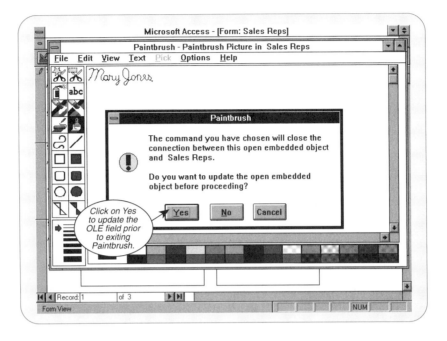

The signature image is placed in the object frame, as Figure M8.11 shows. The image is now embedded as a part of the Access record. If at any time you want to edit the image, you can double click on the signature and the application used to create the image, Paintbrush in this case, is executed so that you can edit the picture while still in Access.

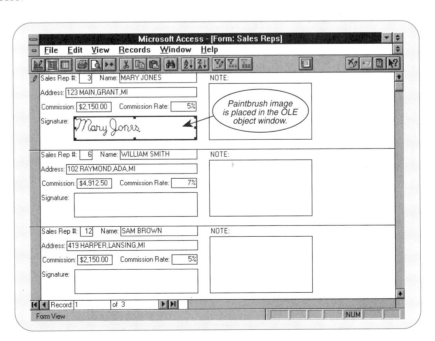

Repeat the above process to create signatures for William Smith and Sam Brown, as shown in Figure M8.12.

Figure M8.12

OLE objects
embedded in an
Access form

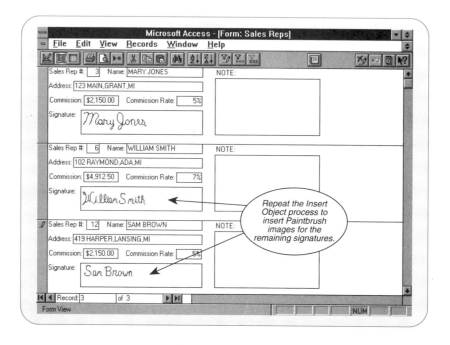

Entering Data in Memo Fields

Entering data in *Memo* fields is similar to entering data in *Text* fields. When you are updating a *Memo* field, you can use Memo view, which gives you some word-processing capabilities. To enter memo text, simply click in a *Memo* field and enter the text. However, if you enter text beyond the display size of the screen memo box, the text scrolls up to allow entering a long string of text. You can then use the scroll bars to move up and down in the box.

For our sales representative example, we will enter information in the *Memo* fields about each representative.

- Click in the *Memo* field for the first record.

You will notice that as you click in a *Memo* field, scroll bars display. These scroll bars are only useful once you enter more text than will fit in the field.

- Type `Fluent in French, German, and Spanish. Likes to travel. Technically strong.`
- Click in the *Memo* field for the second record.
- Type `Somewhat fluent in Spanish. Willing to travel. Works well with new companies. Very innovative.`
- Click in the *Memo* field for the third record.
- Type `Technically excellent. Good reputation among established customers. Advice is well-respected among customers.`

Your screen should now match Figure M8.13.

- Select Save from the File menu to save your updates.
- Close the Form by clicking on the Close button or by selecting close from the File menu.

M8.3   ADVANCED FORM TECHNIQUES

Earlier, you created a form for the *Customer* table that included data from both the *Slsrep* and *Customer* tables. The form displayed customer data as well as the name of the sales rep. In this section, we also create a form that includes data from both the

Figure M8.13

Entering long text
strings in *Memo*
fields

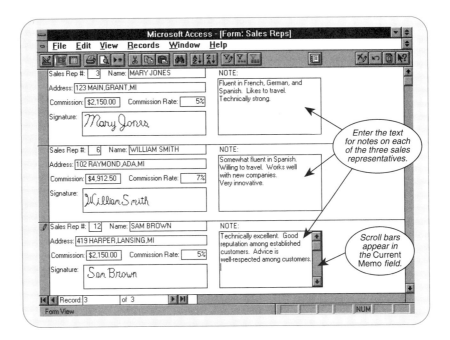

Slsrep and *Customer* tables, but this form displays sales rep data as well as the customers that the sales rep represents.

**Creating a
Form with a
One-To-Many
Relationship**

In this section, we will create a Sales Rep form, and each sales rep can have many customers. There is a one-to-many relationship between sales reps and customers. This can present a problem with the form because, in addition to the basic data about a sales rep (number, name, address, and so on), the form also must include data about the sales rep's many customers. To include the data, we place the customer data in a subform box within the sales rep main form. Access automatically places the data in the subform that matches a link field in the main form.

We begin by creating the customer subform and then include it in the Sales Rep main form just created. You should still have the Forms window on the screen. Click on the New form button and select Field List from the View menu. The customer subform should have the *Customer number, Name, Balance,* and *Credit limit* fields included. You need to set the width, alignment, and format for each field and label, but the position is not as critical since you will be using the report in datasheet view.

- Click on the New button.
- Select the *Customer* table from the drop-down list when prompted for the table or query.
- Click on the Blank Form button to create a custom form.
- Select Field List from the View menu.
- Drag the *Custnumb, Custname, Balance,* and *Credlim* fields to the Detail section of the form.
- Change the labels and position of the fields to match Figure M8.14.

Next, you need to set the properties for the form and each field. Display the Field List and then click in the open portion below the detail section of the form to select the entire form. As you do, the Properties dialog box is set to the entire form. Figure M8.15 shows the Field list with its size enlarged to fill the available space. Set the properties as displayed. (You need to set the Record Source = Customer, Default View = Datasheet, Views Allowed = Datasheet, Scroll Bars = Both, Record Selectors = No, Navigation Buttons = No, Border Style = Thin, Control Box = No, Min. Button = No, Max. Button = No.)

Figure M8.14

Designing the
Customer subform

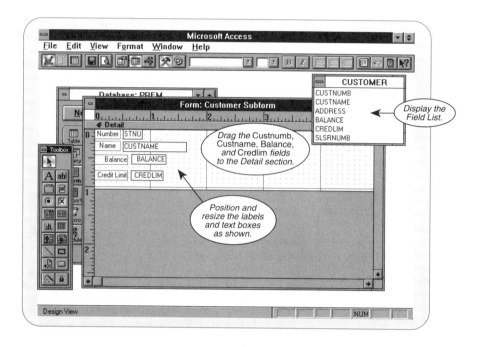

Figure M8.15

Setting Form
properties

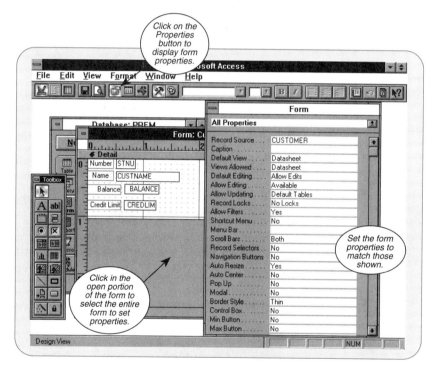

■ Click on the Properties button.
■ Click on the blank area below the Detail section to select the entire form.
■ Set the properties for the form as shown in Figure M8.15.

In addition to the form properties, you need to set field properties for each of the text boxes and labels need to be set (Figure M8.16).

■ Click on the Number label and set the Property Text Align to `Center`.
■ Click on the Name label and set the Property Text Align to `Center`.
■ Click on the Balance label and set the Property Text Align to `Right`.
■ Click on the Credit Limit label and set the Property Text Align to `Right`.

Figure M8.16

Setting Field properties

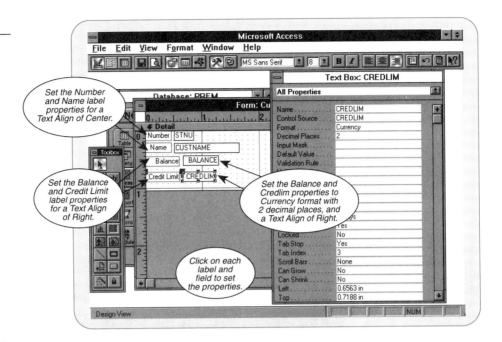

- Click on the Balance text box and set the Property Format to `Currency`, 2 Decimal Places, Text Align `Right`.
- Click on the Credit Limit text box and set the Property Format to `Currency`, 2 Decimal Places, Text Align `Right`.

Once you have set the field properties, close the properties and save the new subform as Customer Subform. Then switch to Form view to view the form to verify that you have created it properly.

- Close the Properties window.
- Save the form as `Customer Subform`.
- Click on the Form View button and compare your form with Figure M8.17.
- When you have finished, close the Customer Subform.

Figure M8.17

Completed Customer subform

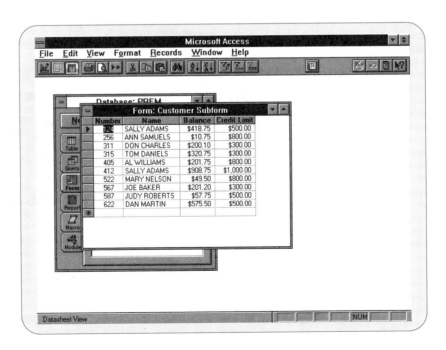

We can now return to the Sales Reps form and insert the subform. Select the Sales Reps form and click on the Design button. You insert a subform by clicking on the Subform button in the Toolbox and then clicking in the form where you want to place the Subform box.

- Select the Sales Reps form in the Forms list.
- Click on the Subform button.
- Click in the space below the Signature box to place the subform.

Your screen should now have a subform box, as Figure M8.18 shows. In order for the one-to-many relationship to be set that will link the sales rep to each customer with the same *Slsrnumb,* you need to open the Properties box and identify the Source Object as the Customer Subform that you just created and *Slsrnumb* as the link field. Resize and position the subform, as Figure M8.19 shows.

Figure M8.18

Adding a subform to the Sales Reps form

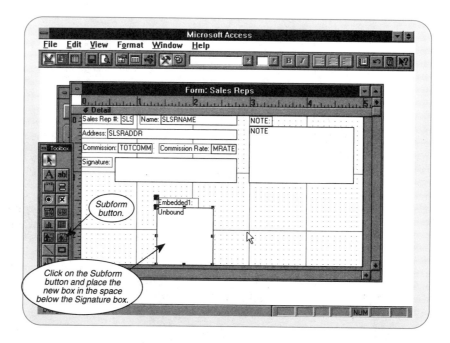

Figure M8.19

Adding a Subform to the Sales Rep form

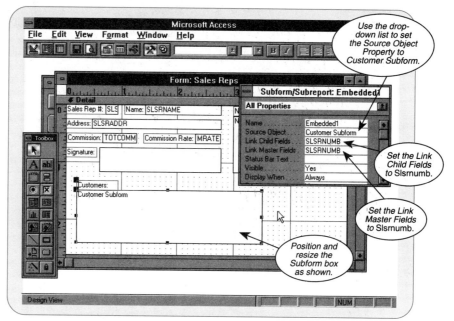

- Open the Properties box.
- Use the Source Object drop-down list to select the Customer Subform.
- Set the Link Child Fields to *Slsrnumb*.
- Set the Link Master Fields to *Slsrnumb*.
- Position and resize the Subform box, as shown in Figure M8.19.

Once your properties have been set and the Subform box positioned, you should save the Sales Reps form and then switch to the Form view.

- Select Save the File menu.
- Click on the Form view button.

Your completed form should now look similar to Figure M8.20 and show the three customers assigned to sales rep 3, Mary Jones. To view each sales rep you can click on the navigation arrows at the bottom of the form. As you move between sales reps, the customers in the subform will change to match their respective sales reps.

When you are finished, close the form.

- Select Close from the File menu.

Figure M8.20

Completed One-to-Many form

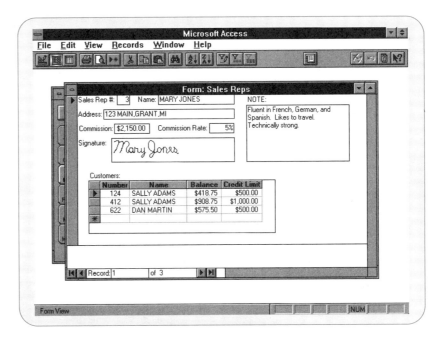

Setting Form Properties

This next section will describe how to make a new form using the `Part` table to demonstrate various advanced form techniques. First you will create a form and set the color and styles of the detail section and each field. Then you will add list boxes and option groups.

First you create a blank form, add the part number, description, units on hand, and price fields, and add a calculated field for the on-hand value. Then you will set the properties of each field.

Your system should still be in the *Prem* database with the Form object selected.

- Click on the Form object.
- Click on the New button.
- Select the *Part* table from the drop-down list.
- Click on the "Blank Form" option to bypass the Form Wizard.
- A blank form appears as Form1.

■ Click on the Field List button of the toolbar to display available fields.
■ Drag the *Partnumb, Partdesc, Unonhand,* and *Unitprce* fields to the Detail section of the form as shown in Figure M8.21.

Figure M8.21

Creating a new form

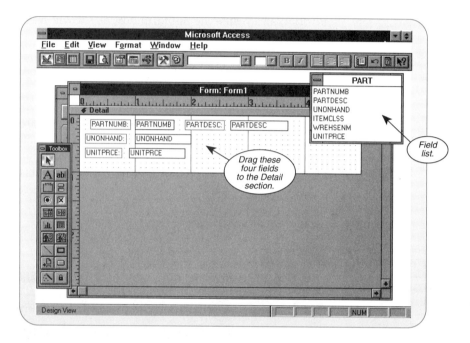

Next, using the same method as in the previous section, change the field labels and size, arranging each field as Figure M8.22 indicates.

■ Change the labels, and arrange each field as indicated in Figure M8.22.

We also want to add a calculated field for the unit value of each part. The unit value is equal to the units on hand times the unit price. Add a textbox, change the label to Unit Value:, and enter the formula =unonhand*unitprce in the textbox.

Figure M8.22

Creating a new form

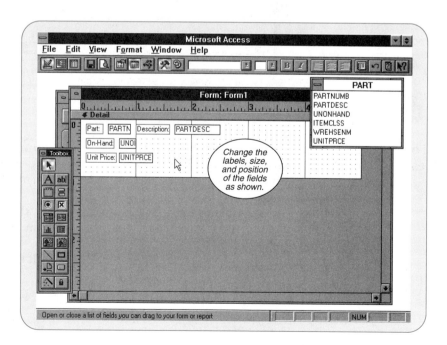

- Click on the Textbox tool of the toolbox.
- Click below the *Unit Price* field to place the new textbox.

Your screen should now have a new textbox, as shown in Figure M8.23.

- Change the label to `Unit Value:`
- Enter the formula `=unonhand*unitprce` in the textbox.
- Click on the Field List to close the Field List window.
- Drag the lower edge of the Detail section down to put some space below the new textbox.

Your screen should now have a new textbox, as Figure M8.24 shows. Our next step will be to change the properties of the detail section and each field. We want to

Figure M8.23

Adding a calculated field to the form

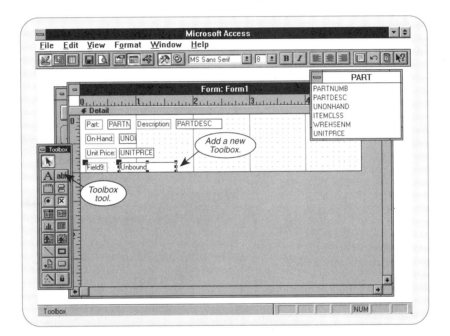

Figure M8.24

Adding a calculated value to a text box

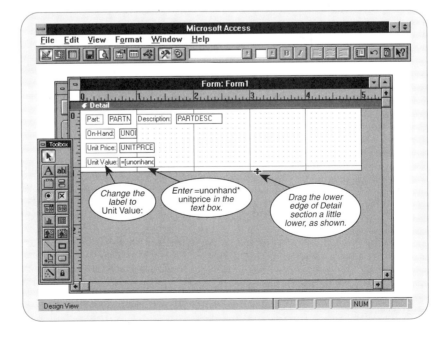

change the color of the detail section and then set the textboxes with a sunken view to set the data fields with a different appearance from the form itself. We can use the palette on the tool bar to set the control appearance.

- Click on the Palette tool to display the palette dialog box.
- Click inside the Detail section.
- Click on the third color option in the background color of the palette.

As you click on the color option, the color of the detail section changes, as Figure M8.25 shows. We also want the labels of each of the textboxes to be the same color as the detail section. When you hold down the Shift key, you can select multiple objects by clicking on each one. Using this method, you could select all of the labels at once and then you would need to select the color option once for all five labels.

- Hold down the Shift key and then click on each of the five labels to select all five.
- With all five labels selected, click on the third color option of the background color in the palette.

Figure M8.25

Changing a form's control properties

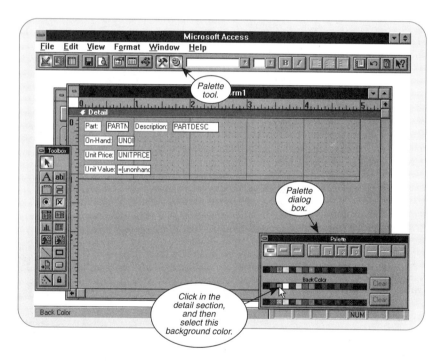

Your screen should now match Figure M8.26, with the labels showing and detail section now having the same shaded color. To see the effects of your actions so far, click on the Form view button. Your screen should now show the records in its Form view, as indicated in Figure M8.27.

- Click on the Form view button to view the form, and then click on the Design view to return.

The next step in changing your form, is to give the *Text Box* fields a sunken look and then change the format of the two *Currency* fields.

Figure M8.26

Changing the properties of labels

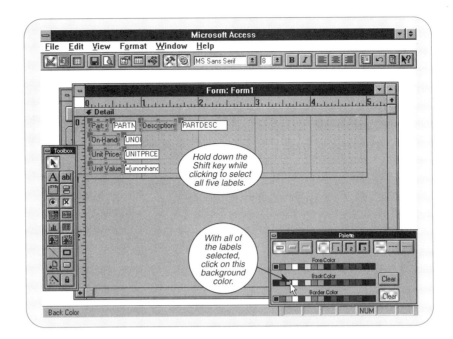

Figure M8.27

Changing a form's property

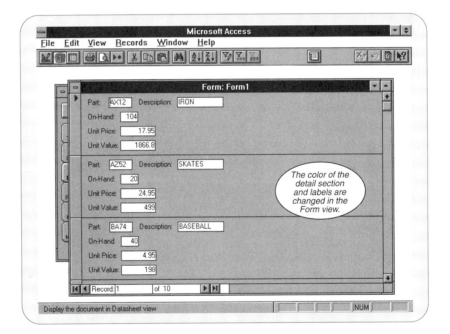

- Hold down the Shift key and click on each of the five Text Boxes to select all five data fields.
- Click on the Sunken View button in the palette. (Figure M8.28)

To set the format property of the two currency fields, close the palette and display the Properties window. Then select the two *Currency* fields and change the format to currency with two decimal places, aligned right.

- Click on the Palette button of the tool bar to close the palette.
- Click on the Properties button of the tool bar to display the properties.
- Hold down the Shift key and click on the Unit Price and Unit Value Text Boxes.

Figure M8.28

Giving the text
boxes a sunken
look

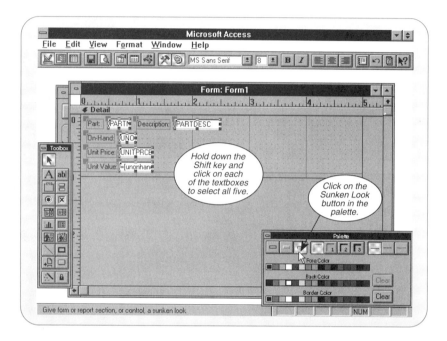

■ Change the properties to a Format of `Currency` with 2 Decimal Places, Text
Aligned `Right`.
■ Click on the Properties button to close that window.
■ Click on the Form View button to check the status of your changes (Figure
M8.29).

Figure M8.29

Form with custom
color and sunken
look

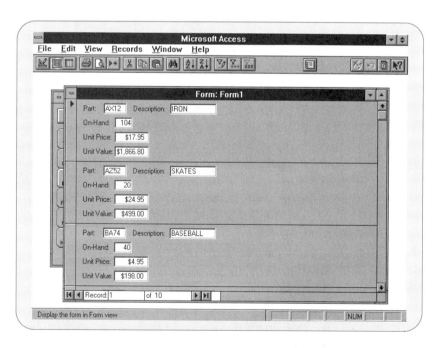

Creating
Option
Groups

Access provides a number of display options that provide a quick way to enter data
into fields that have a limited number of valid values. Earlier in the text, we used a
Combo Box to select from the three valid Item Class values in the Part table. This is an
example of an Option Group. Access allows you to display a number of valid entries
in a controlled group, including Combo boxes, List boxes, Option buttons, and Check

boxes. The Combo box presents a drop-down list in a field from which you can select a valid entry. A List box displays all of the valid entries in a scroll box to select a value. Option buttons present values with radio buttons beside them that can either be on or off, and limit your selection to only one of the items in the group. Check boxes are similar to option boxes, but they present the on/off selection as a small box with an "x" in it to identify the selected item. For this example, you will create two option groups. First, you will create a List box of valid Item Class options for the Part form, followed by option buttons for the three warehouses.

Return to the Design view of the form now. To create the List box for the Item Class field, click on the List Box tool in the Toolbox. Then click in the space under the description to add the new list. Begin by setting the source information in the Properties box.

- Click on the Design view button.
- Click on the List Box tool.
- Click in the Detail section below the Description field.
- Click on the Properties button.

Your screen should now match Figure M8.30. To describe the list box's valid values and what field it is bound to, select *Itemclss* in the field's Control Source drop-down list, and select Value List from the Row Source Type. In the Row Source box, enter HW;AP;SG for the valid values. Change the label to Item Class and then resize and position the List box.

- Change the name of the List box to Item Class.
- Select Itemclss from the Control Source drop-down list.
- Select Value List from the Row Source Type drop-down list.
- Enter HW;AP;SG in the Row Source.
- Change the list box label to Item Class.
- Click on the label and then select the third color option in the Background color of the palette.
- Resize and position the List Box to match Figure M8.31.

Figure M8.30

Adding a List box

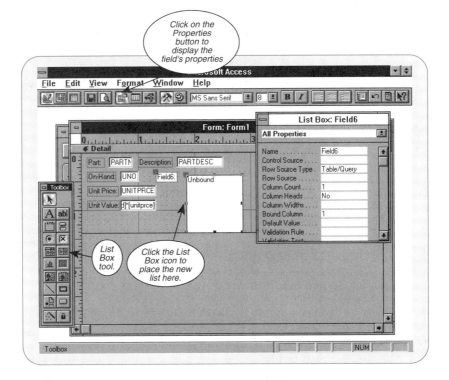

Figure M8.31

Adding a List Box

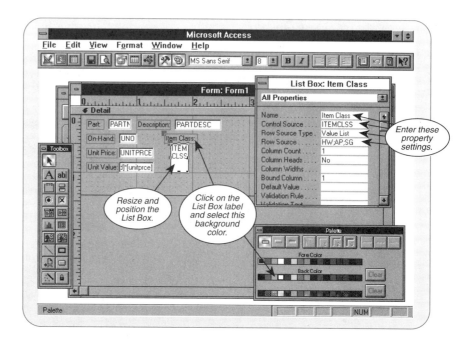

Once your form matches Figure M8.31, click on the Form View button to check your work. Your form should now include the Item Class in a List Box as shown in Figure M8.32.

■ Click on the Form view button to check your form, and then click on the Design view to return to the Design screen.

Figure M8.32

Displaying valid values with List Boxes

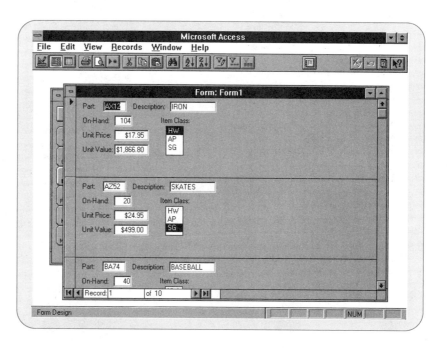

We also want to add an option group of valid warehouse values to select from for the *Wrehsenm* field. Since we will be adding the option group in the upper right side of the form, first drag the Properties box to the lower-left corner of the form. Then click on the Option Group tool and click in the space just below and to the right of the *Description* field to place the Option Group box.

■ Drag the properties box to the lower left of the screen.
■ Click on the Option Group tool.
■ Click below the Description field to place the new Option Group box.

Your should now have an Option Group box as shown in Figure M8.33.

■ Change the label to `Warehouse:`
■ Click on the label and select the third background color option in the palette.

Your screen should now look similar to Figure M8.34. The next step is to add three option buttons for the three warehouse values. Use the Option Button tool to add each

Figure M8.33

Adding Option buttons

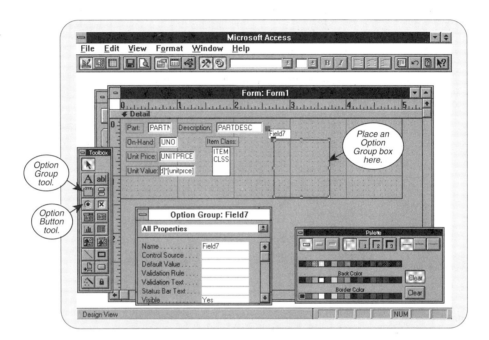

Figure M8.34

Setting Option button properties

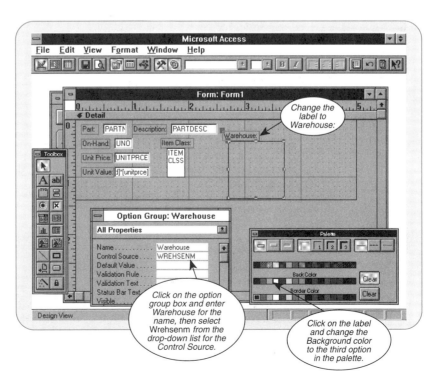

value inside of the Option Group box. Simply click on the tool and then click again inside of the Warehouse to place a new valid button.

■ Add three option buttons to the Option Group box (Figure M8.35).

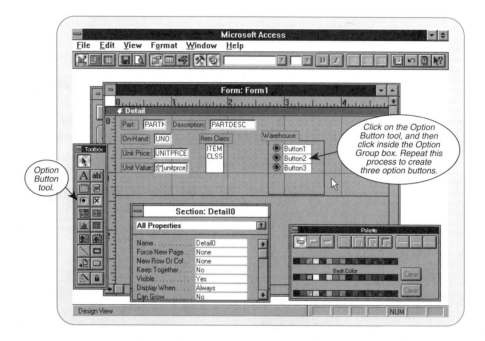

The warehouse numbers in the *Wrehsenm* field indicate where a part is stored, but they aren't very descriptive. You can add a description of the warehouse (for example, the street where the warehouse is located) to our option group.

■ Change the three labels to:
1—Alpine
2—Division
3—Kentwood

Next, hold down the Shift key while clicking on all three labels of the option buttons to select all three. Then select the third background color option in the palette.

■ Hold down the Shift key and select all three option buttons.
■ Click on the third background color option in the palette.
■ Resize and position the warehouse option group and fields, as indicated in Figure M8.36.
■ Drag the lower edge of the detail section so that it is just under the fields.

Your completed form should look similar to Figure M8.37 when you click on the Form View button. Save the form when you are finished.

■ Save the form as Part Listing.

M8.4   USING MACROS

A **macro** automatically performs a series of tasks that had previously been defined. Each task to be performed is called an **action**. Access provides a list of actions that you can select from to create a simple macro. When you run the macro, Access carries out the series of actions in the sequence that they are listed. For example, you could use a macro to search for a selected record, print the current form, or update an entry form.

Figure M8.36

Defining Option
buttons

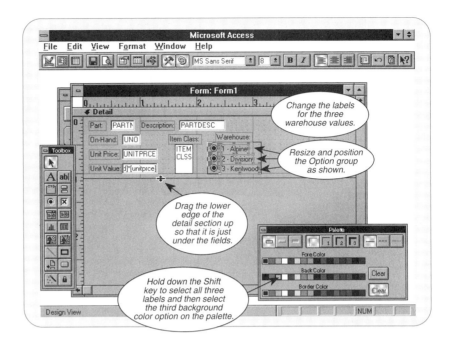

Figure M8.37

Completed form
with option groups

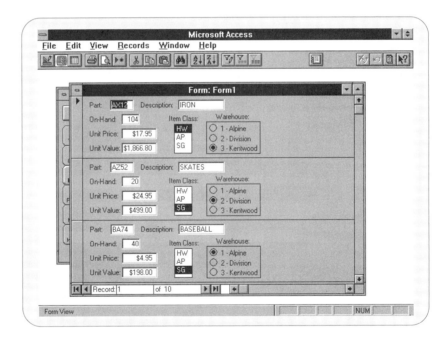

To create a macro, you normally follow the routine of clicking on the Macro button, and then New to create a new macro from scratch. Another method would be to create a macro from within a form or report. In the example, we will add a macro to the Sales Rep form to search for a sales rep chosen from a Combo box. We will create the macro as part of the combo box property.

Macros consist of actions, and each action contains action arguments that define what the action is supposed to do. The actions are selected from drop-down lists of predefined Access actions. Then for each action, you define the specific arguments, or details, of the command. When a macro is run, it carries out each action sequentially.

Modifying the Sales Rep Form

To begin, open the Sales Rep form in the Design view. Add a Form Header to the form, and then add a Combo box that will be used to select the sales rep to search for from the drop-down list.

- Open the Sales Rep form in the Design view.
- Choose Form Header/Footer from the Format menu.
- Click on the Palette button in the tool bar to display the color options.
- Click on the Form Header, and then select the third Background color option from the palette.
- Drag the lower edge of the header down to the 1/2-inch mark.

Your form should now have a shaded header like the one shown in Figure M8.38. You now need to add a Combo box to select the name to search for.

- Click on the Combo Box tool.
- Click in the header to place the Combo box.
- Change the combo box label to `Select Sales Rep name to search for:`
- Click in the Form header, and then select the third background color option in the palette window.

Figure M8.38

Adding a macro to a form

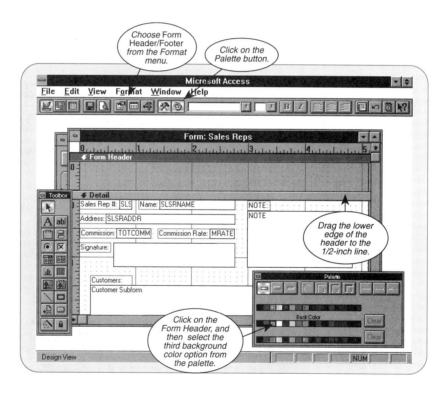

Your screen now should match Figure M8.39, with the bold label in a larger font and the background color matching the header color.

Adding Macro Actions to Form Objects

Now you are ready to define the options in the Combo Box drop-down list and then to create the macro that will be executed when you select one of the options. Refer to Figure M8.40 for the following entries:

- Close the Palette window by clicking on the Palette button.
- Click on the Unbound Combo box.
- Click on the Properties button.
- Enter `SearchName` for the name of the Combo box.

Figure M8.39

Adding a macro to a form

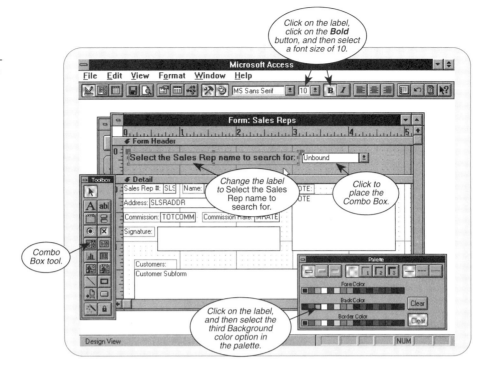

Figure M8.40

Describing the search name Combo Box

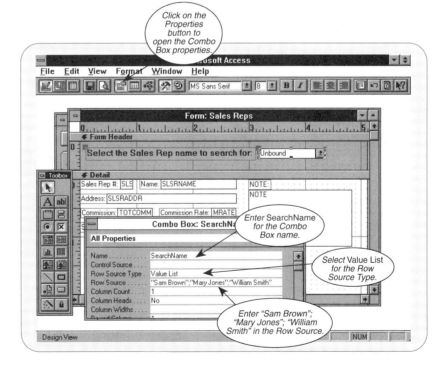

- Select *Value List* from the drop-down list of the Row Source Type.
- Enter "Sam Brown"; "Mary Jones"; "William Smith" in the Row Source.

Then scroll down the Properties window to the After Update box. This entry is where you describe what macro is to be executed when you select an entry from the drop-down list. Since you haven't created a macro yet, you can click on the Build button to create the macro.

- Scroll down to the After Update option.
- Click on the Build option (Figure M8.41).

At this point you may enter a specific expression or, as in this case, use the macro builder (Figure M8.42).

- Select the Macro Builder option.

Figure M8.41

Describing the Macro function for the SearchName combo box

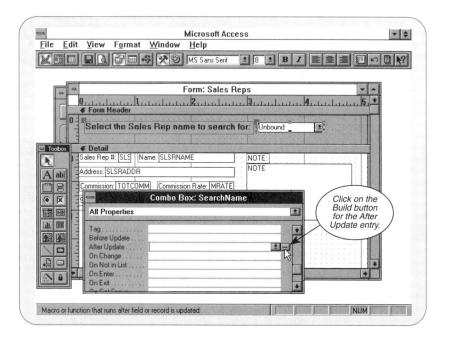

Figure M8.42

Describing the Macro function for the SearchName combo box

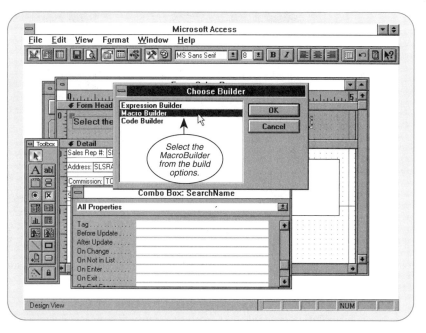

Figure M8.43 shows the Macro window with its two sections. The top half of the window is where you define the actions that the macro is to execute. You can enter comments to explain the action on the right side of the top half. Typically you select the action from the drop-down Macro Action list. Your first action will be to GoTo the

Figure M8.43

Describing actions
for the Name
Search macro

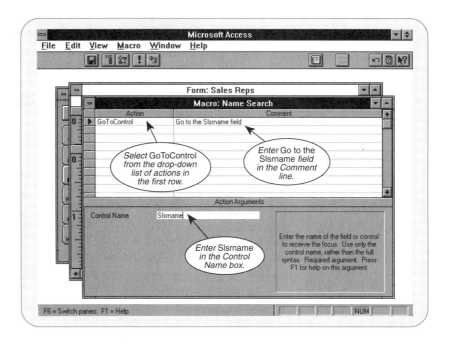

Control field of *Slsrname*. This is the field that you will select from the Combo box and then find a match in the *Slsrname* field.

- Select *GoToControl* for the first action from the drop-down list.
- Enter the comment as `Go to the Slsrname field`.
- Enter *Slsrname* for the field to go to in the Control Name box.

The second action you will take, once you have moved the cursor to the *Slsrname* field, is to find the sales rep record that matches your selection from the SearchName Combo box.

- Move to the second action line.
- Select *FindRecord* from the drop-down list of actions.
- Enter a comment of `Find the record that matches the SearchName selection.`
- Enter `=[SearchName]` in the Find What argument.
- Select *Any Part of Field* from the drop-down list in the Where argument.

Your screen should now match Figure M8.44. This simple macro now consists of only two actions. First, it will place the cursor in the *Slsrname* field and then find the record that has its *Slsrname* match the entry selected in the combo box that you created in our Sales Rep form. Before proceeding you should:

- Save the macro and close the Macro window.

This will return you to the Sales Rep form in the Design view. Click on the Form View button to test the macro. You should now be able to use the drop-down list of the Search Combo box. When you select a sales rep from the SearchName Combo box, the current form will display that record. In Figure M8.45, William Smith had been selected from the Drop-down Combo box. That action caused the macro to run, which found the matching record and displays it.

**Adding a
Help Message
Command
Button**

Another useful action to assign to a macro is to add a command button to a form that will display a message (up to 255 characters) providing hints or help on database use. To demonstrate, we will add a short help button to the Sales Rep form.

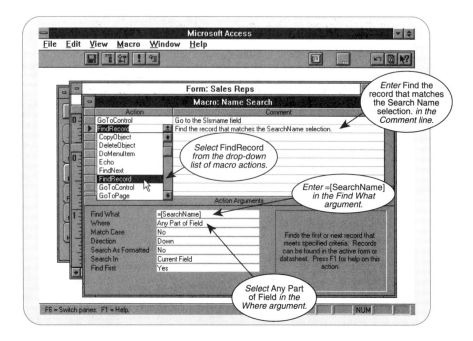

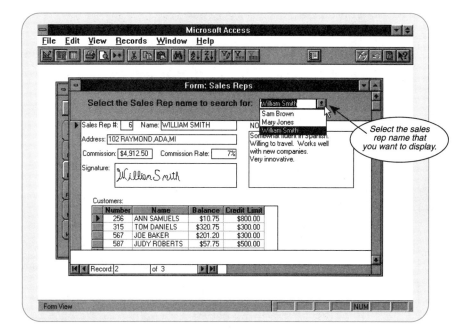

■ Click on the Command Button.
■ Click to the right of the Combo box in the Header to place the Command button.
■ Click on the Properties button.

A Command button has been added to the Header similar to what Figure M8.46 shows. Change the name of the button to Help. Then position the button as shown in Figure M8.47 before clicking on the Build option of the On Click option in the Properties box.

■ Change the name of the button to `Help.`
■ Change the size and position of the command button.
■ Click on the Build option in the On Click property.

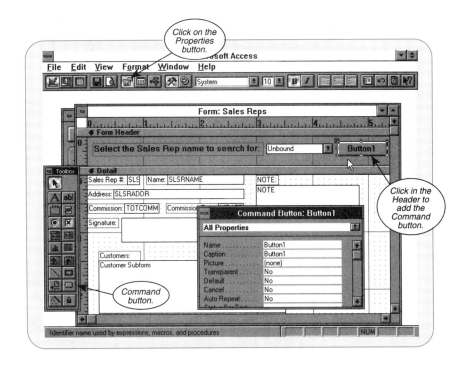

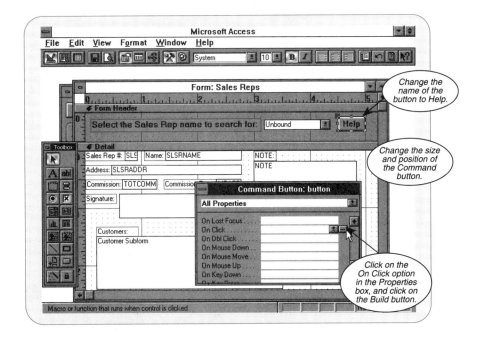

This command button will carry out one action: to display a Message box. Enter your message in the Action Argument Message box.

- Select MsgBox from the Action drop-down list.
- Enter a comment in the first action line: `Provides basic help on searching for a sales rep.`
- Enter the following text in the Message property.
    `To search for a sales representative, click on the drop-down list of sales reps to search for. Then either click on the desired sales rep with the mouse, or use`

the arrows to highlight a representative and press Enter. The matching record is displayed.

■ Change the Beep option to No.

Figure M8.48 shows the Action and arguments for the Help macro. Save and close the Macro. Then Switch to the Form view of the Sales Reps form and click on the Help macro to test the button.

■ Save and Close the Macro.
■ Click on the Form view button.
■ Click on the Help command button to test the macro.

Figure M8.49 shows the completed Help message box. Click on OK to close the message box.

Figure M8.48

Adding a Macro
Command button

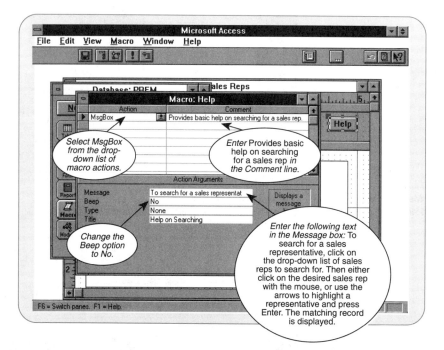

Figure M8.49

Using a Help
Macro Command
button

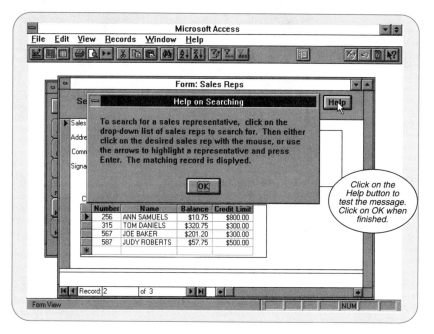

■ Click on OK to close the Help message box.

These two macros illustrate how you could use a macro to automate standard Access actions without using the menus or toolbar buttons. As you become more familiar with Access, you can easily create macros to add to your database.

SUMMARY

The following list summarizes the material covered in this module:

1. Microsoft Access supports full OLE. Object linking and embedding allow you to use other Windows applications to create objects in another application while remaining in Access.
2. Linked objects physically store the object in a separate file in the other application's native format. Embedded objects are created in another application but are stored as part of a table in an Access database.
3. To update an *OLE* field, double click in the field; the application will be launched to create the object. You can then Update or Exit and Update the Access record.
4. Memo fields are long text data types that allow you to enter 64,000 characters.
5. To create forms with a one-to-many relationship, first create a subform for the "many" side of the relationship and then embed the Subform box in the main report for the "one" side of the relationship. Set the link relationship in the Property box of the subform while you are in the Main Reports Design view.
6. You can change the characteristics of a form by setting the properties in the design view.
7. Use the Palette in the Design view to control colors and appearance of a report or form.
8. Option groups tie individual data values together as yes/no values that allow only one of the items to be chosen from the group. You can choose between Option Button or Check Box format.
9. Combo boxes list valid data values in a form or report drop-down list.
10. List boxes list valid data values in a form or report scroll list format.
11. You can add text to option buttons or check boxes to make the entries more descriptive.
12. Macros automatically perform a series of tasks that had previously been defined. Individual tasks are called **actions** and contain arguments describing what the action is supposed to do.
13. You can attach macros to buttons or fields. When you click a button or enter data in a field, the macro actions defined in the object's property are executed.
14. You can create Help messages by adding a Message Command button to a form. The message (up to 255 characters of text) is then displayed in a pop-up window.

EXERCISES

1. What are OLE objects?
2. How do you create or update an *OLE* field?
3. What is a *Memo* field? How do *Memo* fields differ from *Text Data Type* fields?
4. How do you create a form with a one-to-many relationship?
5. What are Option Groups? How do you create an Option Group?
6. What is the difference between Option buttons and check boxes?
7. What is the difference between Combo boxes and List boxes?
8. What is the purpose of a macro?
9. What are actions? How do you assign arguments to actions?
10. How does the MsgBox action work? What are the limitations of these messages?

COMPUTER ASSIGNMENTS

1. Add a *Notes* field and a *Logo* field to the *Publishr* table in the *Books* database. The *Notes* field will store ordering information. The *Logo* field will store a design symbol for each publisher. Use the Paintbrush application to create the logos. (**Hint:** Use abbreviations or first names as the design symbol.) Use the data in the following table to update the *Notes* field:

Pubcode	Notes
AH	Has minimum order requirement of 25 books. Ships weekly.
AP	Will fill single orders. Ships daily.
AW	Has minimum order requirement of 5 books. Ships twice a week.
BB	Will fill single orders and special requests. Ships daily.
BF	Will fill single orders. Ships twice a week.
JT	Will fill single orders and special requests. Ships weekly.
MP	Will fill single orders. Ships orders when needed.
PB	Will fill single orders and special requests. Ships weekly.
RH	Will fill single orders on emergency basis only. Ships orders when needed.
RZ	Has minimum order requirement of 20 books. Ships every other week.
SB	Will fill single orders and special requests. Ships weekly.
SI	Has minimum order requirement of 20 books. Ships every other week.
TH	Will fill single orders on emergency basis only. Ships once a month.
WN	Will fill single orders on emergency basis only. Ships weekly.

2. Create a one-to-many form for the *Publishr* table that lists all the books from the *Book* table published by the publisher. Display the *Pubname* from the *Publshr* table; the *Bktitle, Bktype,* and *Bkprice* from the *Book* table; and the *Authname* from the *Author* table.

3. Create a form for the *Bookcust* table in the *Books* database. Use a combo box to display the *Cstst* field and Option buttons to display the *Brnumb* field. Provide descriptions for the values in the *Brnumb* field.

4. Modify the form created in assignment 2 above. Add a macro to a combo box that will search for a matching *Publishr* value when it is selected from the Combo box.

Graphing in Microsoft Access

OBJECTIVES

1. Describe the process of creating graphs.
2. Examine the ways graphs can be customized.
3. Investigate the use of Crosstab Queries.
4. Modify an existing form to include a graph.

M9.1 INTRODUCTION

This module, examines the power of graphing Access data using the Windows application Microsoft Graph. Instead of building a graphing module into Access, an OLE connection is made with Microsoft Graph though the use of Wizards. This way, you produce professional graphs based on Access data, as well as display data from other Windows applications within forms or reports. Section M9.2 begins by looking at a quick way to create graphs, as well as the ways you can customize graphs and save graph settings for future use. In section M9.3, we will look at different graph types. Section M9.4 addresses the use of crosstabs to format complex data in a tabular form using a crosstab query. We will then create graphs of crosstabs. Finally, section M9.5 investigates how to create and use graphs with existing custom forms.

M9.2 INTRODUCTION TO GRAPHS

We will use the *Slsrep* table as a vehicle for illustrating the graph-creation process. As you will recall, the *Sales Rep* table includes a total commission amount field. We will generate a graph of the three sales reps charting the total commission first as a bar chart and then change the graph type to demonstrate how you can graph data in various formats.

Creating a Graph

The easiest method to use when graphing Access data is to create a form and use the Wizard to create a graph type form. The graph you will create has sales rep names along the X (horizontal) axis. The Y (vertical) axis will represent total commissions.

- Open the *Prem* database.
- Click on the Form object button.
- Click on the New button.
- Select the *Slsrep* table from the drop-down tables/query list.
- Click on the Form Wizard button.

Figure M9.1 shows the various Form Wizard types available. We want to use the Graph Form Wizard.

Figure M9.1

Creating a graph
with the Form
Wizard

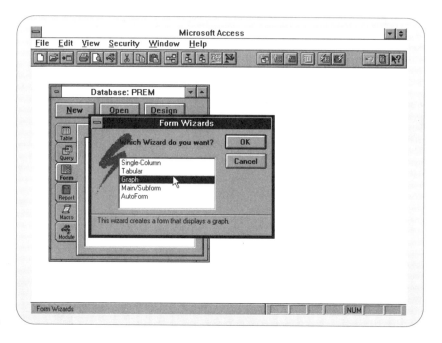

■ Select the Graph Wizard, and click on OK.

The first information that the Graph Wizard needs, is which fields should be
included in the graph. Select *Slsrname* and Totcomm from the Available Fields and, for
each one, click on the ">" button to copy them to the Fields for Graph list. When you
are finished, click on the Next button.

■ Click on *Slsrname* in the Available Fields list.
■ Click on the > button to copy to the Fields for Graph List.
■ Click on *Slsrname* in the Available Fields list.
■ Click on the > button to copy to the Fields for Graph List.

Your screen should match Figure M9.2. You can now move to the next screen.

Figure M9.2

Creating a graph
with the Form
Wizard

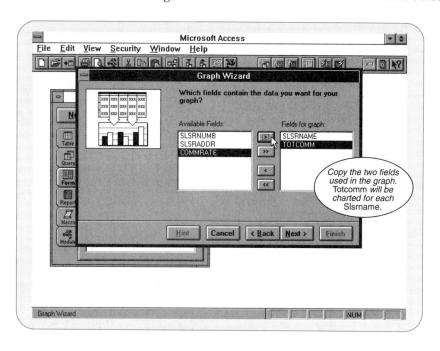

■ Click on the Next button.

The following screen requests a format for calculating totals for each category. Since we will not be using categories in this graph, you can bypass this screen.

■ Click on the Next button.

Next, Graph Wizard presents the twelve basic types of graphs, along with a sample graph window (Figure M9.3). As you select one of the different types of graphs, the sample window indicates what the graph will look like. The default graph is a standard 2-D column graph. To accept the default format, go to the next screen.

■ Click on the Next button.

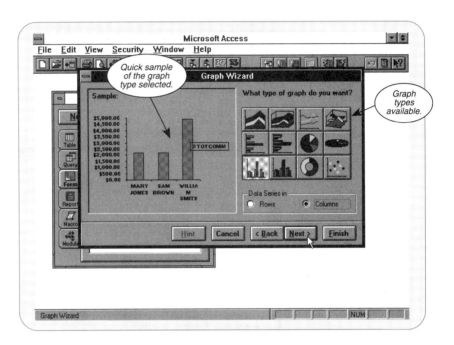

The next screen (Figure M9.4) prompts for a title for the graph and whether a legend should be displayed. Since there is only one data category, it might make more sense to include an explanation in the title instead.

■ Enter `Sales Rep Commissions` in the Title box.
■ Click on No to the Legend display option.
■ Click on the Finish button.

Access completes and displays the resulting graph in a form (Figure M9.5). The graph is considered a special type of form, one that at this time includes only one item, an embedded OLE object. To modify the graph, we will switch to Microsoft Graph and then update and return to the form. Save the form before continuing.

■ Select Save As from the File menu, and save with the name `Commission Graph.`

Adding Axis Titles

To modify the graph and add axis titles and grid lines, for example, you must switch to the Graph application. You could do this by either double clicking on the graph

Figure M9.4

Describing the
graph

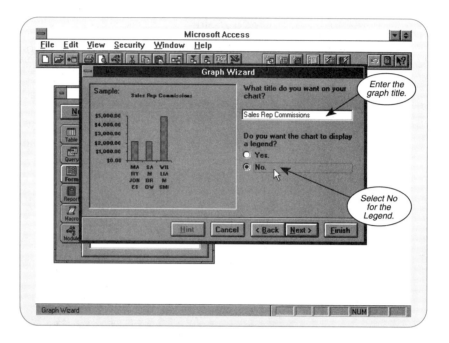

Figure M9.5

Default graph
created by Graph
Wizard

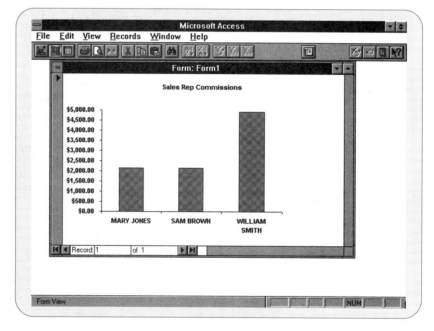

itself, or switching to the Design View, and then selecting Chart Object and Edit from
the Edit menu. This latter method is shown in Figure M9.6.

- Click on the Design View button.
- Select Chart Object and then Edit, from the Edit menu.

You are temporarily placed in Microsoft Graph to modify your chart. At this point,
you will add a few enhancements such as titles on the X and Y axes, horizontal grid
lines, and data labels to show the actual values.

- Select Titles from the Insert menu.

A dialog box as shown in Figure M9.7 displays. You will notice that the check box
for the Chart Title is already checked, because the Graph Wizard inserted it upon

Figure M9.6

Editing the Chart Object

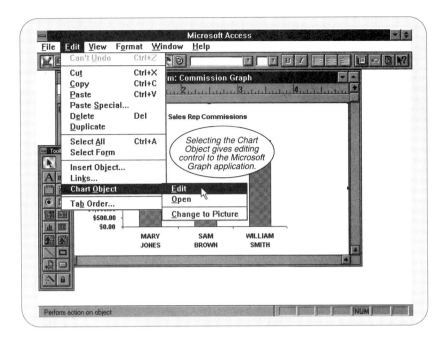

Figure M9.7

Chart Title options

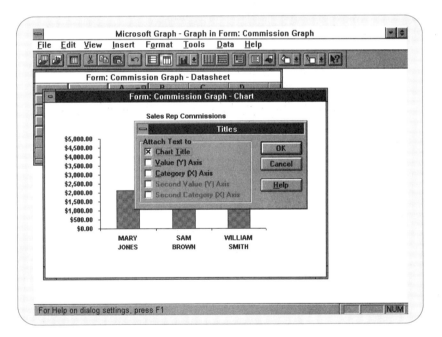

creation. Now you will add the Value title on the Y axis, and a Category title on the X axis.

■ Check the X and Y axis title boxes and then click on OK.

A label box displays with the letter X on the X-axis, and a Y on the Y-axis (Figure M9.8). Use the mouse to select the X and enter Sales Representatives in its place.

■ Select (highlight) the X-axis.
■ Type `Sales Representatives` in the label box.

Your screen should now match Figure M9.9. The next change is not so simple. Although you want to add a title on the Y axis, you will notice that there is not as

Adding titles to a
graph

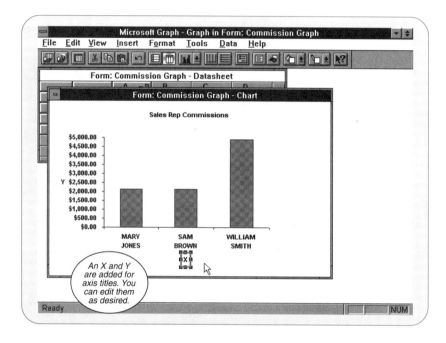

Changing title
labels

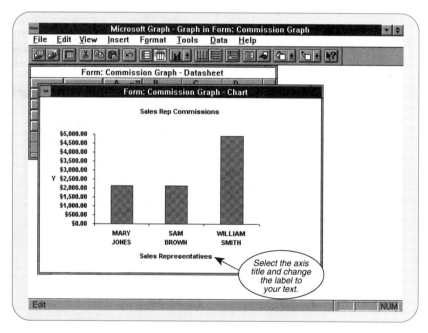

much space for the text to fit in the space to the left of the total commission values.
You will have to insert the title text and then change its alignment.

- Select the Y-axis.
- Type `Total Commission` in the label box.

When you insert a Y-axis title, it uses the default alignment, which unfortunately
overlaps with the values along the Y-axis (Figure M9.10). Microsoft Graph provides a
method for controlling the appearance of each item in the graph.

- Click on the Y-axis title.
- Select Selected Axis Title . . . from the Format menu, or double click on the Y-
  axis.

Figure M9.10

Default title
format

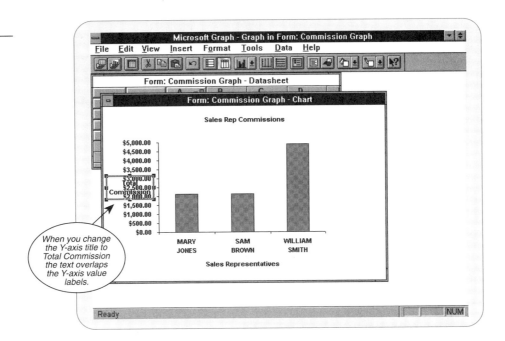

A dialog box displays in the format of index cards. Switch to the Alignment section by clicking on the Alignment tab at the top-right corner of the dialog box.

- Switch to the Alignment view by clicking on the Alignment tab.
- Click on the Orientation view that places the label in the vertical alignment, rotated as Figure M9.11 shows.
- Click on OK. The graph should now match Figure M9.12.

**Changing
Title Fonts**

We can enhance the appearance of the graph by changing the Chart Title font.

- Click on the Chart Title. (Sales Reps Commission should now have a box around it with size/move handles.)

Figure M9.11

Formatting an axis
title

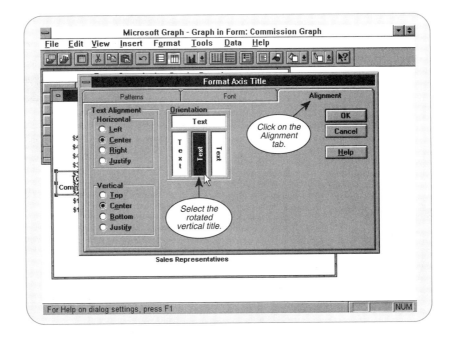

Completed title
format

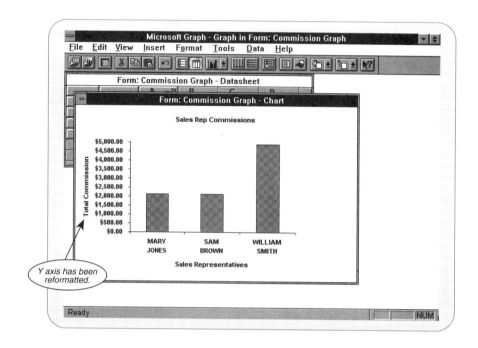

- Select Font from the Format menu.
- Select Times New Roman from the Font list.
- Select Bold Italics from the Font Style list.
- Select 18 from the Size list.

Your screen should now be similar to Figure M9.13.

Note: If your font list is different from this list, select another font similar to the one shown in the Preview window.

- Click on OK to apply the font change.

Your screen should now have the main Chart Title modified, as Figure M9.14 indicates.

Changing Title
fonts

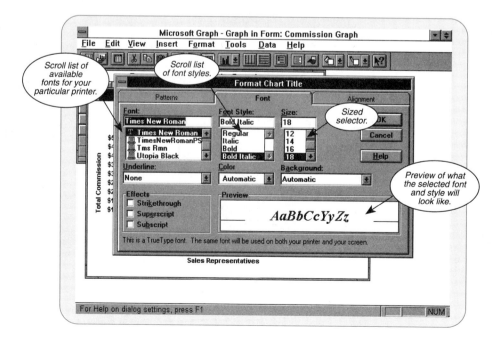

Figure M9.14

Graph with Chart
Title changed

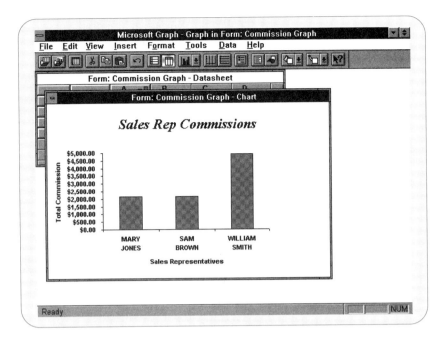

Adding Gridlines

The next change you will make is to add horizontal gridlines to make it easier to follow the graph values to the Y-axis. You could select Gridlines from the Insert menu and then provide the data necessary. However, Graph provides a button on the toolbar that makes it a one-step process.

▪ Click on the Horizontal Gridlines button.

As Figure M9.15 shows, horizontal gridlines are added.

The last addition to the graph that we will make in this section is to add data labels to give the actual values for the three sales rep amounts. Data labels are added above the bar for each of the category items.

Figure M9.15

Adding gridlines to
a graph

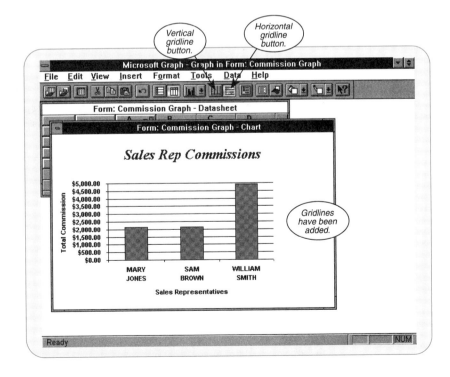

- Select Data Labels from the Insert menu.
- Select Show Values from the Data Labels dialog box.

As Figure M9.16 shows, the commission actual values for the three sales reps are listed above the three columns. Since you are in the Graph application, you can return to the Access form by exiting the application and updating the OLE object in the form.

- Select Exit & Return to Form from the File Menu.

You are returned to the Form window in the Design view.

- Click on the Form View button, or select Form from the View menu.

Your graph is now complete and should match the one shown in Figure M9.17.

- Select Save from the File menu.

Figure M9.16

Adding data labels to a graph

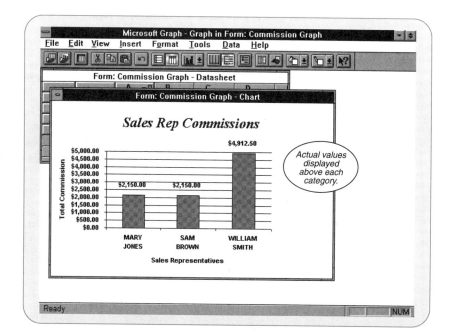

Figure M9.17

Form View including embedded graph

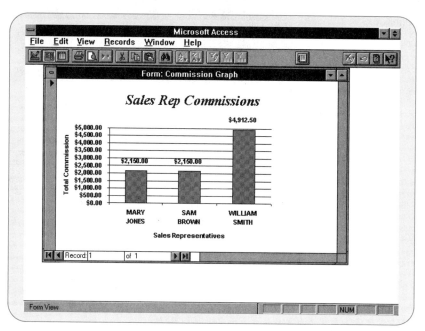

M9.3 GRAPH TYPES

Sometimes, one type of graph will present the data more clearly than another. A pie chart, for example, indicates percentages better than a bar chart does. Fortunately, changing the type of graph is as simple a process as selecting a sample chart from a drop-down list.

- Change to the Design view of the graph form.
- Click on the graph to select it.
- Select Chart Object and then Edit from the Edit menu.

This will give control to the Graph Wizard. You can now edit and modify the format of the graph using the toolbar buttons. In this section, we are looking at a few of the different types of graphs available. There are a number of methods to use when changing graphs. For now, click on the Chart Type drop-down list on the tool bar. Figure M9.18 shows the basic list of chart types.

Figure M9.18

Selecting the chart type

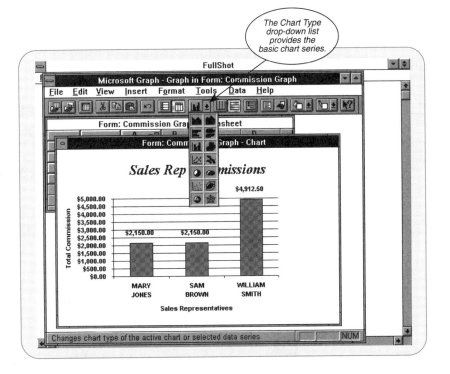

To select one of the chart types, simply click on the desired graph. For example, Figure M9.19 shows the same graph that we produced, but in the 3-D column chart format. Figure M9.20 takes the same chart and rotates it so that it is displayed as a 3-D bar chart. The only difference between the two previous graphs is the orientation of the X and Y axes and the axes titles.

Sometimes the format of a graph doesn't explain the data well. For example, Figure M9.21 presents the same data as the bar and column graphs; however, the line graph is normally used to show trends or changes over time periods. In this example, the bar or column would show the difference between groups better.

The pie graph (Figure M9.22) is a good choice when you want to emphasize the percentage that each item represents. The size of the pie slice for an entry visually indicates the percentage. In the figure, for example, you can see that William Smith's commissions represent over one-half of the total of all commissions. In a pie chart, you can only represent one series of numbers.

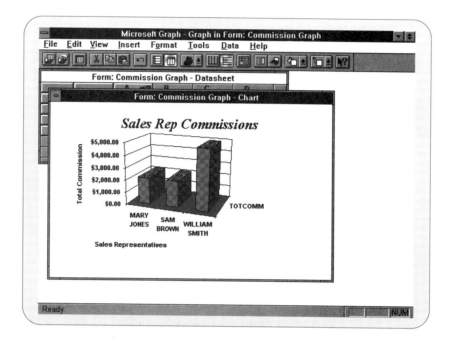

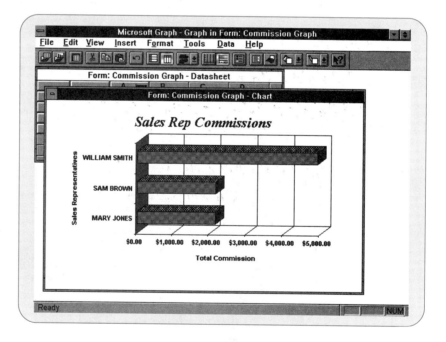

**Customizing
a Graph**

To demonstrate changing the chart type, you will change the appearance of the graph to a 3-D pie chart and then modify the format by "exploding" the slices of the pie. This simply means making the pieces stand out from each other.

■ Select Chart Type from the Format menu.
■ Click on 3-D in the Chart Dimension section.
■ Click on 3-D Pie and then on OK.

Your graph should be similar to Figure M9.22 now, except that the actual commission amount for each item is displayed instead of the name and percentage of the total. To make this change, you only need to change the data label.

Figure M9.21

Line chart

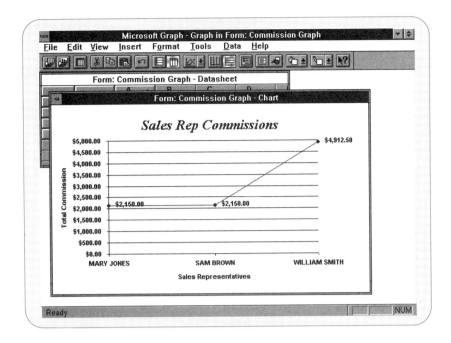

Figure M9.22

3-D Pie chart

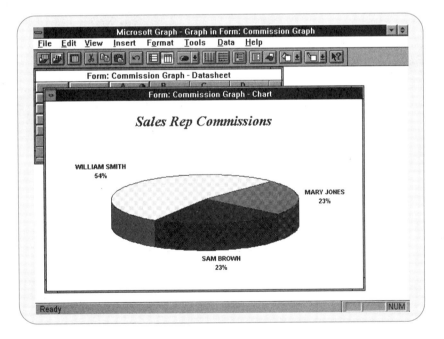

■ Select Data Labels from the Insert menu.

■ Select Show Label and Percent and then click on OK.

Your screen should now match Figure M9.23, showing the sales reps' names and the percentage of their commission amount to the total commissions. If you wanted to make additional changes to the chart, you could select the appropriate format item to change, or you could use the gallery of predefined formats.

■ Select AutoFormat from the Format menu.

■ Select the exploded pie chart format as shown in Figure M9.24, and then click on OK.

Figure M9.23

Changing the
Graph format

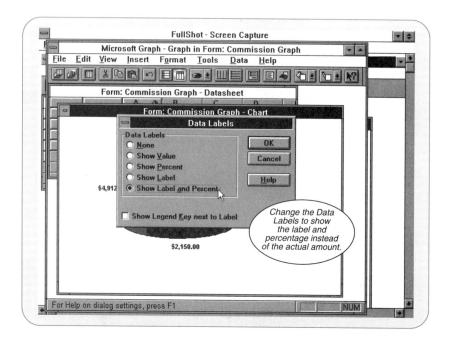

Figure M9.24

The AutoFormat
menu

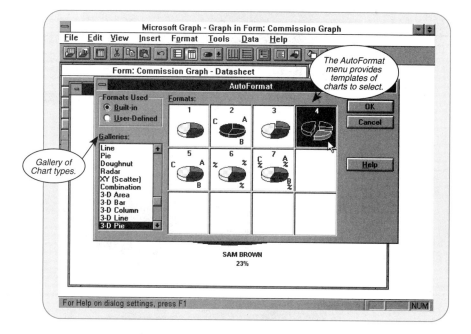

■ Again, select Data Labels from the Insert menu.
■ Select Show Label and Percent and then click on OK.

Your pie chart should now match Figure M9.25, exploded with labels and percentages. There are many other formats and chart types that you could use to present your data. With this short demonstration of changing the type and format of graphs, you should now be able to use any of the graph types.

■ Select Exit & Return to Form from the File menu.
■ Close from the File menu.
■ Answer No when requested whether to save or not. This will leave the form in its original column format.

Figure M9.25

Exploded pie chart

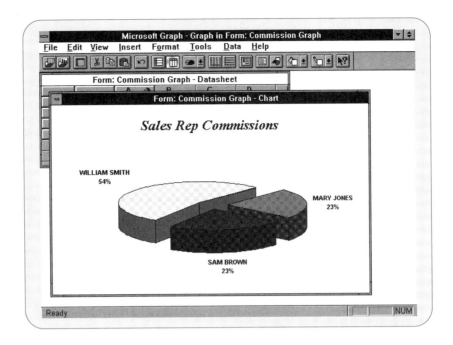

M9.4   USING CROSSTAB

The Need for Crosstabs

Suppose you want to keep sales totals for sales reps over three sales periods (for example, the first three months of the year). During these three periods, you want to break the sales totals down by item class. That is, you want to record the sales amount for each sales rep in housewares, in sporting goods, and in appliances. You also want to represent this information graphically. With all this in mind, how do you structure the data?

One possibility would be to create a table with a single row for each sales rep. In each row, you would have the sales amount in period 1 for housewares, the amount for sporting goods, and the amount for appliances. You would have similar amounts for period 2 and for period 3. Altogether, you would have nine sales amounts.

While this structure might be appropriate for graphing, it is cumbersome. Further, if you decide to extend the number of periods from three to six, the structure is now insufficient. You would have to add several more columns to it. The same thing would happen if you add more item classes.

A better approach is to use the structure shown in Figure M9.26. For each sales amount, add a row to this table consisting of the sales rep number, the item class, the period in which the sale took place, and the amount.

Figure M9.27 shows some sample data for this table. The first row indicates that sales rep 3 sold $4,120.00 of housewares (HW) in period 1. The last row indicates that sales rep 12 sold $1,208.50 of sporting goods (SG) in period 3.

Look at the fourth and fifth rows. The fourth indicates that sales rep 3 sold $5,896.24 of housewares in period 3. The fifth indicates that the same sales rep sold $2,000.00 of the same item class in the same period. To obtain the total amount of housewares sales rep 3 sold in period 3, you need to add these two amounts together, to get $7,896,24.

This form of the data now allows us to prepare a tabular presentation of the data or a graphical representation. Let's look at the tabular presentation first. To do so, we create a crosstab.

Creating a Crosstab

A crosstab summarizes data according to one or more fields and then displays the summary in a tabular format similar to a spreadsheet. A crosstab displays in a query window.

Figure M9.26

Design view for *Slsinfo* table

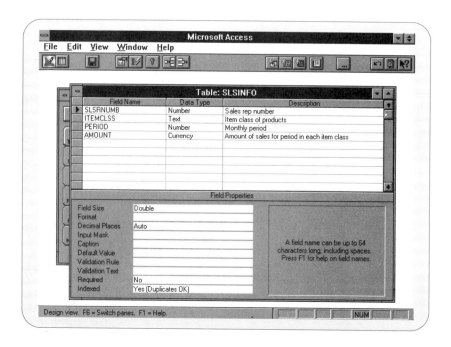

Figure M9.27

Data for *Slsinfo* table

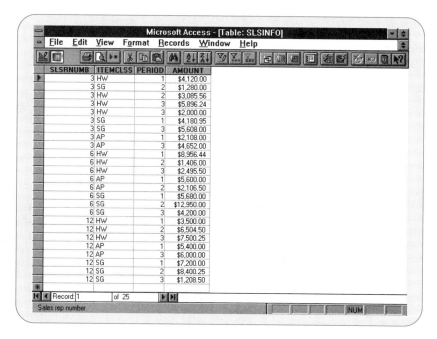

Let's create our first crosstab.

- Click on the Query object.
- Click on New to create the query.
- Click on Query Wizard.

The Query Wizard offers different types of queries (Figure M9.28). Select the Crosstab query.

- Click on Crosstab Query and then click on OK.

You will then be prompted for the data necessary to define the crosstab. Specify the following:

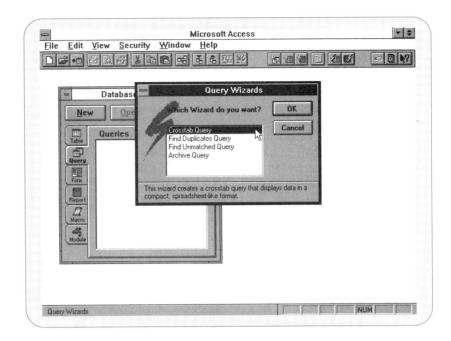

Figure M9.28

Creating a new Crosstab query

1. The field values to use as row headings or categories down the leftmost column of the crosstab.
2. The field values to use as column headings across the top of the crosstab.
3. The field(s) whose values you want to perform the crosstab summary operation on. This provides the data for the crosstab.
4. The type of summary operation to perform.

Let's suppose we want to create a table with a row for each sales rep. Another way of saying this is that the salesrep numbers will be the crosstab row labels. The row labels are the categories.

First, the Query Wizard needs to know what table to use to generate the crosstab. (See Figure M9.29)

Figure M9.29

Creating a Crosstab query

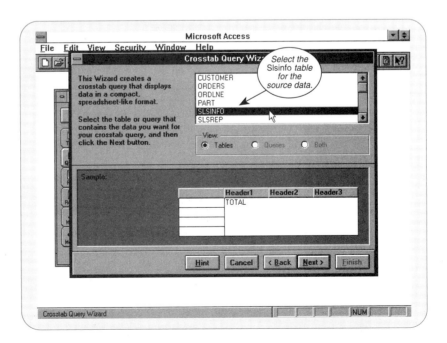

■ Select the *Slsinfo* table.
■ Click on the Next button.

Next, indicate the row headings to use as the categories (Figure M9.30).

■ Select the *Slsrnumb* field for the row heading.
■ Click on the Next button.

Figure M9.30

Creating a
Crosstab query

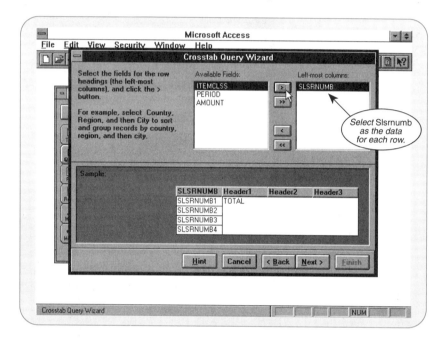

The second part of the crosstab is the *column labels*. Suppose that you want the total sales amount for each item class, without regard for the periods in which the sales took place. In other words, the *Itemclss* field is the one containing the column labels.

■ Select the *Itemclss* field for the column label (Figure M9.31).
■ Click on the Next button.

Figure M9.31

Creating a
Crosstab query

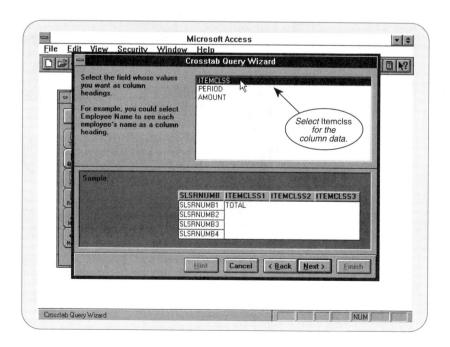

The next step is to indicate the field containing the actual values. In our example, that would be the *Amount* field.

■ Click on Amount in the *Available Fields* list.

The crosstab also provides a number of functions that could be performed to create the crosstab. You could use the Count of the number of items in each category, or the Min, Max, Average, and so forth, of the items, to name a few. The most common use, however, and the one that we wish to use, is the sum of the values in each category.

■ Click on the Sum function.

As Figure M9.32 shows, the crosstab could also calculate a summary for each row. We will not use the summary option at this time.

■ Click on the Summary option to remove the selection. (The X should not be in the check box.)
■ Click on the Next button.

Figure M9.32

Creating a
Crosstab query

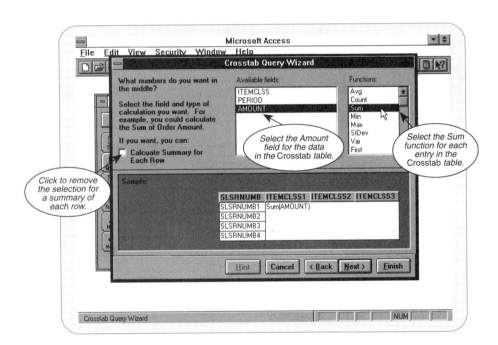

The final step is to provide a name for the query. As Figure M9.33 shows, enter Sales Rep's Sales Amount by Item Class for the query name.

■ Enter `Sales Rep's Sales Amount by Item Class` for the query name.
■ Click on the Finish button to generate the crosstab.

You will then see the crosstab, Figure M9.34 shows. Notice that there is a row for each sales rep. The item classes are the column labels. Finally, notice that Access has automatically combined the three periods, calculated the appropriate totals.

■ Save and Close the Query.

Creating Additional Crosstabs

You can create a variety of crosstabs from the data in the *Slsinfo* table. Let's look at sales rep totals by period rather than by item class.

■ Create a crosstab as you did before, but this time select the *Period* field for your column headings instead of the item class.

Figure M9.33

Creating a
Crosstab query

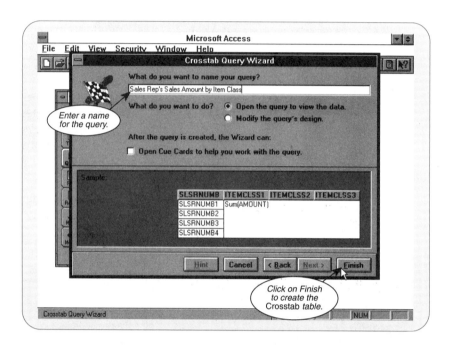

Figure M9.34

Crosstab table

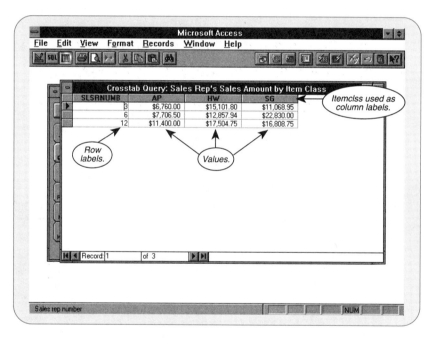

The results are shown in Figure M9.35. Notice that the period numbers are now used as the column labels.

■ Close the crosstab query without saving it.

**Creating
Crosstab
Graphs**

You can depict crosstab information in a graph. Access uses the same procedure to create the graph used in the previous section. We will create a 2-D column graph of sales for each item class by sales rep. Fortunately, the Form Wizard makes it simple to generate this graph because most of the data is already defined in the crosstab query itself. Therefore, we generate a graph using the Form Wizard based on the crosstab query by item class.

Figure M9.35

Crosstab table

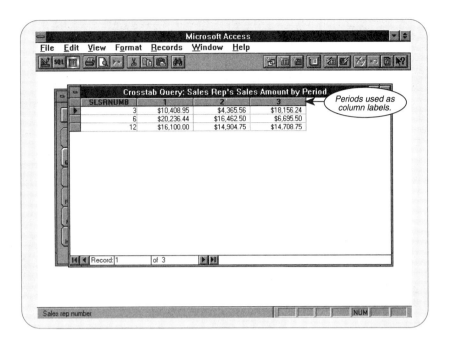

- Click on the Form object button.
- Click on the New button to generate a new form.
- Select the Crosstab query based on Item Class.
- Click on the Form Wizard button.
- Select the Graph Wizard from the Form Wizard options.
- Click on OK.

The Graph Wizard automatically creates a 2-D column graph from the crosstab data. As Figure M9.36 shows, you could select a different type of graph. However, we will use this default type.

- Click on Next to go to the next screen.

Figure M9.36

Graphing Crosstab tables

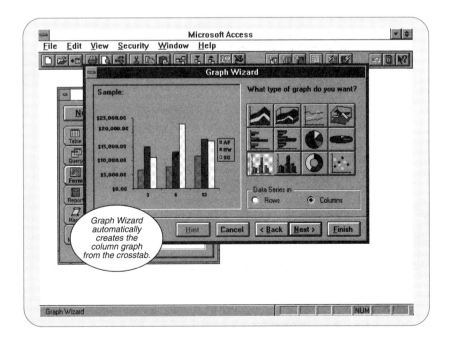

By selecting the default graph, the only information you need to provide is the title for the graph (Figure M9.37).

- Enter `Item class sales by Sales Reps` for the Title.
- Click on Finish to generate the graph.

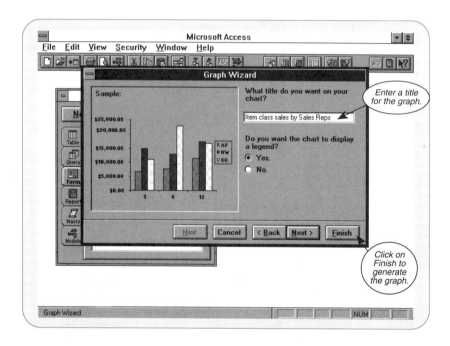

The graph is automatically generated and placed in the Form view, as Figure M9.38 shows.

- Save and Close the Form.

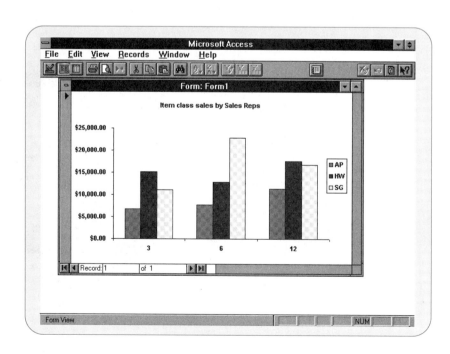

M9.5 ADDING GRAPHS TO A FORM

You can add a graph to an already-existing form by creating a graph object from within the Form Design window. For example, you can replace the customer subform (in a table format) on the Sales Reps form created in Module 8 with a graph (Figure M9.39). Using a graph on a form rather than a table can help show relationships and trends that might be missed with tabular data. You can also include graphs on reports.

In this section, you will create a new Sales Reps form that includes a graph of customer numbers and balances instead of a *Customer* table.

Figure M9.39

Sales Reps form

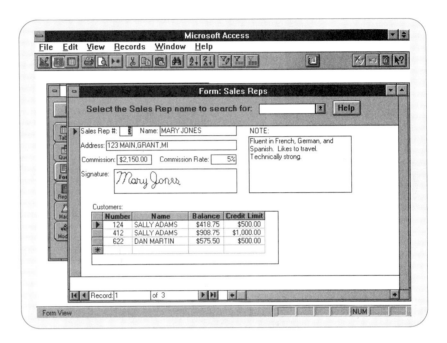

Creating a New Form from an Existing Form

The new form that you will create is similar to the Sales Reps form you created in Module 8. Rather than begin with a totally new form, it is easier to modify the design of an existing form and then save it with a different name. You will modify the design of Sales Reps but save the modified version as Sales Reps Graphs. The first form (Sales Reps) remains unchanged.

To modify a form design:

- Open the *Prem* database.
- Click on the Form object button.
- Select the Sales Reps form.
- Click on the Design button.

The Design window for Sales Reps is displayed (Figure M9.40). Let's save the form with the new name now so that we can take a break whenever necessary.

- Choose Save As from the File menu.
- Enter `Sales Reps Graphs`
- Click on OK.

Adding a Graph Object to a Form

Before adding the graph object, you need to delete the Customers subform and move the *Note* field so that it displays under the *Signature* field.

- Select the customers subform.
- Press the Delete key.

Figure M9.40

Sales Reps Form
Design view

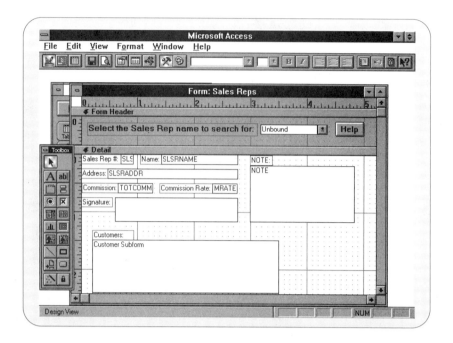

- Select the *Note* field.
- Move the field and adjust its size, Figure M9.41 shows.
- Separate and change the field label as shown.
- Drag the lower edge of the detail section up to the location shown.

Figure M9.41

Modifying the
original form

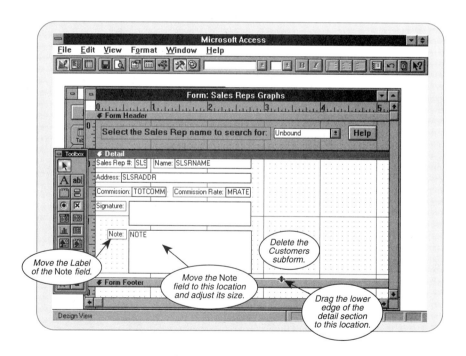

Now, you can add the graph object.

- Click on the Graph tool on the toolbox.
- Move the Graph object pointer to the location shown in Figure M9.42.
- Click to place the graph.

Figure M9.42

Adding a graph to
the form

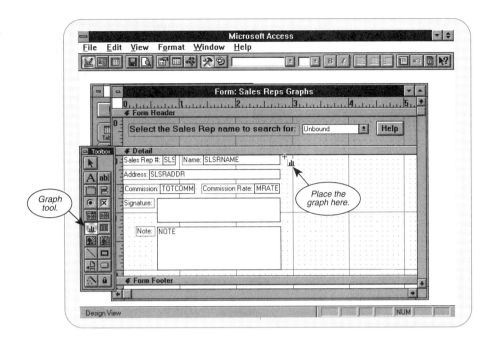

Now that you have added a graph object to the form, the object needs to be defined.
You need to indicate the table that contains the values, the Y-axis values, and the X-
axis values.

The Graph Wizard first needs to know what table to get the data from.

- Select Customer from the list of tables.
- Click on the Next button.

Next you are prompted for the fields to include in the graph (Figure M9.43). As
you select each field that you want, click on the Copy button to copy to the Fields for
Graph list. (The copy button is the one with the ">" symbol.)

Figure M9.43

Selecting fields for
the graph

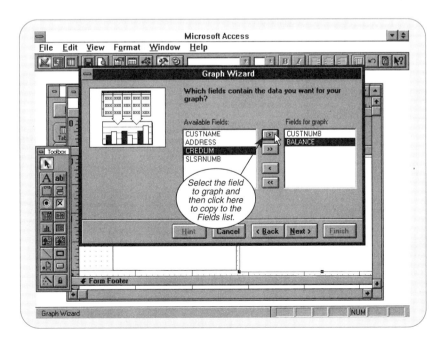

- Select the *Custnumb* field in the Available Fields list.
- Click on the Copy button to copy the field to the Fields for Graph list.
- Select the *Balance* field and copy it to the Fields for Graph list.
- Click on the Next button to move to the next screen.

The Graph Wizard then continues to prompt for the field to be used for axis categories and how to calculate the totals.

When you are prompted for the categories to use for the axis:

- Select *Custnumb.*
- Click on the Copy button.
- Click on the Next button.

When you are prompted on how to calculate totals for the categories, the default should be set for a sum of the item. Therefore:

- Click on the Next button.

The next option of the Graph Wizard is whether to link data in the graph with a field on the form.

- Choose Yes.

When you respond Yes to the link, a dialog box with Fields on the form: and Fields in the graph: displays. The common field between the two is the sales rep number for both the graph and the form. Figure M9.44 shows the two lists with *Slsrnumb* highlighted.

- Select *Slsrnumb* in both the form and the graph lists.
- Click on the "< = >" button.
- Click on the Next button.

Figure M9.44

Linking graph fields with the form

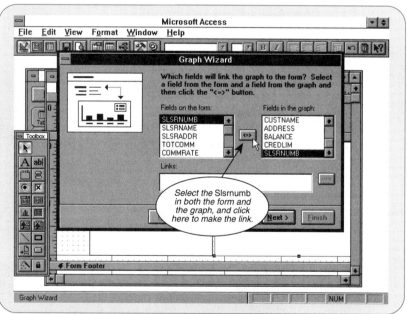

The next set of options is what type of graph to use. The default type is the 2-D column graph, which is the one you want.

- Click on the Next button.

The final set of Wizard prompts asks for a title, and whether to display a legend (Figure M9.45).

- Enter `Customer Balances` for the title.
- Click on No to reject the query to display a legend.
- Click on Finish.

Figure M9.45

Completing the graph

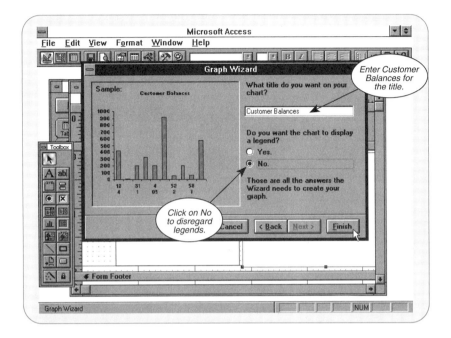

The Wizard now has all of the information it needs to generate the graph on the form. As Figure M9.46 shows, a sample graph is included on the form. You now want to modify the graph to add data labels and an axis title. Recall that to jump to an OLE objects application, you only need to double click on the embedded object.

- Double click on the graph object.

Figure M9.46

Inserting the graph

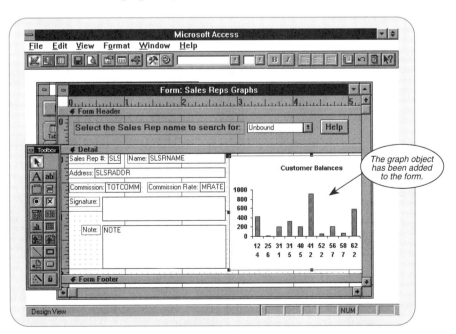

When you double click on the embedded graph (or select Chart Object and Edit from the Edit menu.) you will be switched to the Graph application. There will be two windows on the screen: a datasheet of the fields in the graph, and a sample graph. Before adding the data labels to the chart, change the format of the *Balance* field to show a fixed number of 2 decimal places.

- Click on the Datasheet window.
- Click on the A label of the Balance column (Figure M9.47).

Figure M9.47

Modifying the format of an embedded graph

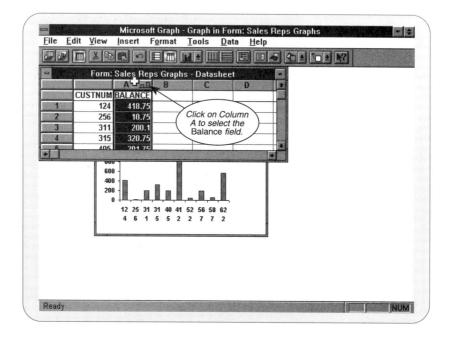

To change the format:

- Select Number from the Format menu.
- Select the 0.00 format, as shown in Figure M9.48.
- Choose OK.

Figure M9.48

Setting the format of graph values

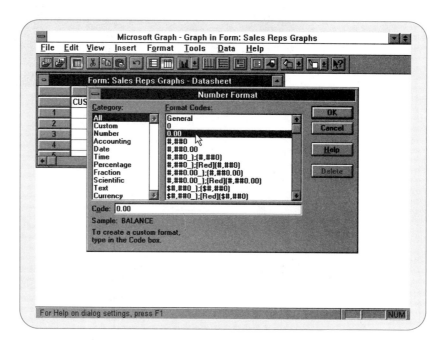

Now that the format is set, you can insert the data labels.

- Select Data Labels from the Insert menu.
- Select Show Value from the Data Labels dialog box (Figure M9.49).
- Choose OK.

Figure M9.49

Setting Data Labels for the graph

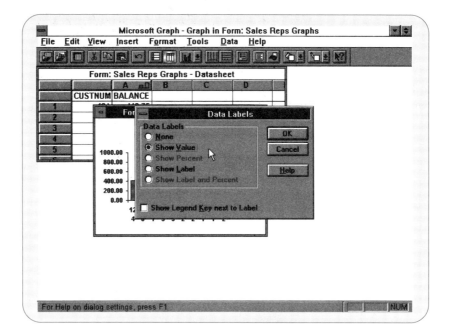

Finally, add a title, Customer Numbers, to the X-axis.

- Select Titles from the Insert menu.

Notice that Chart Title is already selected. You want to add a title to the (Category) X-axis.

- Select Category (X) axis. (This will place an X in the box.)
- Choose OK.

An X has been added to the graph now (Figure M9.50).

Figure M9.50

Adding an axis title

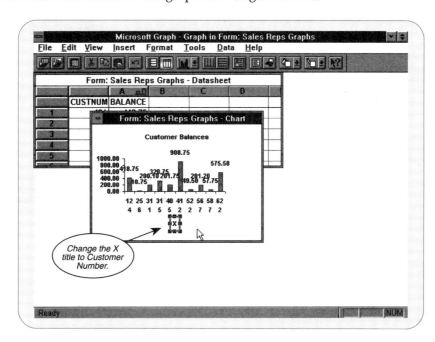

- Highlight the X and change it to `Customer Number`
- Choose OK.

Now that you are finished modifying the graph, you can return to the form.

- Select Exit & Return to Form from the File menu.

The graph has been modified on the form, except that the right edge is wider than the Screen view. Scroll the window to the right so that you can see the right edge of the form. Grab the right handle and resize the graph.

- Scroll the window to the right and resize the graph, as Figure M9.51 shows.
- Click on the Form View button.

The form now includes a graph of the three customers for the first sales rep (Figure M9.52). As you move to the next sales rep, the graph changes to show the linked customers.

Figure M9.51

Form with an embedded graph

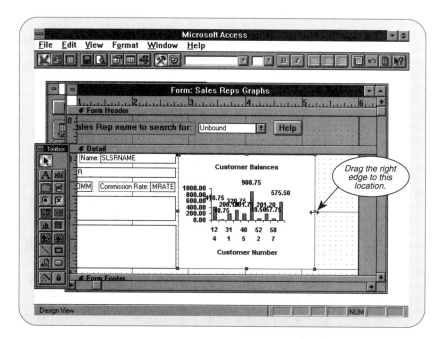

Figure M9.52

Embedded Customer subform has been replaced with graph

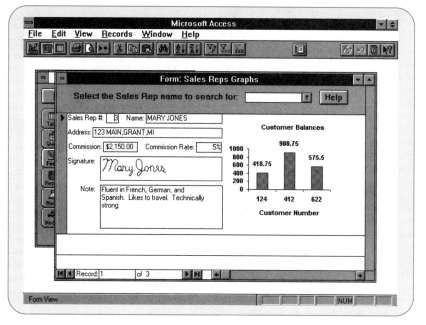

SUMMARY

The following list summarizes the material covered in this module:

1. To create a graph, create a new form and use the Graph Wizard. Define the fields for the X and Y axes as prompted. A graph is an OLE object, created in Microsoft Graph, that displays in a Form window or a report.
2. Access offers several different graph types, including bar, line, column, XY (Scatter), pie, doughnut, surface, and area. Most graphs can display in either 2-D or 3-D.
3. To change the graph type, change to the Design view of the form and double click the graph object. Then choose Chart Type from the Format menu.
4. To modify characteristics of a graph, first select the graph object and double click to enter the Graph application. Then select the appropriate option from the graph's menu.
5. To explode a pie chart, choose AutoFormat from the Format menu. Then select the desired exploded pie type from the Format options.
6. To summarize information from a table, you can create a crosstab. To create a crosstab, create a new Query using the Query Wizard. Select the Crosstab Query Wizard and identify the tables for the data. Then select row and column labels and indicate how amounts are summarized.
7. Crosstabs are considered a special type of query. You can select the crosstab query and change its characteristics using the Design view.
8. To depict crosstab information in a graph form, create a new form using the Graph Wizard. When you identify the Crosstab query that you want to use, Access automatically creates the proper type of format.
9. To customize titles on a graph, select the title and use the Format menu to change the font, style, alignment, color, and so forth. You can change the Chart title and the titles for the X and Y axes.
10. To add legends to a graph, choose Legends from the Insert menu. Data from the Categories column in the datasheet are used as the legends.
11. To change the legends on a graph, select the Datasheet window of the graph object and change the data in the Categories column.

EXERCISES

1. How do you create and display a graph?
2. Are there any restrictions on the fields used for the Y axis?
3. What types of graphs are available in Access?
4. How do you change the chart type?
5. How do you change chart formats?
6. What does it mean to explode a piece of a pie chart? How do you do so?
7. When is a crosstab appropriate? How do you create one? How do you indicate the row and column labels? How do you point out the values to be included in the crosstab?
8. How can you change the titles on a graph? What types of changes can you make?
9. How do you add legends to a graph? How do you change the legend labels?
10. How do place a graph object on a form?
11. How do you define a graph object on a form?

COMPUTER ASSIGNMENTS

The *Premiere Products* Database

1. Create a bar graph representing the unit prices of all parts.

2. Add titles to your graph.
3. Use the *Slsinfo* table to create a column chart of the total sales for each item class in the various periods. Change the titles and legends, and reprint the graph.
4. Create a pie graph that shows totals by item class. Change the titles and legends.
5. Explode the pie and add labels and percent data labels.

The *Books* Database

1. Create a bar graph showing the number of employees in each branch.
2. Create a query using the *Book* table. The query should show the average price for each type of book. Once you have done this, create a bar graph showing the same information. Change the titles on the graph so that they describe the contents of the graph.
3. Create a query joining the *Book* and *Invent* tables. The columns included in the result should be Pubcode, Brnumb, and Oh. Once you have done this, create a crosstab from the results of your query. For row labels, use Pubcode; for column labels, use Brnumb; and for values, use Oh. Select the Sum function.
4. Create a 3-D column graph showing the number of books each publisher has at each branch. Add meaningful titles and legends, and print the graph.
5. Modify the one-to-many form for the *Publishr* table that you created in Module 8. Delete the *Book* table from the form and replace it with a graph showing the price of each book.

Procedure Summary

GENERAL PROCEDURES

Abandoning an Operation
1. Press the Escape key. In some cases, a Cancel button is provided and specific instructions are given.

Selecting a Menu Option
1. If no options on the menu are highlighted, press the Alt key. Once options are highlighted on the menu, simply press the underlined letter for the specific option desired.
2. To move the highlight from one option to another, use the arrow keys or type the first letter of the option.
3. To select the currently highlighted option, press Enter. To use a mouse to select an option, click on the option; that is, move the mouse pointer to the option and press the left button on the mouse.

Viewing a Table
1. Open the Database.
2. Click on the Table Object button.
3. Click on the Table from the list of available tables.
4. Click on the OK button.

    Note: There is a shortcut when using the mouse. Rather than clicking on an item and then clicking on OK, you can double click on the object; that is, rapidly click twice on the object. The second click is recognized as the OK click. If you double clicked but the selection was not activated, you probably took too long for the second click. In this case, you could either try it again, or simply click on OK.

Getting Help
1. Press the function key F1, or click on the Help button on the tool bar.

MANIPULATING WINDOWS

Maximizing or Restoring a Window
1. Select Maximize/Restore from the Control-menu box, or click on the Maximize/Restore button.

Resizing or Moving the Active Window
1. Select Size or Move from the Control-menu box, or:
2. To move, click on the Window Title bar and drag to the new location.
3. To resize, drag the lower-right corner window handle to the new size.

| **Close the Active Window** | 1. Select Close from the Control-menu box,
   or:
2. Double click on the Control-Menu box. |

WORKING WITH A DATABASE

| **Creating an Access Database** | 1. In the Microsoft StartUp window, choose New Database from the File menu.
2. In the File Name box, type in a name for your new database.
3. Select the drive and or directory in which to save the database.
4. Choose OK. |

| **Opening a Database** | 1. To open a database, first start Microsoft Access if necessary.
2. From the File menu, choose Open Database or click on the Open Database button on the toolbar.
3. In the File Name list box, select the database. |

| **Opening a Table** | 1. First open a database.
2. In the Database window, click the Table button to display a list of available tables in the database.
3. Double click on the table name that you want to see. |

| **Viewing Data Using a Form** | 1. In the Database window, click on the Form button to display a list of all forms in the database.
2. Double click on the name of the form that you want to see. |

| **Switching between Datasheet and Form View** | 1. To switch to or from a Datasheet or Form view, click on the Datasheet or the Form button on the toolbar. |

| **Closing a Table or Form** | 1. Choose Close from the file menu,
   or:
2. Double click on the Control-menu box. |

| **Closing a Database** | 1. Choose Close from the file menu,
   or:
2. Double click on the Control-menu box. |

WORKING WITH TABLES

| **Creating a Table** | 1. The first step in creating a database is to create one or more tables. After you have a least one table, then you can create queries, forms, reports, or other objects.
2. In the database window, click on the Table object button.
3. Choose the New button.
4. Choose the Table Wizards button.
5. Follow the prompts for the necessary information. The minimum data for a table are the names of the fields you want to add, and the Data Type for each field.
6. You can either create your own Primary Key by clicking on the Primary Key button, or let the Table Wizards create the primary key for you. |

Modifying a Table

1. In the Database window, click the Table button.
2. Select the name of the Table from the list of tables.
3. Choose the Design button.
4. Make any necessary changes.

Adding Fields to a Table

1. Open the table.
2. Move to the first empty box under Field Name.
3. Enter the name of the field.
4. Press Tab to move to the Data Type box.
5. Press Alt+Down arrow to display the list of data types.
6. Choose the appropriate data type.
7. You could also enter an optional description for each field.

Setting Field Properties

Each field has a set of properties that controls how Access stores or displays data in the field. Changing the format of a field does not change how the data is stored in the database; instead, it affects how the data is displayed in a datasheet, form, or report.

1. Open a table in the Design view.
2. Click in any box of the field that you want to define.
3. Click in any of the Field Properties box in the lower half of the window.
4. Press Alt+Down arrow to display the list of property controls.
5. Select the appropriate format setting, or enter a valid expression.

Adding Records to a Table

You can add records to a table either in the Datasheet view or in the form view.

1. Move to the first available row in a Datasheet (identified with an asterisk "*" in the row handle), or the first available empty form in the Form view.
2. Enter the data for the first field.
3. Press Tab to the next field and enter the data.
4. Repeat for each field.

When you leave a record, either by moving to another record, or closing the table, the new record is automatically saved.

Sorting Records in a Table

1. Open a table in the Datasheet view.
2. Click on the heading of any column that you wish to sort.
3. Click on the Sort Ascending or Sort Descending button on the toolbar.

Deleting Records in a Table

1. Select the record in a table that you wish to delete by clicking on the row handle on the left edge of the table.
2. Press the Delete key.
3. Verify the deletion in the dialog box by choosing OK.

Deleting Unwanted Tables

1. Open a Database window.
2. Click on the Table object button.
3. Select the Table to be deleted.
4. Press the Delete key.
5. Verify the deletion in the dialog box by choosing OK.

WORKING WITH FORMS

Creating a Form

1. In the Database window, click on the Form button.
2. Choose the New button or choose New from the File menu, and then choose Form.
3. Select a table or query from the Table/Query drop-down list.
4. Choose the Form Wizards button.
5. Select AutoForm, or one of the other Form Wizard types available.
6. Choose OK.

   Note: You could also create a form in a table's Datasheet view by clicking on the AutoForm button on the toolbar.

Customizing a Form's Field Controls

1. To change the appearance of a field or object in a form, click to select it.
2. To move an object, drag the object when the pointer changes to a hand icon.
3. To change the size of an object, use the size handles on any of the four sides and drag to the new location.
4. To separate and move labels or fields independently, click on the move handle (the top-left handle) and drag to the new location.

Adding Labels to a Form

1. To draw a freestanding label, click on the Label tool in the Toolbox and click to place the label at the appropriate position.
2. To edit text in an existing label, click to select the label and then click the label again to select the text. You can now edit the text.
3. To change the appearance of text, first click to select the label. Then from the Format menu, choose any of the appropriate controls.

Creating a Form that Contains a Subform

1. In a Database window, click on the Form button.
2. Choose the New button.
3. Select the table/query that will supply data for the form.
4. Click on the Form Wizards button.
5. Select Main/Subform Form Wizard, and choose OK.
6. Make the appropriate choices as you proceed through the Form Wizard, providing sources for the main form and the subform.
7. Choose the Finish button.
8. Save the Form.

Using a Form to View Records

1. Open the table from the Database window.
2. Click on the Form View button on the Toolbar.

Adding Records with a Form

1. From the Records menu, choose GoTo and then choose New, or you could click on the New Record button on the toolbar.
2. Press Tab to move from field to field and enter data as appropriate.
3. To save a record, press Tab to move to the next record. You do not have to take any action to save a new record; simply move to another record or close the table. If you choose to, you could save without leaving the record by choosing Save Record from the File menu.

WORKING WITH QUERIES

Creating a Query

1. Open a Database window.
2. Click on the Query button.
3. Choose the New button.
4. Click the Query Wizards button.
5. Click on the New Query button.
6. Select one or more tables to add to the query.
7. Choose Add for each table added.
8. Choose Close when all tables have been added to the query.
   The Query Wizard displays each table with a list of fields in each table.
9. Drag each field that will be displayed, or searched for to an available Field box.
10. Choose whether the field should be displayed or not by clicking in the Show box.

Sorting records with a Query

1. Open a query in the Design view.
2. Click the Sort cell for the field to be sorted.
3. Select Ascending or Descending from the Sort list.

Joining Tables

Access will automatically show any JOIN lines between fields in tables. If you have not created a relationship between the tables, you can JOIN tables in a query.

1. Drag any field in a table to a matching field in another table. Access will display the JOIN line between the two tables.
2. To delete a JOIN line, click on the JOIN line to select it and then press the Delete key.

   Note: When you draw a JOIN line between tables in a query, the JOIN applies only to that query. Any other query will need to have the JOIN created again.

Creating Crosstab Queries

1. From a Database window, click on the Query button.
2. Choose the New button.
3. Choose the Query Wizards button.
4. Choose the Crosstabs Query Wizard.
5. Select the table or query that will supply data for the query, and then click on the Next button.
6. Select the fields for the categories or row headings, click on the Copy button, and then click on the Next button.
7. Select the fields for the column headings in the crosstab query.
8. Select the field to supply the data in the table and the function to apply to that field. Choose the Next button.
9. Enter a title for the crosstab query and then choose the Finish button.

WORKING WITH REPORTS

Creating a Report with Grouped Data

1. In the Database window, click on the Report button.
2. Choose the New button.
3. Select the table or query to supply data for the report.
4. Choose the ReportWizards button.
5. Select the Groups/Totals Wizard
6. Add all the fields from the query to the report.
7. Group records by a *Category Name* field.
8. Group records on the *Category Name* field by selecting Normal.

9. Sort records on a desired field.
10. Choose an appropriate style for the report.
11. Enter a title for the report.
12. Choose the Finish button.

Creating other Report styles

Follow similar steps as listed above, using the ReportWizard to create Single-Column, Mailing Labels, Tabular, or Summary reports.

Previewing or Printing Reports

1. To see an image of a report on the screen, Click on the Print Preview button on the toolbar.
2. To print a report to the printer, click on the Print button on the toolbar or choose Print from the File menu.

GRAPHING ACCESS DATA

Creating a Graph Using the Graph Wizard

Access graphs are actually special views of a form.

1. In a Database window, click on the Form button.
2. Choose the New button.
3. Choose the Form Wizard button.
4. Choose the table or query to supply data for the graph.
5. Choose the Graph Wizard.
6. Select the fields to supply data for the graph, and then click on the Copy button.
7. Choose the Next button when you have finished selecting the data fields.
8. Choose the calculation for the fields to be displayed.
9. Choose a Chart Type from the dialog box, and then choose the Next button.
10. Enter a Title for the Graph, and then choose Finish.

Modifying a Graph

Graphs are created as an OLE object with Microsoft Graph.

1. In a Database window, click on the Form button.
2. Select the Graph form.
3. Choose the Design view.
4. Double click on the graph to switch control to Microsoft Graph.
5. Select any object or part of the graph (titles, legends, data labels, or chart objects).
6. Use any of the appropriate Format options to make changes to the object.

Answers to Odd-Numbered Exercises

CHAPTER 1 — INTRODUCTION TO DATABASE MANAGEMENT

1. A software package is a collection of programs that is designed to handle some specific task, such as payroll. It is also called a software system, an application system, or an application package.
3. A file is a structure that is used to store data about a single entity; it can be viewed as a table. A record is a row in the table. A field is a column.
5. The number of entities and complex relationships, along with the fact that the entire billing operation of the practice would depend on the system, led Pat to conclude that the problem should be put in the hands of computer professionals.
7. A relationship is an association between entities.
9. A database is a structure that can house information about several types of entities, the attributes of these entities, and the relationships among them.
11. The purchase price of microcomputer DBMSs is relatively low, ranging from $100 to $800. In contrast, the cost of a good mainframe DBMS can run as high as $400,000.
13. Sharing of data means that many users will have access to the same data.
15. Redundancy is the duplication of data. It wastes space, makes updating more difficult, and may lead to inconsistencies in the data.
17. Integrity means that the data in the database follows certain rules (called integrity constraints) that users have established.
19. Data independence is the property by which the structure of a database can be altered without changes having to be made in the programs that access the database. With data independence, it is easy to change the structure of the database when the need arises.
21. The more complex a product is in general (and a DBMS, in particular, is complex), the more difficult it is to understand and correctly apply its features. As a result of this complexity, serious problems may result from mistakes made by the user of the DBMS.
23. Recovery can be more difficult in a database environment, partly because of the greater complexity of the structure. It is also likely that in the database environment several users will be making updates at the same time, which means that recovering the database involves not only restoring it to the last state in which it was known to be correct, but also performing the complex task of redoing all the updates made since that time.

CHAPTER 2 — DATA MODELS

1. A data model is a category of the database management system. It has two components: structure and operations. *Structure* refers to the format in which the DBMS stores data, and *operations* refers to the facilities that are given to users for the purpose of manipulating the data.

3. The three main data models are the relational model, the network model, and the hierarchical model.

5. Order lines in a separate table create a simpler structure, since only one entry is made in any box in the table. Processing that involves finding all the order lines for a given part is much simpler, since a maximum number of order lines does not have to be determined in advance, as it would be if order lines had to be kept in the *Orders* table. Further, no space is wasted by orders that have very few order lines.

7. A relational database is a collection of tables (relations).

9. A tuple is the formal name in the relational model for row; thus, another name for tuple is *row*. It also corresponds to the term *record*.

11. In the shorthand representation, each table is listed, and after each table, all the columns of the table are listed in parentheses. Primary keys are underlined.

```
BRANCH (BRNUMB, BRNAME, BRLOC, NUMEMP)
PUBLSHR (PUBCODE, PUBNAME, PUBCITY)
AUTHOR (AUTHNUMB, AUTHNAME)
BOOK (BKCODE, BKTITLE, PUBCODE, BKTYPE, BKPRICE, PB)
WROTE (BKCODE, AUTHNUMB, SEQNUMB)
INVENT (BKCODE, BRNUMB, OH)
```

13. The primary key is the column or collection of columns that uniquely identifies a given row. The primary key of the *Publshr* table is *Pubcode*. The primary key of the *Author* table is *Authnumb*. The primary key of the *Book* table is *Bkcode*. The primary key of the *Wrote* table is the concatenation (combination) of *Bkcode* and *AUTHNUMB*.

15. The relational model systems are easier to use than systems that follow the other models. Changing the structure of a database is also easier in relational model systems. The disadvantages of relational systems are that they are less efficient than systems that follow the other models, and they provide less support for integrity.

17. The CODASYL model is properly a subset of the network model. To many people, however, the two have become synonymous.

19. Since the network model has a special structure to effect relationships, there is no need for the matching columns that are required by the relational model.

21. Network model systems are more efficient than relational systems, and they provide better support for some types of integrity constraints. The disadvantages are that they are harder to use and they furnish less data independence than relational systems.

23. For any child that has more than one parent, each parent must be in a different physical database. For any parent that is not in the same physical database as the child, the relationship is a logical child relationship.

CHAPTER 3 — THE RELATIONAL MODEL: DATA DEFINITION AND MANIPULATION

(For Exercises 1 through 17, only the SQL formulations are shown.)

1.
```
SELECT PARTNUMB, PARTDESC
    FROM PART
```

3.
```
SELECT CUSTNAME
     FROM CUSTOMER
     WHERE CREDLIM >= 800
```

5.
```
SELECT PARTNUMB, PARTDESC,
     UNONHAND * UNITPRCE
     FROM PART
     WHERE ITEMCLSS = 'AP'
```

7.
```
SELECT *
     FROM PART
     ORDERS BY UNITPRCE
```

9.
```
SELECT SUM(BALANCE)
     FROM CUSTOMER
     WHERE SLSRNUMB = 12
```

11.
```
SELECT ORDNUMB, DATE, CUSTOMER.CUSTNUMB,
     CUSTNAME
     FROM CUSTOMER, ORDERS
     WHERE CUSTOMER.CUSTNUMB = ORDERS.CUSTNUMB
     AND DATE = 90594
```

13.
```
SELECT ORDNUMB, DATE, CUSTOMER.CUSTNUMB,
     CUSTNAME, SLSREP.SLSRNUMB, SLSRNAME
     FROM SLSREP, CUSTOMER, ORDERS
     WHERE CUSTOMER.CUSTNUMB = ORDERS.CUSTNUMB
     AND SLSREP.SLSRNUMB = CUSTOMER.SLSRNUMB
```

15.
```
INSERT INTO ORDERS
     VALUES
     (12600, 90694, 311)
```

17.
```
CREATE TABLE SPGOOD
     (PARTNUMB        CHAR(8),
      PARTDESC        CHAR(25),
      UNITPRCE        DECIMAL (6,2))
INSERT INTO SPGOOD
     SELECT PARTNUMB, PARTDESC, UNITPRCE
          FROM PART
          WHERE ITEMCLSS = 'SG'
```

19.

CUSTOMER	CUSTNUMB	CUSTNAME	ADDRESS	etc.	
P.					

21.

CUSTOMER	CUSTNUMB	CUSTNAME	CREDLIM	SLSRNUMB	
	P.	P.	500	3	

23.

CUSTOMER	CUSTNUMB	CUSTNAME	SLSRNUMB	etc.	
	P.	P.	~3		

25. Relational algebra:

```
SELECT PART WHERE PARTNUMB = 'BT04' GIVING ANSWER
```

SQL:

```
SELECT *
    FROM PART
    WHERE PARTNUMB = 'BT04'
```

27. Relational algebra:

```
JOIN CUSTOMER ORDERS
    WHERE CUSTOMER.CUSTNUMB = ORDER.CUSTNUMB
    GIVING TEMP
PROJECT TEMP OVER (ORDNUMB, ORDDTE, CUSTNUMB,
    CUSTNAME) GIVING ANSWER
```

SQL:

```
SELECT ORDNUMB, DATE, CUSTOMER.CUSTNUMB, CUSTNAME
    FROM CUSTOMER, ORDERS
    WHERE CUSTOMER.CUSTNUMB = ORDERS.CUSTNUMB
```

CHAPTER 4 — RELATIONAL MODEL II: ADVANCED TOPICS

1. A view is an individual user's picture of the database. It is defined through a defining query. The data in the view never actually exists in the form described in the view. Rather, when a user accesses the view, his or her query is merged with the defining query of the view to form a query that pertains to the whole database.

3. a.
```
CREATE VIEW CUSTORD AS
     SELECT CUSTOMER.CUSTNUMB, CUSTNAME,
          BALANCE, ORDNUMB, DATE
          FROM CUSTOMER, ORDERS
          WHERE CUSTOMER.CUSTNUMB =
               ORDERS.CUSTNUMB
```

b.
```
SELECT CUSTNUMB, CUSTNAME, ORDNUMB, DATE
     FROM CUSTORD
     WHERE BALANCE > 100
```

c.
```
SELECT CUSTOMER.CUSTNUMB, CUSTNAME, ORDNUMB,
     DATE
     FROM CUSTOMER, ORDERS
          WHERE CUSTOMER.CUSTNUMB =
               ORDERS.CUSTNUMB
          AND BALANCE > 100
```

5. On relational mainframe DBMSs, the optimizer (a part of the DBMS) makes the decision to use a particular index. On microcomputer DBMSs, the user or programmer makes this decision.

7. If the DBMS updates the catalog automatically, users need not worry about having to update the catalog whenever they make a change in the database structure. If they did have to make such a change, it might be made incorrectly, in which case the data in the catalog wouldn't match the structure of the database.

9. The structure of a table can be changed in SQL through the ALTER command. Columns can be added (ALTER TABLE table-name ADD column-name); columns can be deleted (ALTER TABLE table-name DELETE column-name); and columns can be changed (ALTER TABLE table-name CHANGE COLUMN column-name TO new description). Tables can be deleted (DROP TABLE table-name).

11. A tabular system is one in which users perceive databases as collections of tables. A minimally relational system is one in which users perceive databases as collections of tables and which supports the SELECT, PROJECT, and JOIN commands of the relational algebra. A relationally complete system is one in which users perceive databases as collections of tables and which supports the complete relational algebra. A fully relational system is one in which users perceive databases as collections of tables and which supports the complete relational algebra as well as entity and referential integrity.

CHAPTER 5 — DATABASE DESIGN I: NORMALIZATION

1. Column B is functionally dependent on column A if a value for A determines a unique value for B at any time.

3. The primary key of a table is the column or collection of columns that determines all other columns in the table and for which there is no subcollection that also determines all other columns.

5. A table is in first normal form if it does not contain a repeating group.

7. A table is in third normal form if it is in second normal form and if the only determinants it contains are candidate keys. If a table is not in 3NF, redundant data will cause wasted space and update problems. Inconsistent data may also be a problem.

9. Many answers are possible. See the guidelines in Chapter 5.

11.

```
INVNUMB --> CUSTNUMB, CUSTNAME,
      CUSTADDR, INVDATE
CUSTNUMB --> CUSTNAME, CUSTADDR
PARTNUMB --> PARTDESC, PRICE
INVNUMB, PARTNUMB --> NUMBSHIP


INVOICE (INVNUMB, CUSTNUMB, INVDATE)
CUSTOMER (CUSTNUMB, CUSTNAME, CUSTADDR)
PART (PARTNUMB, PARTDESC, PRICE)
INVLNE (INVNUMB, PARTNUMB, NUMBSHIP)
```

CHAPTER 6 — DATABASE DESIGN II: DESIGN METHODOLOGY

1. A user view is the view of data that is necessary to support the operations of a particular user. By considering individual user views rather than the complete design problem, we greatly simplify the database design process.

3. If the design problem were extremely simple, the overall design might not have to be broken down into a consideration of individual user views.

5. The primary key is the column or columns that uniquely identify a given row and that furnish the main mechanism for directly accessing a row in the table. An alternate key is a column or combination of columns that could have functioned as the primary key but was not chosen to do so. A secondary key is a column or combination of columns that is not any other type of key but is of interest for purposes of retrieval. A foreign key is a column or combination of columns in one table whose values that are required to match the primary key in another table. Foreign keys furnish the mechanism through which relationships are made explicit.

7. a. Include the project number as a foreign key in the employee table.
   b. Include the employee number as a foreign key in the project table.
   c. Create a new table whose primary key is the concatenation of employee number and project number.

9. Instead of the advisor number being included as a foreign key in the student table, there would be an additional table whose primary key was the concatenation of student number and advisor number.

11.

```
PUBLSHR (PUBCODE, PUBNAME, PUBCITY)
    Primary key PUBCODE
    Secondary key PUBNAME
    1.   PUBCODE must be unique.

BRANCH (BRNUMB, BRNAME, BRLOC, NUMEMP)
    Primary key BRNUMB
    Secondary key BRNAME
    1.   BRNUMB must be unique.
```

```
AUTHOR(AUTHNUMB, AUTHNAME)
     Primary key AUTHNUMB
     1.    AUTHNUMB must be unique.

BOOK (BKCODE, BKTITLE, PUBCODE, BKTYPE, BKPRICE, PB)
     Primary key BKCODE
     Foreign key PUBCODE matches PUBLSHR
     1.    BKCODE must be unique.
     2.    PUBCODE must match the code of a publisher in the
           PUBLSHR table.

WROTE(BKCODE, AUTHNUMB, SEQNUMB)
     Primary key BKCODE, AUTHNUMB
     Foreign key BKCODE matches BOOK
     Foreign key AUTHNUMB matches AUTHOR
     1.    The combination of BKCODE and AUTHNUMB must be unique.
     2.    BKCODE must match the code of a book in the
           BOOK table.
     3.    AUTHNUMB must match the number of an author in the
           AUTHOR table.

INVENT (BKCODE, BRNUMB, OH)
     Primary key BRNUMB, BKCODE
     Foreign key BRNUMB matches BRANCH
     Foreign key BKCODE matches BOOK
     1.    The combination of BKCODE and BRNUMB must be unique.
     2.    BRNUMB must match the number of a branch in the
           BRANCH table.
     3.    BKCODE must match the code of a book in the
           BOOK table.
```

CHAPTER 7 — FUNCTIONS OF A DATABASE MANAGEMENT SYSTEM

1. The DBMS must furnish a mechanism for storing data and for enabling users to retrieve data from the database and to update data in the database. This mechanism should not require the users to be aware of the details with respect to how the data is actually stored.

3. Shared update refers to two or more users updating data in a database at the same time.

5. Locking is the process whereby only one user is allowed to access a specific portion of a database at a time. When a user is accessing a portion of the database, it is locked, meaning that it is unavailable to any other user.

7. Deadlock is the circumstance in which user A is waiting for resources that have been locked by user B, and user B is waiting for resources that have been locked by user A; unless action to the contrary is taken, the two will wait for each other forever. It occurs when each of the two users is attempting to access data that is held by the other.

9. a. Each user must attempt to lock all the resources he or she needs before beginning any updates. If any of the resources are already locked by another user, all locks must be released and the process must begin all over again.

   b. Before updating a record, user 1 should make sure that the record has not been updated by user 2 since the time user 1 first read it. To understand why this is necessary, read part c of this answer.

c. After reading a record, a user should immediately release the lock on it.

11. Security is the prevention of unauthorized access to the database.

13. Encryption is the process whereby data is transformed into another form before it is stored in the database. The data is returned to its original form when it is retrieved by a legitimate user. This process prevents a person who bypasses the DBMS and accesses the database directly from seeing the relevant data.

15. A database has integrity when the data in it follows certain established rules, called integrity constraints. Integrity constraints can be handled in four ways: (1) They can be ignored. (2) The responsibility for enforcing them can be assigned to the user (that is, it would be up to the user not to enter invalid data). (3) They can be enforced by programs. (4) They can be enforced by the DBMS. Of these four, the most desirable is the last. When the DBMS enforces the integrity constraints, users don't have to constantly guard against entering incorrect data, and programmers are spared having to build the logic to enforce these constraints into the programs they write.

17. Many examples are possible; if you need help in remembering some, see the list in this chapter.

CHAPTER 8 — DATABASE ADMINISTRATION

1. DBA is database administration, the person or group that is responsible for the database. The responsibilities of DBA are crucial to success in the database environment, especially if the database is to be shared among several users; these responsibilities include determining access privileges; establishing and enforcing security procedures; determining and enforcing policies with respect to the use of a data dictionary; and so on.

3. DBA determines access privileges, uses the DBMS security facilities such as passwords, encryption, and views, and supplements these features, where necessary, with special programs.

   Users often choose passwords that are easy for others to guess, such as the names of family members. Users can also be careless with the paper on which passwords are written. To prevent others from guessing their passwords, users should guard against doing either of these things and should also change their passwords frequently.

5. Certain corporate data, though no longer required in the active database, must be kept for future reference. A data archive is a place for storing this type of data. The use of data archives allows an organization to keep records indefinitely, without causing the database to become unnecessarily large. Data can be removed from the database and placed in the data archive, instead of just being deleted.

7. DBA does some of the training of computer users. Other training, such as that which is provided by a software vendor, is coordinated by DBA.

9. a. What facilities are provided by the system for defining a new database? What data types are supported?

   b. What facilities are present to assist in the restructuring of a database?

   c. What nonprocedural language (a language in which we tell the computer what the task is rather than how to do it) is furnished by the system? How does its functionality compare with that of SQL?

   d. What procedural language (a language in which we tell the computer how to do the task) is provided by the system? How complete is it? How is the integration between the procedural language and the nonprocedural language accomplished?

   e. What data dictionary is included? What types of information can be held in the dictionary? How well is it integrated with the other parts of the system?

    f.   What support does the system provide for shared update? What type of locking is used? Can the system handle deadlock?

    g.   What services does the system provide for backup and recovery? Does recovery consist only of copying a backup over the live database, or does the system support the use of a journal in the recovery process?

    h.   What security features are provided by the system? Does it support passwords, encryption, and/or views? How easy is it for a user to bypass the security controls of the DBMS?

    i.   What type of integrity support is present? What kinds of integrity constraints can be enforced?

    j.   What are the system limitations with respect to the number of tables, columns, rows, and the number of files that can be open at the same time? What hardware limitations exist?

    k.   How good are the manuals? How good is the on-line help facility, if there is one?

    l.   What reputation does the vendor have for support of their products?

    m.   How well does the system perform?

    n.   What is the cost of the DBMS, of additional hardware, and of support?

    o.   What plans does the vendor have for further development of the system?

    p.   This category includes any special requirements an organization might have that do not fit into any of the previous categories.

11.   DBA has primary responsibility for the DBMS once it has been selected. DBA installs the DBMS, makes any changes to its configuration when they are required, determines whether it is appropriate to install a new version of the DBMS when it becomes available, and, if a decision is made to install a new DBMS, coordinates the installation.

CHAPTER 9 — APPLICATION GENERATION

1.   An application system is a collection of programs that is designed to handle some specific task. Application systems can be created by users, by programmers within the organization that will use the system, and by software houses. A DBMS is designed to handle general rather than specific tasks and is thus not an application system. Many DBMSs are used to write application systems, however.

3.   A modern application system is menu-driven and features interactive and automatic updates, reports, some type of query facility, backup and recovery facilities, audit trails, utility services, and administrative services.

5.   Interactive updates are usually achieved through users filling in forms on the screen. Not all updates are accomplished interactively, however. Some occur in response to some other update (for example, a customer's balance is increased automatically as a result of printing an invoice) or in response to a particular request from a user (for example, month-to-date fields are set back to zero when the user chooses the month-end routines).

7.   An audit trail is a record of all the updates that have been made to the database. It is used to ensure that updates have been made correctly and, if any errors have been found, to facilitate restoring the database to a correct state.

9.   A programmer's workbench consists of a collection of tools, such as editors, compilers, and debuggers, that assist the programmer in doing his or her job.

11.   A screen painter is a tool that is used to facilitate the process of developing screen-oriented data entry programs. To use it, a programmer interactively describes what the screen is supposed to look like. A screen painter is also called a screen generator.

13.   A nonprocedural language is one in which a user describes what the task is rather than how it is to be accomplished. SQL is an example of a nonprocedural language. Nonprocedural languages make users more productive, because they do not need to be concerned with specific details with regard to how a task is to be accomplished.

MODULE 1 — INTRODUCTION TO MICROSOFT WINDOWS

1. A GUI uses graphic pictures or icons to replace commands as a means of communication between the user and the computer.

3. The Program Manager is an application that is central to Windows activities. Use the Program Manager to start applications, organize applications into groups, and manage the disk directory.

5. To open an application, open the Group window containing the application and double click the Program Item icon.

   To exit an application, select Exit from the File menu, select Close from the Control-box menu, or double click the application's Control-menu box.

7. *Clicking* means to move the mouse so that the pointer icon moves to an object and then press the left button on the mouse once. *Double clicking* means to move the mouse so that the pointer icon moves to an object and then rapidly press the left mouse button twice. *Dragging* is when you move the mouse so that the mouse pointer icon moves to an object then hold down the left mouse button and move the mouse to another location. The object will be moved with the action.

9. A *dialog box* is a window that displays to communicate an error or request information from the user.

11. Within Windows applications, using the Copy or Cut command from the Edit menu of any application places selected data into a Clipboard. You then can paste the data in the Clipboard into any other application.

13. Choosing Save Settings prior to exiting Windows saves any actions you have taken, such as moving windows, resizing windows, or changing the placement of fields.

MODULE 2 — INTRODUCTION TO MICROSOFT ACCESS

1. To make a selection from an Access menu, press the Alt key and then press the underlined letter of the menu choice. Alternatively, you could press the Alt key and then the left/right or the up/down arrows to move to Menu option and press Enter. To select an option with a mouse, click on the option.

3. To view a table, select the View option of the Main menu and then choose Table. When a list of tables in the database is listed, choose the table to be viewed. Tables can be viewed either in the Datasheet view or Forms view.

5. To create a new table, click on the Table icon and then the New button. Type the name of the table, and then enter the fields that make up the table.

7. You can change the contents of records in a table while in the Datasheet or Form view. To change the contents of records, move to the record and field that you want to change and enter or edit the data.

9. To produce a report without having to specify details about the report layout, select New from the File menu and then select Report. Next, identify the table or query to use as the source of the data, and choose ReportWizard. Then choose AutoReport and answer the questions to the prompts of which fields you want to print.

MODULE 3 — QUERIES

1. A Query-by-Example (QBE) is the default format Access uses to prompt for fields, conditions, and sort data necessary to execute a query. QBE forms ask only that the user identify the field and then the condition to match in the field.

   To produce a QBE, choose the Query object in the Database window and then click on the New button and choose the New Query button.

3. To print the results of a query, switch to the Datasheet view and click on the Print button on the toolbar, or choose Print from the File menu.

5. To enter a condition in a query, enter a logical condition into the Criterion cell of an appropriate column. A compound condition involving an AND is made by entering two or more conditions on the same criterion line. A compound condition involving an OR is made by entering two or more conditions on separate criterion lines.

7. To include a wildcard in a query, enter an asterisk to indicate any number of characters of any type, and a ? to indicate any single characters.

9. Statistical calculations available as built-in functions in Access include: Count, Sum, Average, Max, and Min. To use a built-in function, enter the Design view of a query and in the column that you want to execute one of the functions, click on the Totals button on the toolbar. This will insert a Totals row in the query, in which you can insert the desired function. For example, the Count function would count how many entries are in that particular column and return the result in the query Datasheet view.

11. To change data using a query, create an Update query. Run the Update query and enter the record that you want to update, followed by the update data.

MODULE 4 — ADVANCED RELATIONAL FEATURES

1. To obtain a list of tables in the current directory, open a Database window and click on the Tables Object button. All tables are listed in the Database window.

3. To create a primary index, click on the Primary Key icon on the toolbar, while in the proper field in the Design view. There can be only one primary index, although that primary index can be made up of multiple fields.

5. To add a field to a table, switch to the Design view, move the cursor to the desired location, and insert a row. Type the field name and data type information.

7. To delete a field in a table, switch to the Design view and select the field to be deleted. Press the Delete key and choose OK to verify the deletion.

9. A relationship is a link between to identical fields in two different tables. To set relationships between tables, open a database and select Relationships from the Edit menu. Select the proper table for the primary table and a related table. Then select the matching field between two tables.

MODULE 5 — DATA ENTRY SUPPORT

1. You can create a form in any layout desired. Then you can use the form to display or print records in a single form at a time or in a multi-column format. To create a form, open a database and click on the Form object button, choose New, and select a table or query to supply data for the form. Then use the Form Wizard to create a standard form.

3. To move fields to an different location on a form, enter the Design view of the form and select the field to move. Then place the pointer icon on one of the edges of the field to change the pointer to a small hand icon. You can then press and hold down the left mouse button to drag the field to another location.

5. To add graphical improvements to a form, enter the Design view of a form and select one of the draw tools from the toolbox. To add a box to the form, Select the Box tool and click at the location that you want to place the upper-left corner of the new box. Then resize the box by dragging the size handles on one of the sides of the box.

7. Create a multi-table form by embedding one or more Subforms in a Main form. The Main form contains the primary record, and a Subform then contains data from another table or query related to the record in the Main form.

9. To move between subforms in a Multi-table form, press the Tab key to jump between forms just as you would between Fields. To move to another record in the Master table or Main report of a Multi-table form, use the navigation buttons in the lower-right corner of the form to move to the previous or next record, or enter the record number to jump to.

MODULE 6 — REPORTS AND LABELS

1. To create an AutoReport, select the Report object, choose New, and identify the table to use for the data. Then use the ReportWizard's AutoReport to generate a default single-column report. A single-column report lists each record in a custom layout one after the other. Tabular reports list records on a single row, one after the other.

3. To add a field to a report, open a report in the Design view and either display the Field List and drag a field to a report column or click on the Textbox icon and place it in an available field. To place a new field on the report from another table, you must link the second table to the first table.

5. To change the format and behavior of a field, select the Properties button on the toolbar where the format, alignment, and other characteristics may be set.

7. To construct a report using data from several tables, construct a subreport containing all the required fields from an individual table or query. If you are going to merge numerous tables, it is easier to create a query containing multiple tables. Then, use the new table or query as a section of the report.

9. You can add multiple subgroups to a report in the same manner that you added the first subgroup. Each subgroup is independent, but they all must be linked in some fashion to the record in the Main report.

MODULE 7 — MICROSOFT ACCESS AND THE FUNCTIONS OF A DBMS

1. Access provides a user-accessible catalog through the hierarchical display of a database including tables, queries, forms, reports, and so forth to facilitate finding appropriate objects. The user benefits from having an easy-to-use interface to communicate with the database.

3. Access provides support for shared update through automatic locking and multiuser controls. It also can refresh a user's screen when another user changes the data displayed on the screen. Although automatic locking handles most situations, explicit locking is also available, both from the menu system and within programs.

5. Access provides services to support many different types of integrity constraints through the use of primary indexes and validity checks.

7. Access permits changes to the database structure (adding new tables or columns, deleting tables or columns, or change the characteristics of columns and so on). This is accomplished by adding or deleting tables in a database or adding/deleting columns in the Table Design view.

   You can add or delete table indexes by changing a field's properties settings in the Design view.

   You can Import and Export to and from Access by choosing Import or Export from the File menu while in a database.

   Access provides an easy-to-use edit and query interface in a variety of ways.

   A powerful procedural language, Access Basic is available within the DBMS.

   Most importantly, Access provides an easy-to-use menu-driven interface.

MODULE 8 — ADVANCED TOPICS

1. OLE stands for Object Linking and Embedding. An object is any element created by another Windows application. A linked object is one that was created within another

Windows application and physically stored in a file of the other application. An embedded object is one that was created by another application but has been physically stored as a part of an Access record.

3. *Memo* fields are long text data types that allow you to enter up to 64,000 characters.

5. Option groups tie individual data values together as Yes/No values that allow only one of the items to be chosen from the group. You can choose between Option Button or Check Box format. To create an option group, enter the Design view of a form or report and select the Option Group tool. Place the Option Group box on the form or report. Then any option button or check boxes placed within the box will be treated as a group from which only one may be selected at one time.

7. Combo boxes list valid data values in a drop-down list in a form or report. List boxes list valid data values in a scroll list format in a form or report.

9. Actions are the predefined individual tasks assigned in a macro. Arguments are the specific instructions for each action describing how the action is to be performed.

MODULE 9 — GRAPHING IN MICROSOFT ACCESS

1. To create a graph, create a new form and use the Graph Wizard. Define the fields for the X and Y axes as prompted. A graph is an OLE object, created in Microsoft Graph, that displays in a Form window or a report.

3. Access offers several different graph types, including bar, line, column, XY (Scatter), pie, doughnut, surface, and area. Most graphs can display in either 2-D or 3-D.

5. To modify characteristics of a graph, first select the graph object and double click to enter the Graph application that created the chart. Then select the appropriate part of the graph that you want to modify and choose an option from the Format or Insert menu.

7. Crosstabs summarize information from a table by combining two or more fields in a table format with one of the fields as the row labels and the other field as the column labels. To create a crosstab, create a new Query using the Query Wizard. Select the Crosstab Query Wizard and identify the tables for the data. Then select the fields for row and column labels as well as indicate how amounts are summarized. Crosstabs are a special type of query. You can print out the results of the crosstab query by displaying the query in the datasheet view and selecting Print from the File menu.

9. To add legends to a graph, enter the Design view of the form containing the graph and double click on the graph to enter the Graph application. Then choose Legends from the Insert menu. Data from the Categories column in the datasheet are used as the legends. To change the legend labels, click on the label and edit the text in the textbox.

11. When placing a graph object on a form, you can let the Graph Wizard create the graph. You need only indicate the table or query that contains the fields for the categories to be used for the X and Y axes. Then define how to calculate the totals for the graph, what chart type to use, and finally, enter a chart title.

Glossary

Alias An alternate name for a table; can be used within a query.

ANSI American National Standards Institute.

Application See *application system*.

Application generation The process of developing an application system.

Application generator Software tool that permits the rapid development of an application system. Also called a *fourth-generation environment* or a *fourth-generation language*. The latter term is sometimes applied only to the *nonprocedural language* component of an application generator.

Application package See *application system*.

Application programs The programs that make up an *application system*.

Application system A collection of programs that work together to handle some specific task. Also called an *application*, an *application package*, a *software package*, a *software system*, or sometimes simply a *system*.

Attribute A property of an entity.

Background The permanent part of a screen form, that is, the part that does not change from one transaction to the next. See also *foreground*.

Backup A copy of a database; used to recover the database when it has been damaged or destroyed.

Boyce-Codd Normal Form (BCNF) A relation is in Boyce-Codd normal form if it is in second normal form and the only determinants it contains are candidate keys; also called third normal form in this text.

Candidate key A minimal collection of attributes (columns) in a relation on which all attributes are functionally dependent but which has not necessarily been chosen as the *primary key*.

Catalog A source of information on the types of entities, attributes, and relationships in a database.

Child In the hierarchical model, the record that is the "many" part of the *one-to-many relationship*.

CODASYL COnference on Data SYstems Languages. The group that developed COBOL and proposed the CODASYL model for database management.

CODASYL model The model for database management systems proposed by CODASYL; falls within the general network model of data. Not a standard, although it has been used in the development of many systems.

Concatenation Combination of attributes. To say a key is a concatenation of two attributes, for example, means that a combination of values of both attributes is required to uniquely identify a given tuple.

Concurrent update Several updates taking place to the same file or database at almost the same time; also called *shared update*.

Data archive A place where historical corporate data is kept. Data that is no longer needed in the corporate database but must be retained for future needs is removed from the database and placed in the archive.

Data Base Task Group The group originally appointed by CODASYL to develop specifications for database management systems.

Data definition language A language that is used to communicate the structure of a database to the database management system.

Data dictionary A tool that is used to store descriptions of the entities, attributes, relationships, programs, and so on, that are associated with an organization's database.

Data independence The property that allows the structure of the database to change without requiring changes in programs.

Data manipulation language A language that is used to manipulate the data in the database.

Data model A classification scheme for database management systems. A data model addresses two aspects of database management: *structure* and *operations*.

Data structure diagram A diagram of the records and sets in a network database.

Database A structure that can house information about various types of entities and about the relationships among the entities.

Database administration (DBA) The individual or group that is responsible for the database.

Database administrator The individual who is responsible for the database, or the head of database administration.

Database design The process of determining the content and arrangement of data in the database in order to support some activity on behalf of a user or group of users.

Database Design Language (DBDL) A relational-like language that is used to represent the result of the database design process.

Database management system A software package that is designed to manipulate the data in a database on behalf of a user.

Database navigation The process of finding a path through the relationships in a database in order to satisfy a given request.

Database processing The type of processing in which the data is stored in a *database* and manipulated by a *DBMS*.

DBA See *database administration*. (Sometimes the acronym stands for database administrator.)

DBDL See *database design language*.

DBMS See *database management system*.

DDL See *data definition language*.

Deadlock A state in which two or more users are each waiting to use resources that are held by the other(s).

Deadly embrace Another name for *deadlock*.

Debugging The process of finding and correcting errors in programs.

Defining query The query that is used to define the structure of a view.

Dependency diagram A diagram that indicates the dependencies among the attributes in a relation.

Determinant An attribute that determines at least one other attribute.

DML See *data manipulation language*.

Encryption The transformation of data into another form, for the purpose of security, before it is stored in the database. The data is returned to its original form for any legitimate user who accesses the database.

Entity An object (person, place, or thing) of interest.

Entity integrity The rule that no attribute that participates in the primary key may accept **null** values.

Field The smallest unit of data to which we assign a name; can be thought of as the columns in a table. For example, in a table for customers, the fields (columns) would include such things as the customer's number, the customer's name and address, and so on.

File Technically, a collection of bytes (characters) on a disk; could be data, a program, a document created by a word processor, and so on. Often refers to a data file, which is a structure used to store data about some entity. Such a file can be thought of as a table. The rows in such a table are called records, and the columns are called fields.

First Normal Form (1NF) A relation is in first normal form if it does not contain repeating groups. (Technically, this is part of the definition of a relation.)

Foreground The portion of a screen form that changes from one transaction to the next, that is, the portion of the form into which the user enters data and/or data is displayed. See also *background*.

Foreign key An attribute (or collection of attributes) in a relation whose value is required either to match the value of a primary key in another relation or to be null.

Fourth-generation environment See *application generator*.

Fourth-Generation Language (4GL) Sometimes used to refer to an *application generator* and sometimes to a *nonprocedural language*, which is one of the components of an *application generator*.

Fully relational The expression used to refer to a DBMS in which users perceive data as tables, which supports all the operations of the *relational algebra* and which supports *entity* and *referential integrity*.

Functionally dependent Attribute B is functionally dependent on attribute A (or on a collection of attributes) if a value for A determines a single value for B at any one time.

Functionally determine Attribute A functionally determines attribute B if B is *functionally dependent* on A.

Help facility A facility through which users can receive on-line assistance.

Hierarchical model A data model in which the structure is a *tree*, or hierarchy.

Hierarchy See *tree*.

Index A file that relates key values to records that contain those key values.

Information level of database design The step during *database design* in which the goal is to create a clean, DBMS-independent design that will support user requirements.

Integrity A database has integrity if all *integrity constraints* that have been established for it are currently met.

Integrity constraint A condition that data within a database must satisfy; also, a condition that indicates the types of processing that may or may not take place.

Join In the *relational algebra*, the operation in which two tables are connected on the basis of common data.

Journal A record of all changes in the database; also called a *log*. Used to recover a database that has been damaged or destroyed.

LAN See *local area network*.

Local Area Network (LAN) A configuration of several computers that are all hooked together in a limited geographic area; allows users to share a variety of resources.

Locking The process of placing a lock on a portion of a database, which prevents other users from accessing that portion.

Log A record of all changes in the database; also called a *journal*. Used to recover a database that has been damaged or destroyed.

Logical child relationship A relationship in the *hierarchical model* in which the *parent* is in a different *tree* than the *child*.

Many-to-many relationship A relationship between two entities in which each occurrence of each entity is related to many occurrences of the other entity.

Mapping The process of creating an initial *physical-level design*.

Member In a *CODASYL set*, the record that is the "many" part of the *one-to-many relationship*.

Menu-driven A style of program in which the user selects an action from a list (called a menu) of available options that are displayed on the screen.

Minimally relational A DBMS in which users perceive data as tables and which supports at least the SELECT, PROJECT, and JOIN operations of the *relational algebra* without requiring any predefined access paths.

Natural language A language in which users communicate with the computer through the use of standard English questions and commands.

Navigation See *database navigation*.

Network A structure that contains record types and explicit one-to-many relationships between these record types.

Network model A *data model* in which the structure is a network and the operations involve navigating the network (that is, following the arrows in a data structure diagram).

Nonkey attribute An attribute that is not part of the primary key.

Nonprocedural language A language in which the user specifies the task that is to be accomplished rather than the steps that are required to accomplish it.

Normal form See *first normal form*, *second normal form*, *third normal form*, and *Boyce-Codd Normal Form*.

Normalization Technically, the process of removing repeating groups to produce a *first normal form* relation. Sometimes refers to the process of creating a *third normal form* relation.

Null A special value meaning "unknown" or "not applicable."

One-to-many relationship A relationship between two entities in which each occurrence of the first entity is related to many occurrences of the second entity but each occurrence of the second entity is related to only one occurrence of the first entity.

One-to-one relationship A relationship between two entities in which each occurrence of the first entity is related to one occurrence of the second entity and each occurrence of the second entity is related to one occurrence of the first entity.

Operations One of the two components of a *data model*; the facilities given users of the DBMS to manipulate data within the database.

Optimizer The DBMS component that selects the best way to satisfy a query.

Owner In a CODASYL set, the record that is the "one" part of the one-to-many relationship.

Parent In the hierarchical model, the record that is the "one" part of the one-to-many relationship.

Partial dependency A dependency of an attribute on only a portion of the primary key.

Password A word that must be entered before a user can access certain computer resources.

Physical database In the *hierarchical model*, a single *tree*.

Physical level of database design The step during *database design* in which a design for a given DBMS is produced from the final information-level design.

Primary key A minimal collection of attributes (columns) in a relation on which all attributes are functionally dependent and which is chosen as the main direct-access vehicle to individual tuples (rows). See also *candidate key*.

Procedural language A language in which the user must specify the steps that are required for accomplishing a task instead of merely specifying the task itself.

Program generator A facility that accepts specifications from the user and creates a program that matches these specifications.

Programmer's workbench A collection of tools that simplifies the tasks of programmers; includes editors, compilers, debuggers, and so on.

QBE See *Query-by-Example*.

Qualify To indicate the table (relation) of which a given column (attribute) is a part by preceding the column name with the table name. For example, *CUSTOMER.ADDRESS* indicates the column named *ADDRESS* within the table named *CUSTOMER*.

Query A question, the answer to which is found in the database; also used to refer to a command in a *nonprocedural language* such as SQL that is used to obtain the answer to such a question.

Query-By-Example (QBE) A *data manipulation language* for relational databases in which users indicate the action to be taken by filling in portions of blank tables on the screen.

Query facility A facility that enables users to obtain information easily from the database.

Query language A language that is designed to permit users to obtain information easily from the database.

Record A collection of related fields; can be thought of as a row in a table.

Recovery The process of restoring a database that has been damaged or destroyed.

Redundancy Duplication of data.

Referential integrity The rule that if a relation A contains a *foreign key* that matches the primary key of another relation B, then the value of this foreign key must either match the value of the primary key for some row in relation B or be null.

Relation A two-dimensional table in which all entries are single-valued; each column has a distinct name; all of the values in a column are values of the attribute that is identified by the column name, the order of columns is immaterial; each row is distinct; and the order of rows is immaterial.

Relational algebra A relational **data manipulation language** in which relations are created from existing relations through the use of a set of operations.

Relational database A collection of relations.

Relational model A *data model* in which the structure is the *table* or *relation*.

Relationally complete A term applied to any relational *data manipulation language* that can do whatever can be done through the use of the *relational algebra*; also applied to a DBMS that supplies such a data manipulation language.

Relationship An association between entities.

Repeating group Several entries at a single location in a table.

Report generator See *report writer*.

Report writer A nonprocedural language for producing formatted reports from data in a database; also called a *report generator*.

Save A backup copy.

Screen generator An interactive facility for creating and maintaining display and data-entry formats for screen forms.

Screen painter See *screen generator*.

Second Normal Form (2NF) A relation is in second normal form if it is in first normal form and no non-key attribute is dependent on only a portion of the primary key.

Secondary key An attribute or collection of attributes that is of interest for retrieval purposes (and that is not already designated as some other type of key).

Security The protection of the database against unauthorized access.

Set The CODASYL implementation of a one-to-many relationship.

Shared update Several updates taking place to the same file or database at almost the same time; also called *concurrent update*.

Software package See *application system*.

Software system See *application system*.

SQL See *Structured Query Language*.

Structure One of the two components of a *data model*: the manner in which the system structures data or, at least, the manner in which the users perceive that the data is structured.

Structured Query Language (SQL) A very popular relational *data definition and manipulation language* that is used in many relational DBMSs.

Subquery In SQL, a query that is contained within another query.

System Often used to refer to *application system*; sometimes also used to refer to a *DBMS*.

Table In the database environment, another name for a relation.

Tabular A type of DBMS in which users perceive data as tables but which does not furnish any of the other characteristics of a relational DBMS.

Third Normal Form (3NF) A relation is in third normal form if it is in second normal form and if the only determinants it contains are candidate keys. (Technically, this is the definition of *Boyce-Codd Normal Form*, but in this text the two are used synonymously.)

Tree A network, with an added restriction: no entity can participate as the "many" part of more than one *one-to-many relationship*.

Tuning The process of altering a database design in order to improve performance.

Tuple The formal name for a row in a table.

Unnormalized relation A structure that satisfies the properties required to be a relation with one exception: repeating groups are allowed; that is, the entries in the table do not have to be single-valued.

Update anomaly An update problem that can occur in a database as a result of a faulty design.

User view The view of data that is necessary to support the operations of a particular user.

View An application program's or an individual user's picture of the database.

Wild card A symbol that can be used in place of an unknown character or group of characters in a *query*.

Bibliography

[1] Boar, Bernard H. *Application Prototyping: A Requirements Definition Strategy for the 80s*. John Wiley & Sons, 1984.

[2] Chen, Peter. *The Entity-Relationship Approach to Logical Data Base Design*. QED Monograph Series, 1977.

[3] Codd, E. F. "A Relational Model of Data for Large Shared Databanks." *Communications of the ACM* 13, no. 6 (June 1970).

[4] Codd, E. F. "Further Normalization of the Data Base Relational Model." In *Data Base Systems*, Courant Computer Science Symposia Series, vol. 6, Prentice-Hall, 1972.

[5] Codd, E. F. "Recent Investigations into Relational Data Base Systems." Proceedings of the IFIP Congress, 1974.

[6] Codd, E. F. "Extending the Relational Database Model to Capture More Meaning." *ACM TODS* 4, no. 4 (December 1979).

[7] Codd, E. F. "Relational Database: A Practical Foundation for Productivity." *Communications of the ACM* 25, no. 2 (February 1982).

[8] Date, C. J. *Database: A Primer*. Addison-Wesley, 1983.

[9] Date, C. J. *Introduction to Database Systems: Volume I*, 4th ed. Addison-Wesley, 1986.

[10] Goldstein, Robert C. *Database Technology and Management*. John Wiley & Sons, 1985.

[11] Kroenke, David. *Database Processing*, 2d ed. SRA, 1983.

[12] Kroenke, David M., and Nilson, Donald E. *Database Processing for Microcomputers*. SRA, 1986.

[13] McFadden, Fred R., and Hoffer, Jeffrey A. *Data Base Management*. Benjamin Cummings, 1985.

[14] Pratt, Philip J., and Adamski, Joseph J. *Database Systems: Management and Design*, 3rd ed. boyd & fraser, 1994.

[15] Vasta, Joseph A. *Understanding Data Base Management Systems*. Wadsworth, 1985.

[16] Zloof, M. M. "Query By Example." Proceedings of the NCC 44, May 1975.

SLSREP

SLSRNUMB	SLSRNAME	SLSADDR	TOTCOMM	COMMRATE
3	MARY JONES	123 MAIN, GRANT, MI	2150.00	.05
6	WILLIAM SMITH	102 RAYMOND, ADA, MI	4912.50	.07
12	SAM BROWN	419 HARPER, LANSING, MI	2150.00	.05

CUSTOMER

CUSTNUMB	CUSTNAME	ADDRESS	BALANCE	CREDLIM	SLSRNUMB
124	SALLY ADAMS	481 OAK, LANSING, MI	418.75	500	3
256	ANN SAMUELS	215 PETE, GRANT, MI	10.75	800	6
311	DON CHARLES	48 COLLEGE, IRA, MI	200.10	300	12
315	TOM DANIELS	914 CHERRY, KENT, MI	320.75	300	6
405	AL WILLIAMS	519 WATSON, GRANT, MI	201.75	800	12
412	SALLY ADAMS	16 ELM, LANSING, MI	908.75	1000	3
522	MARY NELSON	108 PINE, ADA, MI	49.50	800	12
567	JOE BAKER	808 RIDGE, HARPER, MI	201.20	300	6
587	JUDY ROBERTS	512 PINE, ADA, MI	57.75	500	6
622	DAN MARTIN	419 CHIP, GRANT, MI	575.50	500	3

ORDERS

ORDNUMB	ORDDTE	CUSTNUMB
12489	90294	124
12491	90294	311
12494	90494	315
12495	90494	256
12498	90594	522
12500	90594	124
12504	90594	522

ORDLNE

ORDNUMB	PARTNUMB	NUMBORD	QUOTPRCE
12489	AX12	11	14.95
12491	BT04	1	402.99
12491	BZ66	1	311.95
12494	CB03	4	175.00
12495	CX11	2	57.95
12498	AZ52	2	22.95
12498	BA74	4	4.95
12500	BT04	1	402.99
12504	CZ81	2	108.99

PART

PARTNUMB	PARTDESC	UNONHAND	ITEMCLSS	WREHSENM	UNITPRCE
AX12	IRON	104	HW	3	17.95
AZ52	SKATES	20	SG	2	24.95
BA74	BASEBALL	40	SG	1	4.95
BH22	TOASTER	95	HW	3	34.95
BT04	STOVE	11	AP	2	402.99
BZ66	WASHER	52	AP	3	311.95
CA14	SKILLET	2	HW	3	19.95
CB03	BIKE	44	SG	1	187.50
CX11	MIXER	112	HW	3	57.95
CZ81	WEIGHTS	208	SG	2	108.99

BRANCH

BRNUMB	BRNAME	BRLOC	NUMEMP
1	Henry's Downtown	16 Riverview	10
2	Henry's On The Hill	1289 Bedford	6
3	Henry's Brentwood	Brentwood Mall	15
4	Henry's Eastshore	Eastshore Mall	9

PUBLSHR

PUBCODE	PUBNAME	PUBCITY
AH	Arkham House Publ.	Sauk City, Wisconsin
AP	Arcade Publishing	New York
AW	Addison-Wesley	Reading, Mass.
BB	Bantam Books	New York
BF	Boyd & Fraser	Boston
JT	Jeremy P. Tarcher	Los Angeles
MP	McPherson and Co.	Kingston
PB	Pocket Books	New York
RH	Random House	New York
RZ	Rizzoli	New York
SB	Schoken Books	New York
SI	Signet	New York
TH	Thames and Hudson	New York
WN	W.W. Norton and Co.	New York

AUTHOR

AUTHNUMB	AUTHNAME
1	Archer, Jeffrey
2	Christie, Agatha
3	Clarke, Arthur C.
4	Francis, Dick
5	Cussler, Clive
6	King, Stephen
7	Pratt, Philip
8	Adamski, Joseph
10	Harmon, Willis
11	Rheingold, Howard
12	Owen, Barbara
13	Williams, Peter
14	Kafka, Franz
15	Novalis
16	Lovecraft, H. P.
17	Paz, Octavio
18	Camus, Albert
19	Castleman, Riva
20	Zinbardo, Philip
21	Gimferrer, Pere
22	Southworth, Rod
23	Wray, Robert

BOOK

BK CODE	BKTITLE	PUB CODE	BK TYPE	BK PRICE	PB
0180	Shyness	BB	PSY	7.65	T
0189	Kane and Abel	PB	FIC	5.55	T
0200	The Stranger	BB	FIC	8.75	T
0378	The Dunwich Horror and Others	PB	HOR	19.75	F
079X	Smoke-screen	PB	MYS	4.55	T
0808	Knockdown	PB	MYS	4.75	T
1351	Cujo	SI	HOR	6.65	T
1382	Marcel Duchamp	PB	ART	11.25	T
138X	Death on the Nile	BB	MYS	3.95	T
2226	Ghost from the Grand Banks	BB	SFI	19.95	F
2281	Prints of the 20th Century	PB	ART	13.25	T
2766	The Prodigal Daughter	PB	FIC	5.45	T
2908	Hymns to the Night	BB	POE	6.75	T
3350	Higher Creativity	PB	PSY	9.75	T
3743	First Among Equals	PB	FIC	3.95	T
3906	Vortex	BB	SUS	5.45	T
5163	The Organ	SI	MUS	16.95	T
5790	Database Systems	BF	CS	54.95	F
6128	Evil Under the Sun	PB	MYS	4.45	T
6328	Vixen 07	BB	SUS	5.55	T
669X	A Guide to SQL	BF	CS	23.95	T
6908	DOS Essentials	BF	CS	20.50	T
7405	Night Probe	BB	SUS	5.65	T
7443	Carrie	SI	HOR	6.75	T
7559	Risk	PB	MYS	3.95	T
7947	dBASE Programming	BF	CS	39.90	T
8092	Magritte	SI	ART	21.95	F
8720	The Castle	BB	FIC	12.15	T
9611	Amerika	BB	FIC	10.95	T

WROTE

BK CODE	AUTH NUMB	SEQ NUMB
0180	20	1
0189	1	1
0200	18	1
0378	16	1
079X	4	1
0808	4	1
1351	6	1
1382	17	1
138X	2	1
2226	3	1
2281	19	1
2766	1	1
2908	15	1
3350	10	1
3350	11	2
3743	1	1
3906	5	1
5163	12	2
5163	13	1
5790	7	1
5790	8	2
6128	2	1
6328	5	1
669X	7	1
6908	22	1
7405	5	1
7443	6	1
7559	4	1
7947	7	1
7947	23	2
8092	21	1
8720	14	1
9611	14	1

INVENT

BK CODE	BR NUMB	OH
0180	1	2
0189	2	2
0200	1	1
0200	2	3
0378	3	2
079X	2	1
079X	3	2
079X	4	3
0808	2	1
1351	2	4
1351	3	2
1382	2	1
138X	2	3
2226	1	3
2226	3	2
2226	4	1
2281	4	3
2766	3	2
2908	1	3
2908	4	1
3350	1	2
3743	2	1
3906	2	1
3906	3	2
5163	1	1
5790	4	2
6128	2	4
6128	3	3
6328	2	2
669X	1	1
6908	2	2
7405	3	2
7443	4	1
7559	2	2
7947	2	2
8092	3	1
8720	1	3
9611	1	2

SLSREP

FIELD NAME	FIELD TYPE	FIELD WIDTH	KEY FIELD	FIELD DESCRIPTION
SLSRNUMB	Number		Yes	Sales rep number
SLSRNAME	Text	15	No	Sales rep name
SLSRADDR	Text	22	No	Sales rep address
TOTCOMM	Currency		No	Total Commission
COMMRATE	Number		No	Commission Rate

CUSTOMER

FIELD NAME	FIELD TYPE	FIELD WIDTH	KEY FIELD	FIELD DESCRIPTION
CUSTNUMB	Number		Yes	Customer number
CUSTNAME	Text	15	No	Customer name
ADDRESS	Text	20	No	Customer address
BALANCE	Currency		No	Customer balance
CREDLIM	Currency		No	Credit limit
SLSRNUMB	Text	2	No	Sales rep number

PART

FIELD NAME	FIELD TYPE	FIELD WIDTH	KEY FIELD	FIELD DESCRIPTION
PARTNUMB	Text	4	Yes	Part number
PARTDESC	Text	10	No	Part description
UNONHAND	Number		No	Units on hand
ITEMCLSS	Text	2	No	Item class
WREHSENM	Number		No	Warehouse number
UNITPRCE	Currency		No	Unit price

ORDERS

FIELD NAME	FIELD TYPE	FIELD WIDTH	KEY FIELD	FIELD DESCRIPTION
ORDNUMB	Number		Yes	Order number
ORDDTE	Date		No	Order date
CUSTNUMB	Text	4	No	Customer number

ORDLNE

FIELD NAME	FIELD TYPE	FIELD WIDTH	KEY FIELD	FIELD DESCRIPTION
ORDNUMB	Number		Yes	Order number
PARTNUMB	Text	4	Yes	Part number
NUMBORD	Number		No	Number ordered
QUOTPRCE	Currency		No	Quoted price

BRANCH

FIELD NAME	FIELD TYPE	FIELD WIDTH	KEY FIELD	FIELD DESCRIPTION
BRNUMB	Number		Yes	Branch Number
BRNAME	Text	20	No	Branch Name
BRLOC	Text	20	No	Branch Location
NUMEMP	Number		No	Number of employees

PUBLSHR

FIELD NAME	FIELD TYPE	FIELD WIDTH	KEY FIELD	FIELD DESCRIPTION
PUBCODE	Text	3	Yes	Publisher code
PUBNAME	Text	20	No	Publisher name
PUBCITY	Text	20	No	Publisher city

AUTHOR

FIELD NAME	FIELD TYPE	FIELD WIDTH	KEY FIELD	FIELD DESCRIPTION
AUTHNUMB	Number		Yes	Author number
AUTHNAME	Text	20	No	Author name

BOOK

FIELD NAME	FIELD TYPE	FIELD WIDTH	KEY FIELD	FIELD DESCRIPTION
BKCODE	Text	4	Yes	Book code
BKTITLE	Text	30	No	Book title
PUBCODE	Text	3	No	Book publisher code
BKTYPE	Text	3	No	Book type
BKPRICE	Number		No	Price
PB	Text	1	No	Paperback (T or F)

WROTE

FIELD NAME	FIELD TYPE	FIELD WIDTH	KEY FIELD	FIELD DESCRIPTION
BKCODE	Text	4	Yes	Book code
AUTHNUMB	Number		No	Author number
SEQNUMB	Number		No	Sequence number

INVITE

FIELD NAME	FIELD TYPE	FIELD WIDTH	KEY FIELD	FIELD DESCRIPTION
BKCODE	Text	4	Yes	Book code
BRNUMB	Number		No	Branch number
OH	Number		No	Number of units on hand

BOOKCUST

CSTNUMB	CSTNAME	CSTADDR	CSTCTY	CSTST	BRNUMB
1	Allen, Donna	21 Wilson	Carson	In	2
2	Peterson, Mark	215 Raymond	Cedar	In	2
3	Sanchez, Miguel	47 Chipwood	Mantin	Il	3
4	Tran, Thanh	108 College	Carson	In	1
5	Roberts, Terry	602 Bridge	Hudson	Mi	4
6	MacDonald, Greg	19 Oak	Carson	In	1
7	VanderJagt, Neal	12 Bishop	Mantin	Il	3
8	Shippers, John	208 Grayton	Cedar	In	2
9	Franklin, Trudy	103 Bedford	Brook	Mi	4
10	Stein, Shelly	82 Harcourt	Hudson	Mi	1

BOOKORD

BKCODE	CSTNUMB	REQDATE
0378	5	05/12/94
0378	10	05/20/94
0808	2	04/28/94
0808	8	05/15/94
1382	1	04/26/94
1382	10	05/20/94
3743	1	04/26/94
3743	5	05/12/94
3743	9	05/18/94
5163	4	04/29/94
7443	3	04/28/94

BOOKCUST

FIELD NAME	FIELD TYPE	FIELD WIDTH	KEY FIELD	FIELD DESCRIPTION
CSTNUMB	Number		Y	Customer number
CSTNAME	Text	16	N	Customer name
CSTADDR	Text	12	N	Customer address (street)
CSTCTY	Text	10	N	Customer city
CSTST	Text	2	N	Customer state
BRNUMB	Number		N	Number of customer's branch

BOOKORD

FIELD NAME	FIELD TYPE	FIELD WIDTH	KEY FIELD	FIELD DESCRIPTION
BKCODE	Number		Y	Book code
CSTNUMB	Number		Y	Number of customer requesting the book
REQDATE	Date/Time		N	Date book was requested